Bill Bryson was b settled in England his English wife He and his family then moved to America for a few years but have now returned to the UK. He is the bestselling author of *The Lost Continent*, *Mother Tongue*, *Neither Here Nor There*, *Made in America*, *Notes From a Small Island*, *A Walk in the Woods*, *Notes From a Big Country*, *Down Under* and, most recently, *A Short History of Nearly Everything*. He is also the author of the bestselling *African Diary* (a charity book for CARE International).

Critical acclaim for
A Short History of Nearly Everything:

'Bryson makes the study of science appear very friendly'
New Scientist

'I defy any reader to be bored or baffled by a single page of its contents. Bryson's "History" is hugely readable' *Sunday Times*

'He attempts his most difficult feat yet – And he succeeds'
Evening Standard

'A carefully constructed trawl, delivered with exemplary clarity, energy and enthusiasm' *Time Out*

'Genuinely enthralling. An utterly absorbing travel guide through our own history' *Irish News*

'An ambitious, sprawling, disciplined journey through the fundamentals of our world' *Irish Times*

'Told with his trademark humour, this is the perfect antidote for all those who find science writing impenetrable' *Woman & Home*

'An impressive testament to Bryson's impulses and powers of assimilation' *Daily Telegraph*

'One of the author's smartest books – one that deserves to be taken seriously' *Baltimore Sun*

'There are lots of laugh-out-loud moments in this book . . . Bryson has crafted a volume that illuminates . . . he has given us a book that might even turn people on to this exciting thing called science'
The Christian Science Monitor

'It moves so fast that it's science on a toboggan . . . Bryson evokes a constant sense of wonder and surprise . . . surprisingly precise, brilliantly encyclopaedic and nicely eloquent . . . In sum, a gifted storyteller who has dared retell the world's biggest story'
Seattle Times

'It's a cornucopia of a book, a joy to read and dawdle through. At its end, we have the sense of time spent in good company'
The Australian

'An ambitious attempt to decipher the mysteries of the universe with levity and in layman's terms' *Herald Sun (Melbourne)*

'Every kind of person would find this book enjoyable and stimulating'
Sydney Morning Herald

'An ambitious attempt to make science come alive' *West Australian*

'A stylish and entertaining armchair guide to nearly everything'
Advertiser (Adelaide)

'In a little more than 400 pages, Bill Bryson does what you always wished your secondary school science teacher would do: make science moreish and fun, even the gross bits' *New Zealand Herald*

'Here's a man who can do the impossible, make the incomprehensible as clear as a polished wine glass, the dullest subjects as lively as a bar room chat . . . It's a delight from start to finish. I doubt there has ever been a better book of its type' *The Timaru Herald*

'Bryson invites us into an exciting world filled with querulous, bombastic, introverted, sly and shy scientists at the forefront of research in their times . . . In short, he brings science alive and makes it interesting and attractive to any layman' *The Daily Post*

'[Bryson] portrays science as a very human activity, and its discoveries and central theories as genuinely awe-inspiring . . . This greatest of all possible guided tours may be his best work yet' *Waikato Times*

'It has been written to entertain, and it does not disappoint . . . It all flows effortlessly with [Bryson's] usual wit and style'
Otago Daily Times

'Bryson is a brilliant writer and this book sparkles with wit and whimsy' *Southland Times*

Also by Bill Bryson

THE LOST CONTINENT

MOTHER TONGUE

NEITHER HERE NOR THERE

MADE IN AMERICA

NOTES FROM A SMALL ISLAND

A WALK IN THE WOODS

NOTES FROM A BIG COUNTRY

DOWN UNDER

AFRICAN DIARY

A Short History *of* Nearly Everything

BILL BRYSON

BLACK SWAN

A SHORT HISTORY OF NEARLY EVERYTHING
A BLACK SWAN BOOK: 0 552 99704 8

Originally published in Great Britain by Doubleday,
a division of Transworld Publishers

PRINTING HISTORY
Doubleday edition published 2003
Black Swan edition published 2004

3 5 7 9 10 8 6 4

Set in 11.5/13pt Bembo by
Falcon Oast Graphic Art Ltd.

Black Swan Books are published by Transworld Publishers,
61–63 Uxbridge Road, London W5 5SA,
a division of The Random House Group Ltd,
in Australia by Random House Australia (Pty) Ltd,
20 Alfred Street, Milsons Point, Sydney, NSW 2061, Australia,
in New Zealand by Random House New Zealand Ltd,
18 Poland Road, Glenfield, Auckland 10, New Zealand
and in South Africa by Random House (Pty) Ltd,
Endulini, 5a Jubilee Road, Parktown 2193, South Africa.

Printed and bound in Great Britain by
Cox & Wyman Ltd, Reading, Berkshire.

Papers used by Transworld Publishers are natural, recyclable products made from wood grown in sustainable forests. The manufacturing processes conform to the environmental regulations of the country of origin.

To Meghan and Chris. Welcome.

The physicist Leo Szilard once announced to his friend Hans Bethe that he was thinking of keeping a diary: 'I don't intend to publish. I am merely going to record the facts for the information of God.' 'Don't you think God knows the facts?' Bethe asked. 'Yes,' said Szilard. 'He knows the facts, but He does not know **this version of the facts**.'

Hans Christian von Baeyer, *Taming the Atom*

CONTENTS

ACKNOWLEDGEMENTS

As I sit here, in early 2003, I have before me several pages of manuscript bearing majestically encouraging and tactful notes from Ian Tattersall of the American Museum of Natural History pointing out, *inter alia*, that Périgueux is not a wine-producing region, that it is inventive but a touch unorthodox of me to italicize taxonomic divisions above the level of genus and species, that I have persistently misspelled Olorgesailie (a place I visited only recently), and so on in similar vein through two chapters of text covering his area of expertise, early humans.

Goodness knows how many other inky embarrassments may lurk in these pages yet, but it is thanks to Dr Tattersall and all of those whom I am about to mention that there aren't many hundreds more. I cannot begin to thank adequately those who helped me in the preparation of this book. I am especially indebted to the following, who were uniformly generous and kindly and showed the most heroic reserves of patience in answering one simple, endlessly repeated question: 'I'm sorry, but can you explain that again?'

In England: David Caplin of Imperial College London; Richard Fortey, Len Ellis and Kathy Way of the Natural

History Museum; Martin Raff of University College London; Rosalind Harding of the Institute of Biological Anthropology in Oxford; Dr Laurence Smaje, formerly of the Wellcome Institute; and Keith Blackmore of *The Times*.

In the United States: Ian Tattersall of the American Museum of Natural History in New York; John Thorstensen, Mary K. Hudson and David Blanchflower of Dartmouth College in Hanover, New Hampshire; Dr William Abdu and Dr Bryan Marsh of Dartmouth–Hitchcock Medical Center in Lebanon, New Hampshire; Ray Anderson and Brian Witzke of the Iowa Department of Natural Resources, Iowa City; Mike Voorhies of the University of Nebraska and Ashfall Fossil Beds State Park near Orchard, Nebraska; Chuck Offenburger of Buena Vista University, Storm Lake, Iowa; Ken Rancourt, director of research, Mount Washington Observatory, Gorham, New Hampshire; Paul Doss, geologist of Yellowstone National Park, and his wife, Heidi, also of the National Park; Frank Asaro of the University of California at Berkeley; Oliver Payne and Lynn Addison of the National Geographic Society; James O. Farlow, Indiana-Purdue University; Roger L. Larson, professor of marine geophysics, University of Rhode Island; Jeff Guinn of the Fort Worth *Star-Telegram* newspaper; Jerry Kasten of Dallas, Texas; and the staff of the Iowa Historical Society in Des Moines.

In Australia: the Reverend Robert Evans of Hazelbrook, New South Wales; Dr Jill Cainey, Australian Bureau of Meteorology; Alan Thorne and Victoria Bennett of the Australian National University in Canberra; Louise Burke and John Hawley of Canberra; Anne Milne of the *Sydney Morning Herald*; Ian Nowak, formerly of the Geological Society of Western Australia; Thomas H. Rich of Museum Victoria; Tim Flannery, director of the South Australian Museum in Adelaide; Natalie Papworth and Alan MacFadyen of the Royal Tasmanian Botanical Gardens,

Hobart; and the very helpful staff of the State Library of New South Wales in Sydney.

And elsewhere: Sue Superville, information centre manager at the Museum of New Zealand in Wellington; and Dr Emma Mbua, Dr Koen Maes and Jillani Ngalla of the Kenya National Museum in Nairobi.

I am also deeply and variously indebted to Patrick Janson-Smith, Gerald Howard, Marianne Velmans, Alison Tulett, Gillian Somerscales, Larry Finlay, Steve Rubin, Jed Mattes, Carol Heaton, Charles Elliott, David Bryson, Felicity Bryson, Dan McLean, Nick Southern, Gerald Engelbretsen, Patrick Gallagher, Larry Ashmead, and the staff of the peerless and ever-cheery Howe Library in Hanover, New Hampshire.

Above all, and as always, my profoundest thanks to my dear, patient, incomparable wife, Cynthia.

INTRODUCTION

Welcome. And congratulations. I am delighted that you could make it. Getting here wasn't easy, I know. In fact, I suspect it was a little tougher than you realize.

To begin with, for you to be here now trillions of drifting atoms had somehow to assemble in an intricate and curiously obliging manner to create you. It's an arrangement so specialized and particular that it has never been tried before and will only exist this once. For the next many years (we hope) these tiny particles will uncomplainingly engage in all the billions of deft, co-operative efforts necessary to keep you intact and let you experience the supremely agreeable but generally under appreciated state known as existence.

Why atoms take this trouble is a bit of a puzzle. Being you is not a gratifying experience at the atomic level. For all their devoted attention, your atoms don't actually care about you – indeed, don't even know that you are there. They don't even know that *they* are there. They are mindless particles, after all, and not even themselves alive. (It is a slightly arresting notion that if you were to pick yourself apart with tweezers, one atom at a time, you would produce a mound of fine atomic dust, none of which had ever been alive but all of which had once been you.) Yet somehow for the period of your existence they will answer to a single rigid impulse: to keep you you.

The bad news is that atoms are fickle and their time of devotion is fleeting – fleeting indeed. Even a long human life adds up to only about 650,000 hours. And when that modest milestone flashes into view, or at some other point thereabouts, for reasons unknown your atoms will close you down, then silently disassemble and go off to be other things. And that's it for you.

Still, you may rejoice that it happens at all. Generally speaking in the universe it doesn't, so far as we can tell. This is decidedly odd because the atoms that so liberally and congenially flock together to form living things on Earth are exactly the same atoms that decline to do it elsewhere. Whatever else it may be, at the level of chemistry life is fantastically mundane: carbon, hydrogen, oxygen and nitrogen, a little calcium, a dash of sulphur, a light dusting of other very ordinary elements – nothing you wouldn't find in any ordinary pharmacy – and that's all you need. The only thing special about the atoms that make you is that they make you. That is, of course, the miracle of life.

Whether or not atoms make life in other corners of the universe, they make plenty else; indeed, they make everything else. Without them there would be no water or air or rocks, no stars and planets, no distant gassy clouds or swirling nebulae or any of the other things that make the universe so agreeably material. Atoms are so numerous and necessary that we easily overlook that they needn't actually exist at all. There is no law that requires the universe to fill itself with small particles of matter or to produce light and gravity and the other properties on which our existence hinges. There needn't actually be a universe at all. For a very long time there wasn't. There were no atoms and no universe for them to float about in. There was nothing – nothing at all anywhere.

So thank goodness for atoms. But the fact that you have atoms and that they assemble in such a willing manner is only part of what got you here. To be here now, alive in the twenty-first century and smart enough to know it, you also had to be the beneficiary of an extraordinary string of biological good fortune. Survival on Earth is a surprisingly tricky business. Of the billions and billions of species of living things that have existed since the dawn of time, most – 99.99 per cent, it has been suggested – are no longer around. Life on Earth, you see, is not only brief but dismayingly tenuous. It is a curious feature of our existence that we come from a planet that is very good at promoting life but even better at extinguishing it.

The average species on Earth lasts for only about four million years, so if you wish to be around for billions of years, you must be as fickle as the atoms that made you. You must be prepared to change everything about yourself – shape, size, colour, species affiliation, everything – and to do so repeatedly. That's much easier said than done, because the process of change is random. To get from 'protoplasmal primordial atomic globule' (as Gilbert and Sullivan put it) to sentient upright modern human has required you to mutate new traits over and over in a precisely timely manner for an exceedingly long while. So at various periods over the last 3.8 billion years you have abhorred oxygen and then doted on it, grown fins and limbs and jaunty sails, laid eggs, flicked the air with a forked tongue, been sleek, been furry, lived underground, lived in trees, been as big as a deer and as small as a mouse, and a million things more. The tiniest deviation from any of these evolutionary imperatives and you might now be licking algae from cave walls or lolling walrus-like on some stony shore or disgorging air through a blowhole in the top of your head before diving sixty feet for a mouthful of delicious sandworms.

Not only have you been lucky enough to be attached

since time immemorial to a favoured evolutionary line, but you have also been extremely – make that miraculously – fortunate in your personal ancestry. Consider the fact that for 3.8 billion years, a period of time older than the Earth's mountains and rivers and oceans, every one of your forebears on both sides has been attractive enough to find a mate, healthy enough to reproduce, and sufficiently blessed by fate and circumstances to live long enough to do so. Not one of your pertinent ancestors was squashed, devoured, drowned, starved, stuck fast, untimely wounded or otherwise deflected from its life's quest of delivering a tiny charge of genetic material to the right partner at the right moment to perpetuate the only possible sequence of hereditary combinations that could result – eventually, astoundingly, and all too briefly – in you.

This is a book about how it happened – in particular, how we went from there being nothing at all to there being something, and then how a little of that something turned into us, and also some of what happened in between and since. That's rather a lot to cover, of course, which is why the book is called *A Short History of Nearly Everything*, even though it isn't really. It couldn't be. But with luck by the time we finish it may feel as if it is.

My own starting point, for what it is worth, was a school science book that I had when I was in fourth or fifth grade. The book was a standard-issue 1950s schoolbook – battered, unloved, grimly hefty – but near the front it had an illustration that just captivated me: a cutaway diagram showing the Earth's interior as it would look if you cut into the planet with a large knife and carefully withdrew a wedge representing about a quarter of its bulk.

It's hard to believe that there was ever a time when I had not seen such an illustration before, but evidently I had not for I clearly remember being transfixed. I suspect, in

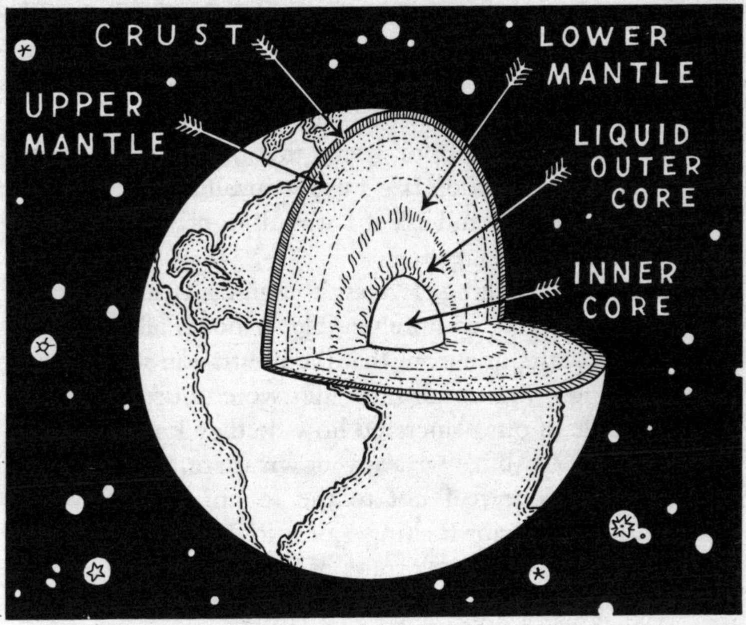

honesty, my initial interest was based on a private image of streams of unsuspecting eastbound motorists in the American plains states plunging over the edge of a sudden four-thousand-mile-high cliff running between Central America and the North Pole, but gradually my attention did turn in a more scholarly manner to the scientific import of the drawing and the realization that the Earth consisted of discrete layers, ending in the centre with a glowing sphere of iron and nickel, which was as hot as the surface of the Sun, according to the caption, and I remember thinking with real wonder: 'How do they know that?'

I didn't doubt the correctness of the information for an instant – I still tend to trust the pronouncements of scientists in the way I trust those of surgeons, plumbers, and other possessors of arcane and privileged information – but I couldn't for the life of me conceive how any human mind could work out what spaces thousands of miles below us, that no eye had ever seen and no X-ray could penetrate, could look like and be made of. To me that was just a miracle. That has been my position with science ever since.

Excited, I took the book home that night and opened it before dinner – an action that I expect prompted my mother to feel my forehead and ask if I was all right – and, starting with the first page, I read.

And here's the thing. It wasn't exciting at all. It wasn't actually altogether comprehensible. Above all, it didn't answer any of the questions that the illustration stirred up in a normal enquiring mind: How did we end up with a Sun in the middle of our planet and how do they know how hot it is? And if it is burning away down there, why isn't the ground under our feet hot to the touch? And why isn't the rest of the interior melting – or is it? And when the core at last burns itself out, will some of the Earth slump into the void, leaving a giant sinkhole on the surface? And how do you *know* this? *How did you figure it out?*

But the author was strangely silent on such details – indeed, silent on everything but anticlines, synclines, axial faults and the like. It was as if he wanted to keep the good stuff secret by making all of it soberly unfathomable. As the years passed, I began to suspect that this was not altogether a private impulse. There seemed to be a mystifying universal conspiracy among textbook authors to make certain the material they dealt with never strayed too near the realm of the mildly interesting and was always at least a long-distance phone call from the frankly interesting.

I now know that there is a happy abundance of science writers who pen the most lucid and thrilling prose – Timothy Ferris, Richard Fortey and Tim Flannery are three that jump out from a single station of the alphabet (and that's not even to mention the late but godlike Richard Feynman) – but, sadly, none of them wrote any textbook I ever used.

All mine were written by men (it was always men) who held the interesting notion that everything became clear when expressed as a formula and the amusingly deluded belief that the children of America would appreciate having chapters end with a section of questions they could mull over in their own time. So I grew up convinced that science was supremely dull, but suspecting that it needn't be, and not really thinking about it at all if I could help it. This, too, became my position for a long time.

Then, much later – about four or five years ago, I suppose – I was on a long flight across the Pacific, staring idly out the window at moonlit ocean, when it occurred to me with a certain uncomfortable forcefulness that I didn't know the first thing about the only planet I was ever going to live on. I had no idea, for example, why the oceans were salty but the Great Lakes weren't. Didn't have the faintest idea. I didn't know if the oceans were growing more salty with time or less, and whether ocean salinity levels was something I should be concerned about or not. (I am very pleased to tell you that until the late 1970s scientists didn't know the answers to these questions either. They just didn't talk about it very audibly.)

And ocean salinity, of course, represented only the merest sliver of my ignorance. I didn't know what a proton was, or a protein, didn't know a quark from a quasar, didn't understand how geologists could look at a layer of rock on a canyon wall and tell you how old it was – didn't know anything, really. I became gripped by a quiet, unwonted but insistent urge to know a little about these matters and to understand above all how people figured them out. That to me remained the greatest of all amazements – how scientists work things out. How does anybody *know* how much the Earth weighs or how old its rocks are or what really is way down there in the centre? How can they know how and when the universe started and what it was like when it did?

How do they know what goes on inside an atom? And how, come to that – or perhaps above all, on reflection – can scientists so often seem to know nearly everything but then still not be able to predict an earthquake or even tell us whether we should take an umbrella with us to the races next Wednesday?

So I decided that I would devote a portion of my life – three years, as it now turns out – to reading books and journals and finding saintly, patient experts prepared to answer a lot of outstandingly dumb questions. The idea was to see if it isn't possible to understand and appreciate – marvel at, enjoy even – the wonder and accomplishments of science at a level that isn't too technical or demanding, but isn't entirely superficial either.

That was my idea and my hope, and that is what the book that follows is intended to do. Anyway, we have a great deal of ground to cover and much less than 650,000 hours in which to do it, so let's begin.

I

LOST IN THE COSMOS

They're all in the same plane. They're all going around in the same direction . . . It's perfect, you know. It's gorgeous. It's almost uncanny.

Astronomer Geoffrey Marcy describing the solar system

1

HOW TO BUILD A UNIVERSE

No matter how hard you try you will never be able to grasp just how tiny, how spatially unassuming, is a proton. It is just way too small.

A proton is an infinitesimal part of an atom, which is itself of course an insubstantial thing. Protons are so small that a little dib of ink like the dot on this 'i' can hold something in the region of 500,000,000,000 of them, or rather more than the number of seconds it takes to make half a million years. So protons are exceedingly microscopic, to say the very least.

Now imagine if you can (and of course you can't) shrinking one of those protons down to a billionth of its normal size into a space so small that it would make a proton look enormous. Now pack into that tiny, tiny space about an ounce of matter. Excellent. You are ready to start a universe.

I'm assuming of course that you wish to build an inflationary universe. If you'd prefer instead to build a more old-fashioned, standard Big Bang universe, you'll need additional materials. In fact, you will need to gather up everything there is – every last mote and particle of matter between here and the edge of creation – and squeeze it into a spot so infinitesimally compact that it has no dimensions at all. It is known as a singularity.

In either case, get ready for a really big bang. Naturally, you will wish to retire to a safe place to observe the spectacle. Unfortunately, there is nowhere to retire to because outside the singularity there is no *where*. When the universe begins to expand, it won't be spreading out to fill a larger emptiness. The only space that exists is the space it creates as it goes.

It is natural but wrong to visualize the singularity as a kind of pregnant dot hanging in a dark, boundless void. But there is no space, no darkness. The singularity has no around around it. There is no space for it to occupy, no place for it to be. We can't even ask how long it has been there – whether it has just lately popped into being, like a good idea, or whether it has been there for ever, quietly awaiting the right moment. Time doesn't exist. There is no past for it to emerge from.

And so, from nothing, our universe begins.

In a single blinding pulse, a moment of glory much too swift and expansive for any form of words, the singularity assumes heavenly dimensions, space beyond conception. The first lively second (a second that many cosmologists will devote careers to shaving into ever-finer wafers) produces gravity and the other forces that govern physics. In less than a minute the universe is a million billion miles across and growing fast. There is a lot of heat now, 10 billion degrees of it, enough to begin the nuclear reactions that create the lighter elements – principally hydrogen and helium, with a dash (about one atom in a hundred million) of lithium. In three minutes, 98 per cent of all the matter there is or will ever be has been produced. We have a universe. It is a place of the most wondrous and gratifying possibility, and beautiful, too. And it was all done in about the time it takes to make a sandwich.

When this moment happened is a matter of some debate. Cosmologists have long argued over whether the moment of creation was ten billion years ago or twice that or something in between. The consensus seems to be heading for a figure of about 13.7 billion years, but these things are notoriously difficult to measure, as we shall see further on. All that can really be said is that at some indeterminate point in the very distant past, for reasons unknown, there came the moment known to science as $t = 0$. We were on our way.

There is of course a great deal we don't know, and much of what we think we know we haven't known, or thought we've known, for long. Even the notion of the Big Bang is quite a recent one. The idea had been kicking around since the 1920s when Georges Lemaître, a Belgian priest–scholar, first tentatively proposed it, but it didn't really become an active notion in cosmology until the mid-1960s, when two young radio astronomers made an extraordinary and inadvertent discovery.

Their names were Arno Penzias and Robert Wilson. In 1965, they were trying to make use of a large communications antenna owned by Bell Laboratories at Holmdel, New Jersey, but they were troubled by a persistent background noise – a steady, steamy hiss that made any experimental work impossible. The noise was unrelenting and unfocused. It came from every point in the sky, day and night, through every season. For a year the young astronomers did everything they could think of to track down and eliminate the noise. They tested every electrical system. They rebuilt instruments, checked circuits, wiggled wires, dusted plugs. They climbed into the dish and placed duct tape over every seam and rivet. They climbed back into the dish with brooms and scrubbing brushes and carefully swept it clean of what they referred to in a later paper as 'white dielectric material', or what is known more commonly as bird shit. Nothing they tried worked.

Unknown to them, just 50 kilometres away at Princeton University a team of scientists led by Robert Dicke was working on how to find the very thing they were trying so diligently to get rid of. The Princeton researchers were pursuing an idea that had been suggested in the 1940s by the Russian-born astrophysicist George Gamow: that if you looked deep enough into space you should find some cosmic background radiation left over from the Big Bang. Gamow calculated that by the time it had crossed the vastness of the cosmos the radiation would reach Earth in the form of microwaves. In a more recent paper he had even suggested an instrument that might do the job: the Bell antenna at Holmdel. Unfortunately, neither Penzias and Wilson, nor any of the Princeton team, had read Gamow's paper.

The noise that Penzias and Wilson were hearing was, of course, the noise that Gamow had postulated. They had found the edge of the universe, or at least the visible part of it, 90 billion trillion miles away. They were 'seeing' the first photons – the most ancient light in the universe – though time and distance had converted them to microwaves, just as Gamow had predicted. In his book *The Inflationary Universe*, Alan Guth provides an analogy that helps to put this finding in perspective. If you think of peering into the depths of the universe as like looking down from the hundredth floor of the Empire State Building (with the hundredth floor representing now and street level representing the moment of the Big Bang), at the time of Wilson and Penzias's discovery the most distant galaxies anyone had ever detected were on about the sixtieth floor and the most distant things – quasars – were on about the twentieth. Penzias and Wilson's finding pushed our acquaintance with the visible universe to within half an inch of the lobby floor.

Still unaware of what caused the noise, Wilson and Penzias phoned Dicke at Princeton and described their problem to him in the hope that he might suggest a solution. Dicke realized at once what the two young men had found. 'Well, boys, we've just been scooped,' he told his colleagues as he hung up the phone.

Soon afterwards the *Astrophysical Journal* published two articles: one by Penzias and Wilson describing their experience with the hiss, the other by Dicke's team explaining its nature. Although Penzias and Wilson had not been looking for cosmic background radiation, didn't know what it was when they had found it, and hadn't described or interpreted its character in any paper, they received the 1978 Nobel Prize in Physics. The Princeton researchers got only sympathy. According to Dennis Overbye in *Lonely Hearts of the Cosmos*, neither Penzias nor Wilson altogether understood the significance of what they had found until they read about it in the *New York Times*.

Incidentally, disturbance from cosmic background radiation is something we have all experienced. Tune your television to any channel it doesn't receive and about 1 per cent of the dancing static you see is accounted for by this ancient remnant of the Big Bang. The next time you complain that there is nothing on, remember that you can always watch the birth of the universe.

Although everyone calls it the Big Bang, many books caution us not to think of it as an explosion in the conventional sense. It was, rather, a vast, sudden expansion on a whopping scale. So what caused it?

One notion is that perhaps the singularity was the relic of an earlier, collapsed universe – that ours is just one of an eternal cycle of expanding and collapsing universes, like the bladder on an oxygen machine. Others attribute the Big Bang to what they call 'a false vacuum' or 'a scalar field' or 'vacuum energy' – some quality or thing, at any rate, that introduced a measure of instability into the nothingness that was. It seems impossible that you could get something from nothing, but the fact that once there was nothing and now there is a universe is evident proof that you can. It may be that our universe is merely part of many larger universes, some in different dimensions, and that Big Bangs are going on all the time all over the place. Or it may be that space and time had some other forms altogether before the Big Bang – forms too alien for us to imagine – and that the Big Bang represents some sort of transition phase, where the universe went from a form we can't understand to one we almost can. 'These are very close to religious questions,' Dr Andrei Linde, a cosmologist at Stanford, told the *New York Times* in 2001.

The Big Bang theory isn't about the bang itself but about what happened after the bang. Not long after, mind you. By doing a lot of maths and watching carefully what goes on in particle accelerators, scientists believe they can look back to 10^{-43} seconds after the moment of creation, when the universe was still so small that you would have needed a microscope to find it. We mustn't swoon over every extraordinary number that comes before us, but it is perhaps worth latching onto one from time to time just to be reminded of their ungraspable and amazing breadth. Thus 10^{-43} is 0.0000000000000000000000000000000000000000001, or one ten million trillion trillion trillionths of a second.*

*A word on scientific notation. Since very large numbers are cumbersome to write and nearly impossible to read, scientists use a shorthand involving powers (or multiples) of ten in which, for instance, 10,000,000,000 is written 10^{10} and 6,500,000 becomes 6.5×10^{6}. The principle is based very simply on multiples of ten: 10×10 (or 100) becomes 10^{2}; $10 \times 10 \times 10$ (or 1,000) is 10^{3}; and so on, obviously and indefinitely. The little superscript number signifies the number of zeroes following the larger principal number. Negative notations provide essentially a mirror image, with the superscript number indicating the number of spaces to the right of the decimal point (so 10^{-4} means 0.0001). Though I salute the principle, it remains an amazement to me that anyone seeing '1.4×10^{9} km^{3}' would see at once that that signifies 1.4 billion cubic kilometres, and no less a wonder that they would choose the former over the latter in print (especially in a book designed for the general reader, where the example was found). On the assumption that many readers are as unmathematical as I am, I will use notations sparingly, though they are occasionally unavoidable, not least in a chapter dealing with things on a cosmic scale.

Most of what we know, or believe we know, about the early moments of the universe is thanks to an idea called inflation theory first propounded in 1979 by a junior particle physicist then at Stanford, now at MIT, named Alan Guth. He was thirty-two years old and, by his own admission, had never done anything much before. He would probably never have had his great theory except that he happened to attend a lecture on the Big Bang given by none other than Robert Dicke. The lecture inspired Guth to take an interest in cosmology, and in particular in the birth of the universe.

The eventual result was the inflation theory, which holds that a fraction of a moment after the dawn of creation, the universe underwent a sudden dramatic expansion. It inflated – in effect ran away with itself, doubling in size every 10^{-34} seconds. The whole episode may have lasted no more than 10^{-30} seconds – that's one million million million million millionths of a second – but it changed the universe from something you could hold in your hand to something at least 10,000,000,000,000,000,000,000,000 times bigger. Inflation theory explains the ripples and eddies that make our universe possible. Without it, there would be no clumps of matter and thus no stars, just drifting gas and ever-lasting darkness.

According to Guth's theory, at one ten-millionth of a trillionth of a trillionth of a trillionth of a second, gravity emerged. After another ludicrously brief interval it was joined by electromagnetism and the strong and weak nuclear forces – the stuff of physics. These were joined an instant later by shoals of elementary particles – the stuff of stuff. From nothing at all, suddenly there were swarms of photons, protons, electrons, neutrons and much else – between 10^{79} and 10^{89} of each, according to the standard Big Bang theory.

Such quantities are of course ungraspable. It is enough to know that in a single cracking instant we were endowed with a universe that was vast – at least a hundred billion light years across, according to the theory, but possibly any size up to infinite – and perfectly arrayed for the creation of stars, galaxies and other complex systems.

What is extraordinary from our point of view is how well it turned out for us. If the universe had formed just a tiny bit differently – if gravity were fractionally stronger or weaker, if the expansion had proceeded just a little more slowly or swiftly – then there might never have been stable elements to make you and me and the ground we stand on. Had gravity been a trifle stronger, the universe itself might have collapsed like a badly erected tent without precisely the right values to give it the necessary dimensions and density and component parts. Had it been weaker, however, nothing would have coalesced. The universe would have remained forever a dull, scattered void.

This is one reason why some experts believe that there may have been many other big bangs, perhaps trillions and trillions of them, spread through the mighty span of eternity, and that the reason we exist in this particular one is that this is one that we could exist in. As Edward P. Tryon of Columbia University once put it: 'In answer to the question of why it happened, I offer the modest proposal that our Universe is simply one of those things which happen from time to time.' To which adds Guth: 'Although the creation of a universe might be very unlikely, Tryon emphasized that no one had counted the failed attempts.'

Martin Rees, Britain's Astronomer Royal, believes that there are many universes, possibly an infinite number, each with different attributes, in different combinations, and that we simply live in one that combines things in the way that allows us to exist. He makes an analogy with a very large clothing store: 'If there is a large stock of clothing, you're not surprised to find a suit that fits. If there are many universes, each governed by a differing set of numbers, there will be one where there is a particular set of numbers suitable to life. We are in that one.'

Rees maintains that six numbers in particular govern our universe, and that if any of these values were changed even very slightly things could not be as they are. For example, for the universe to exist as it does requires that hydrogen be converted to helium in a precise but comparatively stately manner – specifically, in a way that converts seven one-thousandths of its mass to energy. Lower that value very slightly – from 0.07 per cent to 0.06 per cent, say – and no transformation could take place: the universe would consist of hydrogen and nothing else. Raise the value very slightly – to 0.08 per cent – and bonding would be so wildly prolific that the hydrogen would long since have been exhausted. In either case, with the slightest tweaking of the numbers the universe as we know and need it would not be here.

I should say that everything is just right *so far.* In the long term, gravity may turn out to be a little too strong; one day it may halt the expansion of the universe and bring it collapsing in upon itself, until it crushes itself down into another singularity, possibly to start the whole process over again. On the other hand, it may be too weak, in which case the universe will keep racing away for ever until everything is so far apart that there is no chance of material interactions, so that the universe becomes a place that is very roomy, but inert and dead. The third option is that gravity is perfectly pitched – 'critical density' is the cosmologists' term for it – and that it will hold the universe together at just the right dimensions to allow things to go on indefinitely. Cosmologists, in their lighter moments, sometimes call this the 'Goldilocks effect' – that everything is just right. (For the record, these three possible universes are known respectively as closed, open and flat.)

Now, the question that has occurred to all of us at some

point is: what would happen if you travelled out to the edge of the universe and, as it were, put your head through the curtains? Where would your head *be* if it were no longer in the universe? What would you find beyond? The answer, disappointingly, is that you can never get to the edge of the universe. That's not because it would take too long to get there – though of course it would – but because even if you travelled outward and outward in a straight line, indefinitely and pugnaciously, you would never arrive at an outer boundary. Instead, you would come back to where you began (at which point, presumably, you would rather lose heart in the exercise and give up). The reason for this is that the universe bends, in a way we can't adequately imagine, in conformance with Einstein's theory of relativity (which we will get to in due course). For the moment it is enough to know that we are not adrift in some large, ever-expanding bubble. Rather, space curves, in a way that allows it to be boundless but finite. Space cannot even properly be said to be expanding because, as the physicist and Nobel laureate Steven Weinberg notes, 'solar systems and galaxies are not expanding, and space itself is not expanding.' Rather, the galaxies are rushing apart. It is all something of a challenge to intuition. Or, as the biologist J. B. S. Haldane once famously observed: 'The universe is not only queerer than we suppose; it is queerer than we *can* suppose.'

The analogy that is usually given for explaining the curvature of space is to try to imagine someone from a universe of flat surfaces, who had never seen a sphere, being brought to Earth. No matter how far he roamed across the planet's surface, he would never find an edge. He might eventually return to the spot where he had started, and would of course be utterly confounded to explain how that had happened. Well, we are in the same position in space as our puzzled flatlander, only we are flummoxed by a higher dimension.

Just as there is no place where you can find the edge of the universe, so there is no place where you can stand at the centre and say: 'This is where it all began. This is the centre-most point of it all.' We are *all* at the centre of it all. Actually, we don't know that for sure; we can't prove it mathematically. Scientists just assume that we can't really be the centre of the universe – think what that would imply – but that the phenomenon must be the same for all observers in all places. Still, we don't actually know.

For us, the universe goes only as far as light has travelled in the billions of years since the universe was formed. This visible universe – the universe we know and can talk about – is a million million million million (that's 1,000,000,000,000,000,000,000,000) miles across. But according to most theories the universe at large – the meta-universe, as it is sometimes called – is vastly roomier still. According to Rees, the number of light years to the edge of this larger, unseen universe would be written not 'with ten zeroes, not even with a hundred, but with millions'. In short, there's more space than you can imagine already without going to the trouble of trying to envision some additional beyond.

For a long time the Big Bang theory had one gaping hole that troubled a lot of people – namely, that it couldn't begin to explain how we got here. Although 98 per cent of all the matter that exists was created with the Big Bang, that matter consisted exclusively of light gases: the helium, hydrogen and lithium that we mentioned earlier. Not one particle of the heavy stuff so vital to our own being – carbon, nitrogen, oxygen and all the rest – emerged from the gaseous brew of creation. But – and here's the troubling point – to forge these heavy elements, you need the kind of heat and energy thrown off by a Big Bang. Yet there has been only one Big Bang and it didn't produce them. So where did they come from? Interestingly, the man who found the answer to that question was a cosmologist who heartily despised the Big Bang as a theory and coined the term Big Bang sarcastically, as a way of mocking it.

We'll get to him shortly, but before we turn to the question of how we got here, it might be worth taking a few minutes to consider just where exactly 'here' is.

2

WELCOME TO THE SOLAR SYSTEM

Astronomers these days can do the most amazing things. If someone struck a match on the Moon, they could spot the flare. From the tiniest throbs and wobbles of distant stars they can infer the size and character and even potential habitability of planets much too remote to be seen – planets so distant that it would take us half a million years in a spaceship to get there. With their radio telescopes they can capture wisps of radiation so preposterously faint that the total amount of energy collected from outside the solar system by all of them together since collecting began (in 1951) is 'less than the energy of a single snowflake striking the ground', in the words of Carl Sagan.

In short, there isn't a great deal that goes on in the universe that astronomers can't find when they have a mind to. Which is why it is all the more remarkable to reflect that until 1978 no-one had ever noticed that Pluto has a moon. In the summer of that year, a young astronomer named James Christy at the Lowell Observatory in Flagstaff, Arizona, was making a routine examination of photographic images of Pluto when he saw that there was something there – something blurry and uncertain but definitely other than Pluto. Consulting a colleague named Robert Harrington, he concluded that what he was looking at was a moon. And it wasn't just any moon. Relative to the planet, it was the biggest moon in the solar system.

This was actually something of a blow to Pluto's status as a planet, which had never been terribly robust anyway. Since previously the space occupied by the moon and the space occupied by Pluto were thought to be one and the same, it meant that Pluto was much smaller than anyone had supposed – smaller even than Mercury. Indeed, seven moons in the solar system, including our own, are larger.

Now, a natural question is why it took so long for anyone to find a moon in our own solar system. The answer is that it is partly a matter of where astronomers point their instruments and partly a matter of what their instruments are designed to detect and partly it's just Pluto. Mostly it's where they point their instruments. In the words of the astronomer Clark Chapman: 'Most people think that astronomers get out at night in observatories and scan the skies. That's not true. Almost all the telescopes we have in the world are designed to peer at very tiny little pieces of the sky way off in the distance to see a quasar or hunt for black holes or look at a distant galaxy. The only real network of telescopes that scans the skies has been designed and built by the military.'

We have been spoiled by artists' renderings into imagining a clarity of resolution that doesn't exist in actual astronomy. Pluto in Christy's photograph is faint and fuzzy – a piece of cosmic lint – and its moon is not the romantically backlit, crisply delineated companion orb you would get in a *National Geographic* painting, but rather just a tiny and extremely indistinct hint of additional fuzziness. Such was the fuzziness, in fact, that it took seven years for anyone to spot the moon again and thus independently confirm its existence.

One nice touch about Christy's discovery was that it happened in Flagstaff, for it was there in 1930 that Pluto had been found in the first place. That seminal event in astronomy was largely to the credit of the astronomer Percival Lowell. Lowell, who came from one of the oldest and wealthiest Boston families (the one in the famous ditty about Boston being the home of the bean and the cod, where Lowells spoke only to Cabots, while Cabots spoke only to God), endowed the famous observatory that bears his name, but is most indelibly remembered for his belief that Mars was covered with canals built by industrious Martians for purposes of conveying water from polar regions to the dry but productive lands nearer the equator.

Lowell's other abiding conviction was that there existed, somewhere out beyond Neptune, an undiscovered ninth planet, dubbed Planet X. Lowell based this belief on irregularities he detected in the orbits of Uranus and Neptune, and devoted the last years of his life to trying to find the gassy giant he was certain was out there. Unfortunately, he died suddenly in 1916, at least partly exhausted by his quest, and the search fell into abeyance while Lowell's heirs squabbled over his estate. However, in 1929, partly as a way of deflecting attention away from the Mars canal saga (which by now had become a serious embarrassment) the Lowell Observatory directors decided to resume the search and to that end hired a young man from Kansas named Clyde Tombaugh.

Tombaugh had no formal training as an astronomer, but he was diligent and he was astute, and after a year's patient searching he somehow spotted Pluto, a faint point of light in a glittery firmament. It was a miraculous find, and what made it all the more striking was that the observations on which Lowell had predicted the existence of a planet beyond Neptune proved to be comprehensively erroneous. Tombaugh could see at once that the new planet was nothing like the massive gasball Lowell had postulated – but any reservations he or anyone else had about the character of the new planet were soon swept aside in the delirium that attended almost any big news story in that easily excited age. This was the first American-discovered planet, and no-one was going to be distracted by the thought that it was really just a distant icy dot. It was named Pluto, at least partly because the first two letters made a monogram from Lowell's initials. Lowell was posthumously hailed everywhere as a genius of the first order and Tombaugh was largely forgotten, except among planetary astronomers, who tend to revere him.

A few astronomers continue to think there may yet be a Planet X out there – a real whopper, perhaps as much as ten times the size of Jupiter, but so far out as to be invisible to us. (It would receive so little sunlight that it would have almost none to reflect.) The idea is that it wouldn't be a conventional planet like Jupiter or Saturn – it's much too far away for that; we're talking perhaps 4.5 trillion miles – but more like a sun that never quite made it. Most star systems in the cosmos are binary (double-starred), which makes our solitary sun a slight oddity.

As for Pluto itself, nobody is quite sure how big it is, what it is made of, what kind of atmosphere it has, or even what it really is. A lot of astronomers believe it isn't a planet at all, but merely the largest object so far found in a zone of galactic debris known as the Kuiper belt. The Kuiper belt was actually theorized by an astronomer named F. C. Leonard in 1930, but the name honours Gerard Kuiper, a Dutch native working in America, who expanded the idea. The Kuiper belt is the source of what are known as short-period comets – those that come past pretty regularly – of which the most famous is Halley's comet. The more reclusive long-period comets (among them the recent visitors Hale–Bopp and Hyakutake) come from the much more distant Oort cloud, about which more presently.

It is certainly true that Pluto doesn't act much like the

other planets. Not only is it runty and obscure, it is so variable in its motions that no-one can tell you exactly where Pluto will be a century hence. Whereas the other planets orbit on more or less the same plane, Pluto's orbital path is tipped (as it were) out of alignment at an angle of 17 degrees, like the brim of a hat tilted rakishly on someone's head. Its orbit is so irregular that for substantial periods on each of its lonely circuits around the Sun it is closer to us than Neptune is. For most of the 1980s and 1990s, Neptune was in fact the solar system's most far-flung planet. Only on 11 February 1999 did Pluto return to the outside lane, there to remain for the next 228 years.

So if Pluto really is a planet, it is certainly an odd one. It is very tiny: just one quarter of 1 per cent as massive as Earth. If you set it down on top of the United States, it would cover not quite half the lower forty-eight states. This alone makes it extremely anomalous; it means that our planetary system consists of four rocky inner planets, four gassy outer giants, and a tiny, solitary iceball. Moreover, there is every reason to suppose that we may soon begin to find other, even larger icy spheres in the same portion of space. Then we *will* have problems. After Christy spotted Pluto's moon, astronomers began to regard that section of the cosmos more attentively, and as of early December 2002 had found over six hundred additional Trans-Neptunian Objects or Plutinos as they are alternatively called. One, dubbed Varuna, is nearly as big as Pluto's moon. Astronomers now think there may be billions of these objects. The difficulty is that many of them are awfully dark. Typically they have an albedo, or reflectiveness, of just 4 per cent, about the same as a lump of charcoal – and of course these lumps of charcoal are over six billion kilometres away.

And how far is that, exactly? It's almost beyond imagining. Space, you see, is just enormous – just enormous. Let's imagine, for purposes of edification and entertainment, that we are about to go on a journey by rocketship. We won't go terribly far – just to the edge of our own solar system – but we need to get a fix on how big a place space is and what a small part of it we occupy.

Now the bad news, I'm afraid, is that we won't be home for supper. Even at the speed of light (300,000 kilometres per second) it would take seven hours to get to Pluto. But of course we can't travel at anything like that speed. We'll have to go at the speed of a spaceship, and these are rather more lumbering. The best speeds yet achieved by any human object are those of the Voyager 1 and 2 spacecrafts, which are now flying away from us at about 56,000 kilometres an hour.

The reason the Voyager craft were launched when they were (in August and September 1977) was that Jupiter, Saturn, Uranus and Neptune were aligned in a way that happens only once every 175 years. This enabled the two Voyagers to use a 'gravity assist' technique in which the craft were successively flung from one gassy giant to the next in a kind of cosmic version of crack the whip. Even so, it took them nine years to reach Uranus and a dozen to cross the orbit of Pluto. The good news is that if we wait until January 2006 (which is when NASA's New Horizons spacecraft is tentatively scheduled to depart for Pluto) we can take advantage of favourable Jovian positioning, plus some advances in technology, and get there in only a decade or so – though getting home again will take rather longer, I'm afraid. At all events, it's going to be a long trip.

Now, the first thing you are likely to realize is that space is extremely well named and rather dismayingly uneventful. Our solar system may be the liveliest thing for trillions of miles, but all the visible stuff in it – the Sun, the planets and their moons, the billion or so tumbling rocks of the asteroid belt, comets and other miscellaneous drifting detritus – fills less than a trillionth of the available space. You also quickly realize that none of the maps you have ever seen of the solar system was drawn remotely to scale. Most schoolroom charts show the planets coming one after the other at neighbourly intervals – the outer giants actually cast shadows over each other in many illustrations – but this is a necessary deceit to get them all on the same piece of paper. Neptune in reality isn't just a little bit beyond Jupiter, it's *way* beyond Jupiter – five times further from Jupiter than Jupiter is from us, so far out that it receives only 3 per cent as much sunlight as Jupiter.

Such are the distances, in fact, that it isn't possible, in any practical terms, to draw the solar system to scale. Even if you added lots of fold-out pages to your textbooks or used a really long sheet of poster paper, you wouldn't come close. On a diagram of the solar system to scale, with the Earth reduced to about the diameter of a pea, Jupiter would be over 300 metres away and Pluto would be two and a half kilometres distant (and about the size of a bacterium, so you wouldn't be able to see it anyway). On the same scale, Proxima Centauri, our nearest star, would be 16,000 kilometres away. Even if you shrank down everything so that Jupiter was as small as the full stop at the end of this sentence, and Pluto was no bigger than a molecule, Pluto would still be over 10 metres away.

So the solar system is really quite enormous. By the time we reach Pluto, we have come so far that the Sun – our dear, warm, skin-tanning, life-giving Sun – has shrunk to the size of a pinhead. It is little more than a bright star. In such a lonely void you can begin to understand how even the most significant objects – Pluto's moon, for example – have escaped attention. In this respect, Pluto has hardly been alone. Until the Voyager expeditions, Neptune was thought to have two moons; Voyager found six more. When I was a boy, the solar system was thought to contain thirty moons. The total now is at least ninety, about a third of which have been found in just the last ten years. The point to remember, of course, when considering the universe at large is that we don't actually know what is in our own solar system.

Now, the other thing you will notice as we speed past Pluto is that we are speeding past Pluto. If you check your itinerary, you will see that this is a trip to the edge of our solar system, and I'm afraid we're not there yet. Pluto may be the last object marked on schoolroom charts, but the sytem doesn't end there. In fact, it isn't even close to ending there. We won't get to the solar system's edge until we have passed through the Oort cloud, a vast celestial realm of drifting comets, and we won't reach the Oort cloud for another – I'm so sorry about this – ten thousand years. Far from marking the outer edge of the solar system, as those schoolroom maps so cavalierly imply, Pluto is barely one-fifty-thousandth of the way.

Of course we have no prospect of such a journey. A trip of 386,000 kilometres to the Moon still represents a very big undertaking for us. A manned mission to Mars, called for by the first President Bush in a moment of passing giddiness, was quietly dropped when someone worked out that it would cost $450 billion and probably result in the deaths of all the crew (their DNA torn to tatters by high-energy solar particles from which they could not be shielded).

Based on what we know now and can reasonably imagine, there is absolutely no prospect that any human being will ever visit the edge of our own solar system – ever. It is just too far. As it is, even with the Hubble telescope we can't see even into the Oort cloud, so we don't actually know that it is there. Its existence is probable but entirely hypothetical.*

*Properly called the Öpik–Oort cloud, it is named for the Estonian astronomer Ernst Öpik, who hypothesized its existence in 1932, and for the Dutch astronomer Jan Oort, who refined the calculations eighteen years later.

About all that can be said with confidence about the Oort cloud is that it starts somewhere beyond Pluto and stretches some two light years out into the cosmos. The basic unit of measure in the solar system is the Astronomical Unit, or AU, representing the distance from the Sun to the Earth. Pluto is about 40 AUs from us, the heart of the Oort cloud about fifty thousand. In a word, it is remote.

But let's pretend again that we have made it to the Oort cloud. The first thing you might notice is how very peaceful it is out here. We're a long way from anywhere now – so far from our own Sun that it's not even the brightest star in the sky. It is a remarkable thought that that distant tiny twinkle has enough gravity to hold all these comets in orbit. It's not a very strong bond, so the comets drift in a stately manner, moving at only about 220 miles an hour. From time to time one of these lonely comets is nudged out of its normal orbit by some slight gravitational perturbation – a passing star, perhaps. Sometimes they are ejected into the emptiness of space, never to be seen again, but sometimes they fall into a long orbit around the Sun. About three or four of these a year, known as long-period comets, pass through the inner solar system. Just occasionally these stray visitors smack into something solid, like Earth. That's why we've come out here now – because the comet we have come to see has just begun a long fall towards the centre of the solar system. It is headed for, of all places, Manson, Iowa. It is going to take a long time to get there – three or four million years at least – so we'll leave it for now, and return to it much later in the story.

So that's your solar system. And what else is out there, beyond the solar system? Well, nothing and a great deal, depending on how you look at it.

In the short term, it's nothing. The most perfect vacuum ever created by humans is not as empty as the emptiness of interstellar space. And there is a great deal of this nothingness until you get to the next bit of something. Our nearest neighbour in the cosmos, Proxima Centauri, which is part of the three-star cluster known as Alpha Centauri, is 4.3 light years away, a sissy skip in galactic terms, but still a hundred million times further than a trip to the Moon. To reach it by spaceship would take at least twenty-five thousand years, and even if you made the trip you still wouldn't be anywhere except at a lonely clutch of stars in the middle of a vast nowhere. To reach the next landmark of consequence, Sirius, would involve another 4.6 light years of travel. And so it would go if you tried to star-hop your way across the cosmos. Just reaching the centre of our own galaxy would take far longer than we have existed as beings.

Space, let me repeat, is enormous. The average distance between stars out there is over 30 million million kilometres. Even at speeds approaching those of light, these are fantastically challenging distances for any travelling individual. Of course, it is *possible* that alien beings travel billions of miles to amuse themselves by planting crop circles in Wiltshire or frightening the daylights out of some poor guy in a pickup truck on a lonely road in Arizona (they must have teenagers, after all), but it does seem unlikely.

Still, statistically the probability that there are other thinking beings out there is good. Nobody knows how many stars there are in the Milky Way – estimates range from a hundred billion or so to perhaps four hundred billion – and the Milky Way is just one of a hundred and forty billion or so other galaxies, many of them even larger than ours. In the 1960s, a professor at Cornell named Frank Drake, excited by such whopping numbers, worked out a famous equation designed to calculate the chances of advanced life existing in the cosmos, based on a series of diminishing probabilities.

Under Drake's equation you divide the number of stars in

a selected portion of the universe by the number of stars that are likely to have planetary systems; divide that by the number of planetary systems that could theoretically support life; divide that by the number on which life, having arisen, advances to a state of intelligence; and so on. At each such division, the number shrinks colossally – yet even with the most conservative inputs the number of advanced civilizations just in the Milky Way always works out to be somewhere in the millions.

What an interesting and exciting thought. We may be only one of millions of advanced civilizations. Unfortunately, space being spacious, the average distance between any two of these civilizations is reckoned to be at least two hundred light years, which is a great deal more than merely saying it makes it sound. It means, for a start, that even if these beings know we are here and are somehow able to see us in their telescopes, they're watching light that left Earth two hundred years ago. So they're not seeing you and me. They're watching the French Revolution and Thomas Jefferson and people in silk stockings and powdered wigs – people who don't know what an atom is, or a gene, and who make their electricity by rubbing a rod of amber with a piece of fur and think that's quite a trick. Any message we receive from these observers is likely to begin 'Dear Sire', and congratulate us on the handsomeness of our horses and our mastery of whale oil. Two hundred light years is a distance so far beyond us as to be, well, just beyond us.

So even if we are not really alone, in all practical terms we are. Carl Sagan calculated the number of probable planets in the universe at as many as ten billion trillion – a number vastly beyond imagining. But what is equally beyond imagining is the amount of space through which they are lightly scattered. 'If we were randomly inserted into the universe,' Sagan wrote, 'the chances that you would be on or near a planet would be less than one in a billion trillion

trillion.' (That's 10^{33}, or 1 followed by 33 zeroes.) 'Worlds are precious.'

Which is why perhaps it is good news that in February 1999 the International Astronomical Union ruled officially that Pluto is a planet. The universe is a big and lonely place. We can do with all the neighbours we can get.

3

THE REVEREND EVANS'S UNIVERSE

When the skies are clear and the Moon is not too bright, the Reverend Robert Evans, a quiet and cheerful man, lugs a bulky telescope onto the back sun-deck of his home in the Blue Mountains of Australia, about 80 kilometres west of Sydney, and does an extraordinary thing. He looks deep into the past and finds dying stars.

Looking into the past is, of course, the easy part. Glance at the night sky and what you see is history and lots of it – not the stars as they are now but as they were when their light left them. For all we know, the North Star, our faithful companion, might actually have burned out last January or in 1854 or at any time since the early fourteenth century and news of it just hasn't reached us yet. The best we can say – can ever say – is that it was still burning on this date 680 years ago. Stars die all the time. What Bob Evans does better than anyone else who has ever tried is spot these moments of celestial farewell.

By day, Evans is a kindly and now semi-retired minister in the Uniting Church in Australia, who does a bit of locum work and researches the history of nineteenth-century religious movements. But by night he is, in his unassuming way, a titan of the skies. He hunts supernovae.

A supernova occurs when a giant star, one much bigger than our own Sun, collapses and then spectacularly explodes, releasing in an instant the energy of a hundred billion suns, burning for a time more brightly than all the stars in its galaxy. 'It's like a trillion hydrogen bombs going off at once,' says Evans. If a supernova explosion happened within five hundred light years of us, we would be goners, according to Evans – 'it would wreck the show,' as he cheerfully puts it. But the universe is vast and supernovae are normally much too far away to harm us. In fact, most are so unimaginably distant that their light reaches us as no more than the faintest twinkle. For the month or so that they are visible, all that distinguishes them from the other stars in the sky is that they occupy a point of space that wasn't filled before. It is these anomalous, very occasional pricks in the crowded dome of the night sky that the Reverend Evans finds.

To understand what a feat this is, imagine a standard dining-room table covered in a black tablecloth and throwing a handful of salt across it. The scattered grains can be thought of as a galaxy. Now imagine fifteen hundred more tables like the first one – enough to make a single line two miles long – each with a random array of salt across it. Now add one grain of salt to any table and let Bob Evans walk among them. At a glance he will spot it. That grain of salt is the supernova.

Evans's is a talent so exceptional that Oliver Sacks, in *An Anthropologist on Mars*, devotes a passage to him in a chapter on autistic savants – quickly adding that 'there is no suggestion that he is autistic.' Evans, who has not met Sacks, laughs at the suggestion that he might be either autistic or a savant, but he is powerless to explain quite where his talent comes from.

'I just seem to have a knack for memorizing star fields,' he told me, with a frankly apologetic look, when I visited him and his wife, Elaine, in their picture-book bungalow on a tranquil edge of the village of Hazelbrook, out where Sydney finally ends and the boundless Australian bush begins. 'I'm not particularly good at other things,' he added. 'I don't remember names well.'

'Or where he's put things,' called Elaine from the kitchen.

He nodded frankly again and grinned, then asked me if I'd like to see his telescope. I had imagined that Evans would have a proper observatory in his back yard – a scaled-down version of a Mount Wilson or Palomar, with a sliding domed roof and a mechanized chair that would be a pleasure to manoeuvre. In fact, he led me not outside but to a crowded storeroom off the kitchen where he keeps his books and papers and where his telescope – a white cylinder that is about the size and shape of a household hot-water tank – rests in a home-made, swivelling plywood mount. When he wishes to observe, he carries them, in two trips, to a small sun-deck off the kitchen. Between the overhang of the roof and the feathery tops of eucalyptus trees growing up from the slope below, he has only a letterbox view of the sky, but he says it is more than good enough for his purposes. And there, when the skies are clear and the Moon is not too bright, he finds his supernovae.

The term supernova was coined in the 1930s by a memorably odd astrophysicist named Fritz Zwicky. Born in Bulgaria and raised in Switzerland, Zwicky came to the California Institute of Technology in the 1920s and there at once distinguished himself by his abrasive personality and erratic talents. He didn't seem to be outstandingly bright, and many of his colleagues considered him little more than 'an irritating buffoon'. A fitness fanatic, he would often drop to the floor of the Caltech dining hall or some other public area and do one-armed push-ups to demonstrate his virility to anyone who seemed inclined to doubt it. He was notoriously aggressive, his manner eventually becoming so intimidating that his closest collaborator, a gentle man named Walter Baade, refused to be left alone with him. Among other things, Zwicky accused Baade, who was German, of being a Nazi, which he was not. On at least one occasion Zwicky threatened to kill Baade, who worked up the hill at the Mount Wilson Observatory, if he saw him on the Caltech campus.

But Zwicky was also capable of insights of the most startling brilliance. In the early 1930s he turned his attention to a question that had long troubled astronomers: the appearance in the sky of occasional unexplained points of light, new stars. Improbably, he wondered if the neutron – the subatomic particle that had just been discovered in England by James Chadwick, and was thus both novel and rather fashionable – might be at the heart of things. It occurred to him that if a star collapsed to the sort of densities found in the core of atoms, the result would be an unimaginably compacted core. Atoms would literally be crushed together, their electrons forced into the nucleus, forming neutrons. You would have a neutron star. Imagine a million really weighty cannonballs squeezed down to the size of a marble and – well, you're still not even close. The core of a neutron star is so dense that a single spoonful of matter from it would weigh 90 billion kilograms. A spoonful! But there was more. Zwicky realized that after the collapse of such a star there would be a huge amount of energy left over – enough to make the biggest bang in the universe. He called these resultant explosions supernovae. They would be – they are – the biggest events in creation.

On 15 January 1934 the journal *Physical Review* published a very concise abstract of a presentation that had been conducted by Zwicky and Baade the previous month at Stanford University. Despite its extreme brevity – one paragraph of twenty-four lines – the abstract contained an enormous amount of new science: it provided the first reference to supernovae and to neutron stars; convincingly explained their method of formation; correctly calculated the scale of their explosiveness; and, as a kind of concluding bonus, connected supernova explosions to the production of a mysterious new phenomenon called cosmic rays, which had recently been found swarming through the universe. These ideas were revolutionary, to say the least. The existence of neutron stars wouldn't be confirmed for thirty-four years. The cosmic rays notion, though considered plausible, hasn't been verified yet. Altogether, the abstract was, in the words of Caltech astrophysicist Kip S. Thorne, 'one of the most prescient documents in the history of physics and astronomy'.

Interestingly, Zwicky had almost no understanding of why any of this would happen. According to Thorne, 'he did not understand the laws of physics well enough to be able to substantiate his ideas.' Zwicky's talent was for big ideas. Others – Baade mostly – were left to do the mathematical sweeping up.

Zwicky was also the first to recognize that there wasn't nearly enough visible mass in the universe to hold galaxies together, and that there must be some other gravitational influence – what we now call dark matter. One thing he failed to see was that if a neutron star shrank enough it would become so dense that even light couldn't escape its immense gravitational pull. You would have a black hole. Unfortunately, Zwicky was held in such disdain by most of his colleagues that his ideas attracted almost no notice. When, five years later, the great Robert Oppenheimer turned his attention to neutron stars in a landmark paper, he made not a single reference to any of Zwicky's work, even though Zwicky had been working for years on the same problem in an office just down the corridor. Zwicky's deductions concerning dark matter wouldn't attract serious attention for nearly four decades. We can only assume that he did a lot of push-ups in this period.

Surprisingly little of the universe is visible to us when we incline our heads to the sky. Only about six thousand stars are visible to the naked eye from Earth, and only about two thousand can be seen from any one spot. With binoculars the number of stars you can see from a single location rises to about fifty thousand, and with a small 2-inch telescope it leaps to three hundred thousand. With a 16-inch telescope, such as Evans uses, you begin to count not in stars but in galaxies. From his deck, Evans supposes he can see between fifty thousand and one hundred thousand galaxies, each containing tens of billions of stars. These are of course respectable numbers, but even with so much to take in, supernovae are extremely rare. A star can burn for billions of years, but it dies just once and quickly, and only a few dying stars explode. Most expire quietly, like a camp fire at dawn. In a typical galaxy, consisting of a hundred billion stars, a supernova will occur on average once every two or three hundred years. Looking for a supernova, therefore, was a little like standing on the observation platform of the Empire State Building with a telescope and searching windows around Manhattan in the hope of finding, let us say, someone lighting a twenty-first birthday cake.

So when a hopeful and softly spoken minister got in touch to ask if they had any usable field charts for hunting supernovae, the astronomical community thought he was out of his mind. At the time Evans had a 10-inch telescope – a very respectable size for amateur star-gazing, but hardly the sort of thing with which to do serious cosmology – and he was proposing to find one of the universe's rarer phenomena. In the whole of astronomical history before Evans started looking in 1980, fewer than sixty supernovae had been found. (At the time I visited him, in August 2001, he had just recorded his thirty-fourth visual discovery; a thirty-fifth followed three months later, and a thirty-sixth in early 2003.)

Evans, however, had certain advantages. Most observers, like most people generally, are in the northern hemisphere, so he had a lot of sky largely to himself, especially at first. He also had speed and his uncanny memory. Large telescopes are cumbersome things, and much of their operational time is consumed in being manoeuvred into position. Evans could swing his little 16-inch telescope around like a tail-gunner in a dogfight, spending no more than a couple of seconds on any particular point in the sky. In consequence, he could observe perhaps four hundred galaxies in an evening while a large professional telescope would be lucky to do fifty or sixty.

Looking for supernovae is mostly a matter of not finding them. From 1980 to 1996 he averaged two discoveries a year – not a huge payoff for hundreds of nights of peering and peering. Once he found three in fifteen days, but another time he went three years without finding any at all.

'There is actually a certain value in not finding anything,' he said. 'It helps cosmologists to work out the rate at which galaxies are evolving. It's one of those rare areas where the absence of evidence *is* evidence.'

On a table beside the telescope were stacks of photos and papers relevant to his pursuits, and he showed me some of them now. If you have ever looked through popular astronomical publications, and at some time you must have, you will know that they are generally full of richly luminous colour photos of distant nebulae and the like – fairy-lit clouds of celestial light of the most delicate and moving splendour. Evans's working images are nothing like that. They are just blurry black-and-white photos with little points of haloed brightness. One he showed me depicted a swarm of stars in which lurked a trifling flare that I had to put close to my face to discern. This, Evans told me, was a star in a constellation called Fornax from a galaxy known to astronomy as NGC1365. (NGC stands for New General Catalogue, where these things are recorded. Once it was a heavy book on someone's desk in Dublin; today, needless to say, it's a database.) For sixty million years, the light from this star's spectacular demise travelled unceasingly through space until one night in August 2001 it arrived at Earth in the form of a puff of radiance, the tiniest brightening, in the night sky. It was, of course, Robert Evans on his eucalypt-scented hill-side who spotted it.

'There's something satisfying, I think,' Evans said, 'about the idea of light travelling for millions of years through space and *just* at the right moment as it reaches Earth someone looks at the right bit of sky and sees it. It just seems right that an event of that magnitude should be witnessed.'

Supernovae do much more than simply impart a sense of wonder. They come in several types (one of them discovered by Evans), and of these, one in particular, known as the Ia supernova, is important to astronomy because these supernovae always explode in the same way, with the same critical mass. For this reason they can be used as 'standard candles' – benchmarks by which to measure the brightness (and hence relative distance) of other stars, and thus to measure the expansion rate of the universe.

In 1987 Saul Perlmutter at the Lawrence Berkeley Laboratory in California, needing more Ia supernovae than visual sightings were providing, set out to find a more systematic method of searching for them. Perlmutter devised a nifty system using sophisticated computers and charge-coupled devices – in essence, really good digital cameras. It automated supernova hunting. Telescopes could now take thousands of pictures and let a computer detect the tell-tale bright spots that marked a supernova explosion. In five years, with the new technique, Perlmutter and his colleagues at Berkeley found forty-two supernovae. Now even amateurs are finding supernovae with charge-coupled devices. 'With CCDs you can aim a telescope at the sky and go watch television,' Evans said with a touch of dismay. 'It took all the romance out of it.'

I asked him if he was tempted to adopt the new technology. 'Oh, no,' he said, 'I enjoy my way too much. Besides' – he gave a nod at the photo of his latest supernova and smiled – 'I can still beat them sometimes.'

The question that naturally occurs is: what would it be like if a star exploded nearby? Our nearest stellar neighbour, as we have seen, is Alpha Centauri, 4.3 light years away. I had imagined that if there were an explosion there we would have 4.3 years to watch the light of this magnificent event spreading across the sky, as if tipped from a giant can. What would it be like if we had four years and four months to watch an inescapable doom advancing towards us, knowing that when it finally arrived it would blow the skin right off our bones? Would people still go to work? Would farmers plant crops? Would anyone deliver them to the shops?

Weeks later, back in the town in New Hampshire where I then lived, I put these questions to John Thorstensen, an astronomer at Dartmouth College. 'Oh no,' he said, laughing. 'The news of such an event travels out at the speed of light, but so does the destructiveness, so you'd learn about it and die from it in the same instant. But don't worry, because it's not going to happen.'

For the blast of a supernova explosion to kill you, he explained, you would have to be 'ridiculously close' – probably within ten light years or so. 'The danger would be various types of radiation – cosmic rays and so on.' These would produce fabulous auroras, shimmering curtains of spooky light that would fill the whole sky. This would not be a good thing. Anything potent enough to put on such a show could well blow away the magnetosphere, the magnetic zone high above the Earth that normally protects us from ultraviolet rays and other cosmic assaults. Without the magnetosphere anyone unfortunate enough to step into sunlight would pretty quickly take on the appearance of, let us say, an overcooked pizza.

The reason we can be reasonably confident that such an event won't happen in our corner of the galaxy, Thorstensen said, is that it takes a particular kind of star to make a supernova in the first place. A candidate star must be ten to twenty times as massive as our own Sun, and 'we don't have anything of the requisite size that's that close. The universe is a mercifully big place.' The nearest likely candidate, he added, is Betelgeuse, whose various sputterings have for years suggested that something interestingly unstable is going on there. But Betelgeuse is fifty thousand light years away.

Only half a dozen times in recorded history have supernovae been close enough to be visible to the naked eye. One was a blast in 1054 that created the Crab Nebula. Another, in 1604, made a star bright enough to be seen during the day for over three weeks. The most recent was in 1987, when a supernova flared in a zone of the cosmos known as the Large Magellanic Cloud, but that was only barely visible and only in the southern hemisphere – and it was a comfortably safe 169,000 light years away.

Supernovae are significant to us in one other decidedly central way. Without them we wouldn't be here. You will recall the cosmological conundrum with which we ended the first chapter – that the Big Bang created lots of light gases but no heavy elements. Those came later, but for a very long time nobody could figure out *how* they came later. The problem was that you needed something really hot – hotter even than the middle of the hottest stars – to forge carbon and iron and the other elements without which we would be distressingly immaterial. Supernovae provided the explanation, and it was an English cosmologist almost as singular in manner as Fritz Zwicky who worked it out.

He was a Yorkshireman named Fred Hoyle. Hoyle, who died in 2001, was described in an obituary in *Nature* as a 'cosmologist and controversialist', and both of those he most certainly was. He was, according to *Nature*'s obituary, 'embroiled in controversy for most of his life' and 'put his name to much rubbish'. He claimed, for instance, and without evidence, that the Natural History Museum's treasured fossil of an archaeopteryx was a forgery along the lines of the Piltdown hoax, causing much exasperation to the museum's palaeontologists, who had to spend days fielding phone calls from journalists all over the world. He also believed that the Earth was seeded from space not only by life but also by many of its diseases, such as influenza and bubonic plague, and suggested at one point that humans evolved projecting noses with the nostrils underneath as a way of keeping cosmic pathogens from falling into them.

It was he who coined the term Big Bang, in a moment of facetiousness, for a radio broadcast in 1952. He pointed out that nothing in our understanding of physics could account for why everything, gathered to a point, would suddenly and dramatically begin to expand. Hoyle favoured a steady-state theory in which the universe was constantly expanding and continually creating new matter as it went. Hoyle also realized that if stars imploded they would liberate huge amounts of heat – 100 million degrees or more, enough to begin to generate the heavier elements in a process known as nucleosynthesis. In 1957, working with others, Hoyle showed how the heavier elements were formed in supernova explosions. For this work, W. A. Fowler, one of his collaborators, received a Nobel Prize. Hoyle, shamefully, did not.

According to Hoyle's theory, an exploding star would

generate enough heat to create all the new elements and spray them into the cosmos where they would form gaseous clouds – the interstellar medium, as it is known – that could eventually coalesce into new solar systems. With the new theories it became possible at last to construct plausible scenarios for how we got here. What we now think we know is this:

About 4.6 billion years ago, a great swirl of gas and dust some 24 billion kilometres across accumulated in space where we are now and began to aggregate. Virtually all of it – 99.9 per cent of the mass of the solar system – went to make the Sun. Out of the floating material that was left over, two microscopic grains floated close enough together to be joined by electrostatic forces. This was the moment of conception for our planet. All over the inchoate solar system, the same was happening. Colliding dust grains formed larger and larger clumps. Eventually the clumps grew large enough to be called planetesimals. As these endlessly bumped and collided, they fractured or split or recombined in endless random permutations, but in every encounter there was a winner, and some of the winners grew big enough to dominate the orbit around which they travelled.

It all happened remarkably quickly. To grow from a tiny cluster of grains to a baby planet some hundreds of kilometres across is thought to have taken only a few tens of thousands of years. In just 200 million years, possibly less, the Earth was essentially formed, though still molten and subject to constant bombardment from all the debris that remained floating about.

At this point, about 4.4 billion years ago, an object the size of Mars crashed into the Earth, blowing out enough material to form a companion sphere, the Moon. Within weeks, it is thought, the flung material had reassembled itself into a single clump, and within a year it had formed into the spherical rock that companions us yet. Most of the lunar material, it is thought, came from the Earth's crust, not its core, which is why the Moon has so little iron while we have a lot. The theory, incidentally, is almost always presented as a recent one, but in fact it was first proposed in the 1940s by Reginald Daly of Harvard. The only recent thing about it is people paying any attention to it.

When the Earth was only about a third of its eventual size, it was probably already beginning to form an atmosphere, mostly of carbon dioxide, nitrogen, methane and sulphur. Hardly the sort of stuff that we would associate with life, and yet from this noxious stew life formed. Carbon dioxide is a powerful greenhouse gas. This was a good thing, because the Sun was significantly dimmer back then. Had we not had the benefit of a greenhouse effect, the Earth might well have frozen over permanently, and life might never have got a toehold. But somehow life did.

For the next 500 million years the young Earth continued to be pelted relentlessly by comets, meteorites and other galactic debris, which brought water to fill the oceans and the components necessary for the successful formation of life. It was a singularly hostile environment, and yet somehow life got going. Some tiny bag of chemicals twitched and became animate. We were on our way.

Four billion years later, people began to wonder how it had all happened. And it is there that our story next takes us.

II

THE SIZE OF THE EARTH

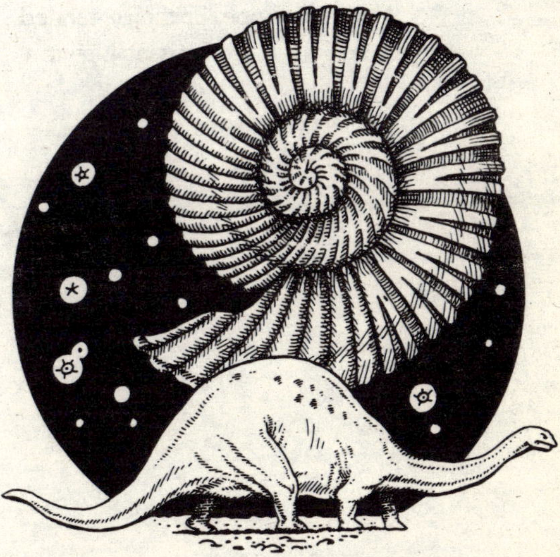

Nature and Nature's laws lay hid in night;
God said, Let Newton be! and all was light.
Alexander Pope, 'Epitaph: Intended for Sir Isaac Newton'

4

THE MEASURE OF THINGS

If you had to select the least convivial scientific field trip of all time, you could certainly do worse than the French Royal Academy of Sciences' Peruvian expedition of 1735. Led by a hydrologist named Pierre Bouguer and a soldier–mathematician named Charles Marie de La Condamine, it was a party of scientists and adventurers who travelled to Peru with the purpose of triangulating distances through the Andes.

At the time people had lately become infected with a powerful desire to understand the Earth – to determine how old it was, and how massive, where it hung in space, and how it had come to be. The French party's goal was to help settle the question of the circumference of the planet by measuring the length of one degree of meridian (or one-360th of the distance around the planet) along a line reaching from Yarouqui, near Quito, to just beyond Cuenca in what is now Ecuador, a distance of about 320 kilometres.*

Almost at once things began to go wrong, sometimes spectacularly so. In Quito, the visitors somehow provoked the locals and were chased out of town by a mob armed with stones. Soon after, the expedition's doctor was murdered in a misunderstanding over a woman. The botanist became

* see footnote overpage

deranged. Others died of fevers and falls. The third most senior member of the party, a man named Jean Godin, ran off with a thirteen-year-old girl and could not be induced to return.

At one point the group had to suspend work for eight months while La Condamine rode off to Lima to sort out a problem with their permits. Eventually he and Bouguer stopped speaking and refused to work together. Everywhere the dwindling party went it was met with the deepest suspicions from officials who found it difficult to believe that a group of French scientists would travel halfway around the world to measure the world. That made no sense at all. Two and a half centuries later, it still seems a reasonable question. Why didn't the French make their measurements in France and save themselves all the bother and discomfort of their Andean adventure?

The answer lies partly with the fact that eighteenth-century scientists, the French in particular, seldom did things

*Triangulation, their chosen method, was a popular technique based on the geometric fact that if you know the length of one side of a triangle and the angles of two corners, you can work out all its other dimensions without leaving your chair. Suppose, by way of example, that you and I decided we wished to know how far it is to the Moon. Using triangulation, the first thing we must do is put some distance between us, so let's say for argument that you stay in Paris and I go to Moscow and we both look at the Moon at the same time. Now, if you imagine a line connecting the three principals of this exercise – that is, you and me and the Moon – it forms a triangle. Measure the length of the baseline between you and me and the angles of our two corners and the rest can be simply calculated. (Because the interior angles of a triangle always add up to 180 degrees, if you know the sum of two of the angles you can instantly calculate the third; and knowing the precise shape of a triangle and the length of one side tells you the lengths of the other sides.) This was in fact the method used by a Greek astronomer. Hipparchus of Nicaea, in 150BC to work out the Moon's distance from the Earth. At ground level, the principles of triangulation are the same, except that the triangles don't reach into space but rather are laid side to side on a map. In measuring a degree of meridian, the surveyors would create a sort of chain of triangles marching across the landscape.

simply if an absurdly demanding alternative was available, and partly with a practical problem that had first arisen with the English astronomer Edmond Halley many years before – long before Bouguer and La Condamine dreamed of going to South America, much less had a reason for doing so.

Halley was an exceptional figure. In the course of a long and productive career, he was a sea captain, a cartographer, a professor of geometry at the University of Oxford, deputy controller of the Royal Mint, Astronomer Royal, and inventor of the deep-sea diving bell. He wrote authoritatively on magnetism, tides and the motions of the planets, and fondly on the effects of opium. He invented the weather map and actuarial table, proposed methods for working out the age of the Earth and its distance from the Sun, even devised a practical method for keeping fish fresh out of season. The one thing he didn't do was discover the comet that bears his name. He merely recognized that the comet he saw in 1682 was the same one that had been seen by others in 1456, 1531 and 1607. It didn't become Halley's comet until 1758, some sixteen years after his death.

For all his achievements, however, Halley's greatest contribution to human knowledge may simply have been to take part in a modest scientific wager with two other worthies of his day: Robert Hooke, who is perhaps best remembered now as the first person to describe a cell, and the great and stately Sir Christopher Wren, who was actually an astronomer first and an architect second, though that is not often generally remembered now. In 1683, Halley, Hooke and Wren were dining in London when the conversation turned to the motions of celestial objects. It was known that planets were inclined to orbit in a particular kind of oval known as an ellipse – 'a very specific and precise curve', to quote Richard Feynman – but it wasn't understood why. Wren generously offered a prize worth 40 shillings (equivalent to a couple of weeks' pay) to whichever of the men could provide a solution.

Hooke, who was well known for taking credit for ideas that weren't necessarily his own, claimed that he had solved the problem already but declined now to share it on the interesting and inventive grounds that it would rob others of the satisfaction of discovering the answer for themselves. He would instead 'conceal it for some time, that others might know how to value it'. If he thought any more on the matter, he left no evidence of it. Halley, however, became consumed with finding the answer, to the point that the following year he travelled to Cambridge and boldly called upon the university's Lucasian Professor of Mathematics, Isaac Newton, in the hope that he could help.

Newton was a decidedly odd figure – brilliant beyond measure, but solitary, joyless, prickly to the point of paranoia, famously distracted (upon swinging his feet out of bed in the morning he would reportedly sometimes sit for hours, immobilized by the sudden rush of thoughts to his head), and capable of the most riveting strangeness. He built his own laboratory, the first at Cambridge, but then engaged in the most bizarre experiments. Once he inserted a bodkin – a long needle of the sort used for sewing leather – into his eye socket and rubbed it around 'betwixt my eye and the bone as near to [the] backside of my eye as I could' just to see what would happen. What happened, miraculously, was nothing – at least, nothing lasting. On another occasion, he stared at the Sun for as long as he could bear, to determine what effect it would have upon his vision. Again he escaped lasting damage, though he had to spend some days in a darkened room before his eyes forgave him.

Set atop these odd beliefs and quirky traits, however, was the mind of a supreme genius – though even when working in conventional channels he often showed a tendency to peculiarity. As a student, frustrated by the limitations of conventional mathematics, he invented an entirely new form, the calculus, but then told no-one about it for twenty-seven years. In like manner, he did work in optics that transformed our understanding of light and laid the foundation for the science of spectroscopy, and again chose not to share the results for three decades.

For all his brilliance, real science accounted for only a part of his interests. At least half his working life was given over to alchemy and wayward religious pursuits. These were not mere dabblings but wholehearted devotions. He was a secret adherent of a dangerously heretical sect called Arianism, whose principal tenet was the belief that there had been no Holy Trinity (slightly ironic, since Newton's college at Cambridge was Trinity). He spent endless hours studying the floor plan of the lost Temple of King Solomon in Jerusalem (teaching himself Hebrew in the process, the better to scan original texts) in the belief that it held mathematical clues to the dates of the second coming of Christ and the end of the world. His attachment to alchemy was no less ardent. In 1936, the economist John Maynard Keynes bought a trunk of Newton's papers at auction and discovered with astonishment that they were overwhelmingly preoccupied not with optics or planetary motions, but with a single-minded quest to turn base metals into precious ones. An analysis of a strand of Newton's hair in the 1970s found it contained mercury – an element of interest to alchemists, hatters and thermometer-makers but almost no-one else – at a concentration some forty times the natural level. It is perhaps little wonder that he had trouble remembering to get up in the morning.

Quite what Halley expected to get from him when he made his unannounced visit in August 1684 we can only guess. But thanks to the later account of a Newton confidant, Abraham DeMoivre, we do have a record of one of science's most historic encounters:

> In 1684 Dr Halley came to visit at Cambridge [and] after they had some time together the Dr asked him what he thought the curve would be that would be described by the Planets supposing the force of attraction towards the Sun to be reciprocal to the square of their distance from it.

This was a reference to a piece of mathematics known as the inverse square law, which Halley was convinced lay at the heart of the explanation, though he wasn't sure exactly how.

> Sr Isaac replied immediately that it would be an [ellipse]. The Doctor, struck with joy & amazement, asked him how he knew it. 'Why,' saith he, 'I have calculated it,' whereupon Dr Halley asked him for his calculation without farther delay. Sr Isaac looked among his papers but could not find it.

This was astounding – like someone saying he had found a cure for cancer but couldn't remember where he had put the formula. Pressed by Halley, Newton agreed to redo the calculations and produce a paper. He did as promised, but then did much more. He retired for two years of intensive reflection and scribbling, and at length produced his masterwork: the *Philosophiae Naturalis Principia Mathematica* or *Mathematical Principles of Natural Philosophy*, better known as the *Principia*.

Once in a great while, a few times in history, a human mind produces an observation so acute and unexpected that people can't quite decide which is the more amazing – the fact or the thinking of it. The appearance of the *Principia* was one of those moments. It made Newton instantly famous. For the rest of his life he would be draped with plaudits and honours, becoming, among much else, the first person in Britain knighted for scientific achievement. Even the great German mathematician Gottfried von Leibniz, with whom

Newton had a long, bitter fight over priority for the invention of the calculus, thought his contributions to mathematics equal to all the accumulated work that had preceded him. 'Nearer the gods no mortal may approach,' wrote Halley in a sentiment that was endlessly echoed by his contemporaries and by many others since.

Although the *Principia* has been called 'one of the most inaccessible books ever written' (Newton intentionally made it difficult so that he wouldn't be pestered by mathematical 'smatterers', as he called them), it was a beacon to those who could follow it. It not only explained mathematically the orbits of heavenly bodies, but also identified the attractive force that got them moving in the first place – gravity. Suddenly every motion in the universe made sense.

At the *Principia*'s heart were Newton's three laws of motion (which state, very baldly, that a thing moves in the direction in which it is pushed; that it will keep moving in a straight line until some other force acts to slow or deflect it; and that every action has an opposite and equal reaction) and his universal law of gravitation. This states that every object in the universe exerts a tug on every other. It may not seem like it, but as you sit here now you are pulling everything around you – walls, ceiling, lamp, pet cat – towards you with your own little (indeed, very little) gravitational field. And these things are also pulling on you. It was Newton who realized that the pull of any two objects is, to quote Feynman again, 'proportional to the mass of each and varies inversely as the square of the distance between them'. Put another way, if you double the distance between two objects, the attraction between them becomes four times weaker. This can be expressed with the formula

$$F = G \frac{mm'}{r^2}$$

which is of course way beyond anything that most of us could make practical use of, but at least we can appreciate that it is elegantly compact. A couple of brief multiplications, a simple division and, bingo, you know your gravitational position wherever you go. It was the first really universal law of nature ever propounded by a human mind, which is why Newton is everywhere regarded with such profound esteem.

The *Principia*'s production was not without drama. To Halley's horror, just as work was nearing completion Newton and Hooke fell into dispute over the priority for the inverse square law and Newton refused to release the crucial third volume, without which the first two made little sense. Only with some frantic shuttle diplomacy and the most liberal applications of flattery did Halley manage finally to extract the concluding volume from the erratic professor.

Halley's traumas were not yet quite over. The Royal Society had promised to publish the work, but now pulled out, citing financial embarrassment. The year before, the society had backed a costly flop called *The History of Fishes*, and suspected that the market for a book on mathematical principles would be less than clamorous. Halley, whose means were not great, paid for the book's publication out of his own pocket. Newton, as was his custom, contributed nothing. To make matters worse, Halley at this time had just accepted a position as the society's clerk, and he was informed that the society could no longer afford to provide him with a promised salary of £50 per annum. He was to be paid instead in copies of *The History of Fishes*.

Newton's laws explained so many things – the slosh and roll of ocean tides, the motions of planets, why cannonballs trace a particular trajectory before thudding back to earth, why we aren't flung into space as the planet spins beneath us at hundreds of kilometres an hour* – that it took a while for all their implications to seep in. But one revelation became almost immediately controversial.

This was the suggestion that the Earth is not quite round. According to Newton's theory, the centrifugal force of the Earth's spin should result in a slight flattening at the poles and a bulging at the equator, which would make the planet slightly oblate. That meant that the length of a degree of meridian wouldn't be the same in Italy as it was in Scotland. Specifically, the length would shorten as you moved away from the poles. This was not good news for those people whose measurements of the planet were based on the assumption that it was a perfect sphere, which was everyone.

For half a century people had been trying to work out the size of the Earth, mostly by making very exacting measurements. One of the first such attempts was by an English mathematician named Richard Norwood. As a young man Norwood had travelled to Bermuda with a diving bell modelled on Halley's device, intending to make a fortune scooping pearls from the seabed. The scheme failed because there were no pearls and anyway Norwood's bell didn't work, but Norwood was not one to waste an experience. In the early seventeenth century Bermuda was well known among ships' captains for being hard to locate. The problem was that the ocean was big, Bermuda small and the navigational tools for dealing with this disparity hopelessly inadequate. There wasn't even yet an agreed length for a nautical mile. Over the breadth of an ocean the smallest miscalculations would become magnified so that ships often missed Bermuda-sized targets by dismayingly large margins. Norwood, whose first love was trigonometry and thus

*How fast you are spinning depends on where you are. The speed of the Earth's spin varies from something over 1,600 kilometres an hour at the equator to zero at the poles. In London the speed is 998 kilometres an hour.

angles, decided to bring a little mathematical rigour to navigation, and to that end he determined to calculate the length of a degree.

Starting with his back against the Tower of London, Norwood spent two devoted years marching 208 miles north to York, repeatedly stretching and measuring a length of chain as he went, all the while making the most meticulous adjustments for the rise and fall of the land and the meanderings of the road. The final step was to measure the angle of the sun at York at the same time of day and on the same day of the year as he had made his first measurement in London. From this, he reasoned he could determine the length of one degree of the Earth's meridian and thus calculate the distance around the whole. It was an almost ludicrously ambitious undertaking – a mistake of the slightest fraction of a degree would throw the whole thing out by miles – but in fact, as Norwood proudly declaimed, he was accurate to 'within a scantling' – or, more precisely, to within about six hundred yards. In metric terms, his figure worked out at 110.72 kilometres per degree of arc.

In 1637, Norwood's masterwork of navigation, *The Seaman's Practice*, was published and found an immediate following. It went through seventeen editions and was still in print twenty-five years after his death. Norwood returned to Bermuda with his family, where he became a successful planter and devoted his leisure hours to his first love, trigonometry. He survived there for thirty-eight years and it would be pleasing to report that he passed this span in happiness and adulation. In fact, he didn't. On the crossing from England, his two young sons were placed in a cabin with the Reverend Nathaniel White, and somehow so successfully traumatized the young vicar that he devoted much of the rest of his career to persecuting Norwood in any small way he could think of.

Norwood's two daughters brought their father additional

pain by making poor marriages. One of the husbands, possibly incited by the vicar, continually laid small charges against Norwood in court, causing him much exasperation and necessitating repeated trips across Bermuda to defend himself. Finally, in the 1650s witchcraft trials came to Bermuda and Norwood spent his final years in severe unease that his papers on trigonometry, with their arcane symbols, would be taken as communications with the devil and that he would be treated to a dreadful execution. So little is known of Norwood that it may in fact be that he deserved his unhappy declining years. What is certainly true is that he got them.

Meanwhile, the momentum for determining the Earth's circumference passed to France. There, the astronomer Jean Picard devised an impressively complicated method of triangulation involving quadrants, pendulum clocks, zenith sectors and telescopes (for observing the motions of the moons of Jupiter). After two years of trundling and triangulating his way across France, in 1669 he announced a more accurate measure of 110.46 kilometres for one degree of arc. This was a great source of pride for the French but it was predicated on the assumption that the Earth was a perfect sphere – which Newton now said it was not.

To complicate matters, after Picard's death the father and son team of Giovanni and Jacques Cassini repeated Picard's experiments over a larger area and came up with results that suggested that the Earth was fatter not at the equator but at the poles – that Newton, in other words, was exactly wrong. It was this that prompted the Academy of Sciences to dispatch Bouguer and La Condamine to South America to take new measurements.

They chose the Andes because they needed to measure near the equator, to determine if there really was a difference in sphericity there, and because they reasoned that mountains would give them good sightlines. In fact, the mountains of Peru were so constantly lost in cloud that the team often had to wait weeks for an hour's clear surveying. On top of that, they had selected one of the most nearly impossible terrains on Earth. Peruvians refer to their landscape as *muy accidentado* – 'much accidented' – and this it most certainly is. Not only did the French have to scale some of the world's most challenging mountains – mountains that defeated even their mules – but to reach the mountains they had to ford wild rivers, hack their way through jungles, and cross miles of high, stony desert, nearly all of it uncharted and far from any source of supplies. But Bouguer and La Condamine were nothing if not tenacious, and they stuck to the task for nine and a half long, grim, sun-blistered years. Shortly before concluding the project, word reached them that a second French team, taking measurements in northern Scandinavia (and facing notable discomforts of their own, from squelching bogs to dangerous ice floes), had found that a degree was in fact longer near the poles, as Newton had promised. The Earth was 43 kilometres stouter when measured equatorially than when measured from top to bottom around the poles.

Bouguer and La Condamine thus had spent nearly a decade working towards a result they didn't wish to find only to learn now that they weren't even the first to find it. Listlessly, they completed their survey, which confirmed that the first French team was correct. Then, still not speaking, they returned to the coast and took separate ships home.

Something else conjectured by Newton in the *Principia* was that a plumb line hung near a mountain would incline very slightly towards the mountain, affected by the mountain's gravitational mass as well as by the Earth's. This was more than a curious fact. If you measured the deflection accurately and worked out the mass of the mountain, you could calculate the universal gravitational constant – that is, the basic value of gravity, known as *G* – and along with it the mass of the Earth.

Bouguer and La Condamine had tried this on Peru's Mount Chimborazo, but had been defeated by both the technical difficulties and their own squabbling, and so the notion lay dormant for another thirty years until resurrected in England by Nevil Maskelyne, the Astronomer Royal. In Dava Sobel's popular book *Longitude*, Maskelyne is presented as a ninny and villain for failing to appreciate the brilliance of the clockmaker John Harrison, and this may be so; but we are indebted to him in other ways not mentioned in her book, not least for his successful scheme to weigh the Earth.

Maskelyne realized that the nub of the problem lay with finding a mountain of sufficiently regular shape to judge its mass. At his urging, the Royal Society agreed to engage a reliable figure to tour the British Isles to see if such a mountain could be found. Maskelyne knew just such a person – the astronomer and surveyor Charles Mason. Maskelyne and Mason had become friends eleven years earlier while engaged in a project to measure an astronomical event of great importance: the passage of the planet Venus across the face of the Sun. The tireless Edmond Halley had suggested years before that if you measured one of these passages from selected points on the Earth, you could use the principles of triangulation to work out the distance from the Earth to the Sun, and thence to calibrate the distances to all the other bodies in the solar system.

Unfortunately, transits of Venus, as they are known, are an irregular occurrence. They come in pairs eight years apart, but then are absent for a century or more, and there were none in Halley's lifetime.* But the idea simmered and when

*The next transit will be on 8 June 2004, with a second in 2012. There were none in the twentieth century.

the next transit fell due in 1761, nearly two decades after Halley's death, the scientific world was ready – indeed, more ready than it had been for an astronomical event before.

With the instinct for ordeal that characterized the age, scientists set off for more than a hundred locations around the globe – to Siberia, China, South Africa, Indonesia and the woods of Wisconsin, among many others. France dispatched thirty-two observers, Britain eighteen more, and still others set out from Sweden, Russia, Italy, Germany, Ireland and elsewhere.

It was history's first co-operative international scientific venture, and almost everywhere it ran into problems. Many observers were waylaid by war, sickness or shipwreck. Others made their destinations but opened their crates to find equipment broken or warped by tropical heat. Once again the French seemed fated to provide the most memorably unlucky participants. Jean Chappe spent months travelling to Siberia by coach, boat and sleigh, nursing his delicate instruments over every perilous bump, only to find the last vital stretch blocked by swollen rivers, the result of unusually heavy spring rains, which the locals were swift to blame on him after they saw him pointing strange instruments at the sky. Chappe managed to escape with his life, but with no useful measurements.

Unluckier still was Guillaume le Gentil, whose experiences are wonderfully summarized by Timothy Ferris in *Coming of Age in the Milky Way*. Le Gentil set off from France a year ahead of time to observe the transit from India, but various setbacks left him still at sea on the day of the transit – just about the worst place to be, since steady measurements were impossible on a pitching ship.

Undaunted, Le Gentil continued on to India to await the next transit in 1769. With eight years to prepare, he erected a first-rate viewing station, tested and retested his instruments and had everything in a state of perfect readiness. On the morning of the second transit, 4 June 1769, he awoke to a fine day; but, just as Venus began its pass, a cloud slid in front of the Sun and remained there for almost exactly the duration of the transit of three hours, fourteen minutes and seven seconds.

Stoically, Le Gentil packed up his instruments and set off for the nearest port, but en route he contracted dysentery and was laid up for nearly a year. Still weakened, he finally made it onto a ship. It was nearly wrecked in a hurricane off the African coast. When at last he reached home, eleven and a half years after setting off, and having achieved nothing, he discovered that his relatives had had him declared dead in his absence and had enthusiastically plundered his estate.

In comparison, the disappointments experienced by Britain's eighteen scattered observers were mild. Mason found himself paired with a young surveyor named Jeremiah Dixon and apparently they got along well, for they formed a lasting partnership. Their instructions were to travel to Sumatra and chart the transit there, but after just one night at sea their ship was attacked by a French frigate. (Although scientists were in an internationally co-operative mood, nations weren't.) Mason and Dixon sent a note to the Royal Society observing that it seemed awfully dangerous on the high seas and wondering if perhaps the whole thing oughtn't to be called off. In reply they received a swift and chilly rebuke, noting that they had already been paid, that the nation and scientific community were counting on them, and that their failure to proceed would result in the irretrievable loss of their reputations. Chastened, they sailed on, but en route word reached them that Sumatra had fallen to the French and so they observed the transit inconclusively from the Cape of Good Hope. On the way home they stopped on the lonely Atlantic outcrop of St Helena, where they met Maskelyne, whose observations had been thwarted by cloud cover. Mason and Maskelyne formed a solid friendship and spent several happy, and possibly even mildly useful, weeks charting tidal flows.

Soon afterwards Maskelyne returned to England, where he became Astronomer Royal, and Mason and Dixon – now evidently more seasoned – set off for four long and often perilous years surveying their way through 244 miles of dangerous American wilderness to settle a boundary dispute between the estates of William Penn and Lord Baltimore and their respective colonies of Pennsylvania and Maryland. The result was the famous Mason–Dixon line, which later took on symbolic importance as the dividing line between the slave and free states. (Although the line was their principal task, they also contributed several astronomical surveys, including one of the century's most accurate measurements of a degree of meridian – an achievement that brought them far more acclaim in England than the settling of a boundary dispute between spoiled aristocrats.)

Back in Europe, Maskelyne and his counterparts in Germany and France were forced to the conclusion that the transit measurements of 1761 were essentially a failure. One of the problems, ironically, was that there were too many observations, which when brought together often proved contradictory and impossible to resolve. The successful charting of a Venusian transit fell instead to a little-known Yorkshire-born sea captain named James Cook, who watched the 1769 transit from a sunny hilltop in Tahiti, and then went on to chart and claim Australia for the British crown. Upon his return there was now enough information for the French astronomer Joseph Lalande to calculate that the mean distance from the Earth to the Sun was a little over 150 million kilometres. (Two further transits in the nineteenth century allowed astronomers to put the figure at 149.59 million kilometres, where it has remained ever since. The precise distance, we now know, is 149.597870691 million kilometres.) The Earth at last had a position in space.

* * *

As for Mason and Dixon, they returned to England as scientific heroes and, for reasons unknown, dissolved their partnership. Considering the frequency with which they turn up at seminal events in eighteenth-century science, remarkably little is known about either man. No likenesses exist and few written references. Of Dixon, the *Dictionary of National Biography* notes intriguingly that he was 'said to have been born in a coal mine', but then leaves it to the reader's imagination to supply a plausible explanatory circumstance, and adds that he died at Durham in 1777. Apart from his name and long association with Mason, nothing more is known.

Mason is only slightly less shadowy. We know that in 1772, at Maskelyne's behest, he accepted the commission to find a suitable mountain for the gravitational deflection experiment, at length reporting back that the mountain they needed was in the central Scottish Highlands, just above Loch Tay, and was called Schiehallion. Nothing, however, would induce him to spend a summer surveying it. He never returned to the field again. His next known movement was in 1786 when, abruptly and mysteriously, he turned up in Philadelphia with his wife and eight children, apparently on the verge of destitution. He had not been back to America since completing his survey there eighteen years earlier and had no known reason for being there, nor any friends or patrons to greet him. A few weeks later he was dead.

With Mason refusing to survey the mountain, the job fell to Maskelyne. So, for four months in the summer of 1774, Maskelyne lived in a tent in a remote Scottish glen and spent his days directing a team of surveyors, who took hundreds of measurements from every possible position. To find the mass of the mountain from all these numbers required a great deal of tedious calculating, for which a mathematician named Charles Hutton was engaged. The surveyors had covered a map with scores of figures, each marking an elevation at some point on or around the mountain. It was essentially just a confusing mass of numbers, but Hutton noticed that if he used a pencil to connect points of equal height, it all became much more orderly. Indeed, one could instantly get a sense of the overall shape and slope of the mountain. He had invented contour lines.

Extrapolating from his Schiehallion measurements, Hutton calculated the mass of the Earth at 5,000 million million tons, from which could reasonably be deduced the masses of all the other major bodies in the solar system, including the Sun. So from this one experiment we learned the masses of the Earth, the Sun, the Moon, the other planets and *their* moons, and got contour lines into the bargain – not bad for a summer's work.

Not everyone was satisfied with the results, however. The shortcoming of the Schiehallion experiment was that it was not possible to get a truly accurate figure without knowing the actual density of the mountain. For convenience, Hutton had assumed that the mountain had the same density as ordinary stone, about 2.5 times that of water, but this was little more than an educated guess.

One improbable-seeming person who turned his mind to the matter was a country parson named John Michell, who resided in the lonely Yorkshire village of Thornhill. Despite his remote and comparatively humble situation, Michell was one of the great scientific thinkers of the eighteenth century and much esteemed for it.

Among a great deal else, he perceived the wavelike nature of earthquakes, conducted much original research into magnetism and gravity, and, quite extraordinarily, envisioned the possibility of black holes two hundred years before anyone else – a leap that not even Newton could make. When the German-born musician William Herschel decided his real interest in life was astronomy, it was Michell to whom he turned for instruction in making telescopes, a kindness for which planetary science has been in his debt ever since.*

But of all that Michell accomplished, nothing was more ingenious or had greater impact than a machine he designed and built for measuring the mass of the Earth. Unfortunately, he died before he could conduct the experiments, and both the idea and the necessary equipment were passed on to a brilliant but magnificently retiring London scientist named Henry Cavendish.

Cavendish is a book in himself. Born into a life of sumptuous privilege – his grandfathers were dukes, respectively, of Devonshire and Kent – he was the most gifted English scientist of his age, but also the strangest. He suffered, in the words of one of his few biographers, from shyness to a 'degree bordering on disease'. Any human contact was for him a source of the deepest discomfort.

Once he opened his door to find an Austrian admirer, freshly arrived from Vienna, on the front step. Excitedly, the Austrian began to babble out praise. For a few moments Cavendish received the compliments as if they were blows from a blunt object and then, unable to take any more, fled down the path and out the gate, leaving the front door wide open. It was some hours before he could be coaxed back to the property. Even his housekeeper communicated with him by letter.

Although he did sometimes venture into society – he was particularly devoted to the weekly scientific soirées of the great naturalist Sir Joseph Banks – it was always made clear to the other guests that Cavendish was on no account to be approached or even looked at. Those who sought his views

*In 1781 Herschel became the first person in the modern era to discover a planet. He wanted to call it George, after the British monarch, but was overruled. Instead it became Uranus.

were advised to wander into his vicinity as if by accident and to 'talk as it were into vacancy'. If their remarks were scientifically worthy they might receive a mumbled reply, but more often than not they would hear a peeved squeak (his voice appears to have been high-pitched) and turn to find an actual vacancy and the sight of Cavendish fleeing for a more peaceful corner.

His wealth and solitary inclinations allowed him to turn his house in Clapham into a large laboratory where he could range undisturbed through every corner of the physical sciences – electricity, heat, gravity, gases, anything to do with the composition of matter. The second half of the eighteenth century was a time when people of a scientific bent grew intensely interested in the physical properties of fundamental things – gases and electricity in particular – and began seeing what they could do with them, often with more enthusiasm than sense. In America, Benjamin Franklin famously risked his life by flying a kite in an electrical storm. In France, a chemist named Pilatre de Rozier tested the flammability of hydrogen by gulping a mouthful and blowing across an open flame, proving at a stroke that hydrogen is indeed explosively combustible and that eyebrows are not necessarily a permanent feature of one's face. Cavendish, for his part, conducted experiments in which he subjected himself to graduated jolts of electrical current, diligently noting the increasing levels of agony until he could keep hold of his quill, and sometimes his consciousness, no longer.

In the course of a long life Cavendish made a string of signal discoveries – among much else, he was the first person to isolate hydrogen and the first to combine hydrogen and oxygen to form water – but almost nothing he did was entirely divorced from strangeness. To the continuing exasperation of his fellow scientists, he often alluded in published work to the results of experiments that he had not told anyone about. In his secretiveness he didn't merely resemble Newton, but actively exceeded him. His experiments with electrical conductivity were a century ahead of their time, but unfortunately remained undiscovered until that century had passed. Indeed, the greater part of what he did wasn't known until the late nineteenth century, when the Cambridge physicist James Clerk Maxwell took on the task of editing Cavendish's papers, by which time credit for his discoveries had nearly always been given to others.

Among much else, and without telling anyone, Cavendish discovered or anticipated the law of the conservation of energy, Ohm's Law, Dalton's Law of Partial Pressures, Richter's Law of Reciprocal Proportions, Charles's Law of Gases, and the principles of electrical conductivity. That's just some of it. According to the science historian J. G. Crowther, he also foreshadowed 'the work of Kelvin and G. H. Darwin on the effect of tidal friction on slowing the rotation of the earth, and Larmor's discovery, published in 1915, on the effect of local atmospheric cooling . . . the work of Pickering on freezing mixtures, and some of the work of Rooseboom on heterogeneous equilibria'. Finally, he left clues that led directly to the discovery of the group of elements known as the noble gases, some of which are so elusive that the last of them wasn't found until 1962. But our interest here is in Cavendish's last known experiment when, in the late summer of 1797, at the age of sixty-seven, he turned his attention to the crates of equipment that had been left to him – evidently out of simple scientific respect – by John Michell.

When assembled, Michell's apparatus looked like nothing so much as an eighteenth-century version of a Nautilus weight-training machine. It incorporated weights, counterweights, pendulums, shafts and torsion wires. At the heart of the machine were two 350-pound lead balls, which were suspended beside two smaller spheres. The idea was to measure the gravitational deflection of the smaller spheres by the larger ones, which would allow the first measurement of the elusive force known as the gravitational constant, and from which the weight (strictly speaking the mass)* of the Earth could be deduced.

Because gravity holds planets in orbit and makes falling objects land with a bang, we tend to think of it as a powerful force, but it isn't really. It is only powerful in a kind of collective sense, when one massive object, like the Sun, holds onto another massive object, like the Earth. At an elemental level gravity is extraordinarily unrobust. Each time you pick up a book from a table or a coin from the floor you effortlessly overcome the gravitational exertion of an entire planet. What Cavendish was trying to do was measure gravity at this extremely featherweight level.

Delicacy was the keyword. Not a whisper of disturbance could be allowed into the room containing the apparatus, so Cavendish took up a position in an adjoining room and made his observations with a telescope aimed through a peephole. The work was incredibly exacting, involving seventeen delicate, interconnected measurements, which together took nearly a year to complete. When at last he had finished his calculations, Cavendish announced that the Earth weighed a little over 13,000,000,000,000,000,000,000 pounds, or six billion trillion metric tons, to use the modern measure. (A metric ton, or tonne, is 1,000 kilograms or 2,205 pounds.)

Today, scientists have at their disposal machines so precise they can detect the weight of a single bacterium and so sensitive that readings can be disturbed by someone yawning

*To a physicist, mass and weight are two quite different things. Your mass stays the same wherever you go, but your weight varies depending on how far you are from the centre of some other massive object like a planet. Travel to the Moon and you will be much lighter but no less massive. On Earth, for all practical purposes, mass and weight are the same and so the terms can be treated as synonymous, at least outside the classroom.

seventy-five feet away, but they have not significantly improved on Cavendish's measurements of 1797. The current best estimate for the Earth's weight is 5.9725 billion trillion tonnes, a difference of only about 1 per cent from Cavendish's finding. Interestingly, all of this merely confirmed estimates made by Newton 110 years before Cavendish without any experimental evidence at all.

At all events, by the late eighteenth century scientists knew very precisely the shape and dimensions of the Earth and its distance from the Sun and planets; and now Cavendish, without even leaving home, had given them its weight. So you might think that determining the age of the Earth would be relatively straightforward. After all, the necessary materials were literally at their feet. But no. Human beings would split the atom and invent television, nylon and instant coffee before they could figure out the age of their own planet.

To understand why, we must travel north to Scotland and begin with a brilliant and genial man, of whom few have ever heard, who had just invented a new science called geology.

5

THE STONE-BREAKERS

At just the time that Henry Cavendish was completing his experiments in London, four hundred miles away in Edinburgh another kind of concluding moment was about to take place with the death of James Hutton. This was bad news for Hutton, of course, but good news for science as it cleared the way for a man named John Playfair to rewrite Hutton's work without fear of embarrassment.

Hutton was by all accounts a man of the keenest insights and liveliest conversation, a delight in company, and without rival when it came to understanding the mysterious slow processes that shaped the Earth. Unfortunately, it was beyond him to set down his notions in a form that anyone could begin to understand. He was, as one biographer observed with an all but audible sigh, 'almost entirely innocent of rhetorical accomplishments'. Nearly every line he penned was an invitation to slumber. Here he is in his 1795 masterwork, *A Theory of the Earth with Proofs and Illustrations*, discussing . . . well, something:

> The world which we inhabit is composed of the materials, not of the earth which was the immediate predecessor of the present, but of the earth which, in ascending from the present,

> we consider as the third, and which had preceded the land that was above the surface of the sea, while our present land was yet beneath the water of the ocean.

Yet almost singlehandedly, and quite brilliantly, he created the science of geology and transformed our understanding of the Earth.

Hutton was born in 1726 into a prosperous Scottish family, and enjoyed the sort of material comfort that allowed him to pass much of his life in a genially expansive round of light work and intellectual betterment. He studied medicine, but found it not to his liking and turned instead to farming, which he followed in a relaxed and scientific way on the family estate in Berwickshire. Tiring of field and flock, in 1768 he moved to Edinburgh, where he founded a successful business producing sal ammoniac from coal soot, and busied himself with various scientific pursuits. Edinburgh at that time was a centre of intellectual vigour and Hutton luxuriated in its enriching possibilities. He became a leading member of a society called the Oyster Club, where he passed his evenings in the company of men such as the economist Adam Smith, the chemist Joseph Black and the philosopher David Hume, as well as such occasional visiting sparks as Benjamin Franklin and James Watt.

In the tradition of the day, Hutton took an interest in nearly everything, from mineralogy to metaphysics. He conducted experiments with chemicals, investigated methods of coal mining and canal building, toured salt mines, speculated on the mechanisms of heredity, collected fossils and propounded theories on rain, the composition of air and the laws of motion, among much else. But his particular interest was geology.

Among the questions that attracted interest in that fanatically inquisitive age was one that had puzzled people for a very long time – namely, why ancient clam shells and other marine fossils were so often found on mountaintops. How on earth did they get there? Those who thought they had a solution fell into two opposing camps. One group, known as the Neptunists, were convinced that everything on the Earth, including sea shells in improbably lofty places, could be explained by rising and falling sea levels. They believed that mountains, hills and other features were as old as the Earth itself, and were changed only when water sloshed over them during periods of global flooding.

Opposing them were the Plutonists, who noted that volcanoes and earthquakes, among other enlivening agents, continually changed the face of the planet but clearly owed nothing to wayward seas. The Plutonists also raised awkward questions about where all the water went when it wasn't in flood. If there was enough of it at times to cover the Alps, then where, pray, was it during times of tranquillity, such as now? Their belief was that the Earth was subject to profound internal forces as well as surface ones. However, they couldn't convincingly explain how all those clam shells got up there.

It was while puzzling over these matters that Hutton had a series of exceptional insights. From looking at his own farmland, he could see that soil was created by the erosion of rocks and that particles of this soil were continually washed away and carried off by streams and rivers and re-deposited elsewhere. He realized that if such a process were carried to its natural conclusion then the Earth would eventually be worn quite smooth. Yet everywhere around him there were hills. Clearly there had to be some additional process, some form of renewal and uplift, that created new hills and mountains to keep the cycle going. The marine fossils on mountaintops, he decided, had not been deposited during floods, but had risen along with the mountains themselves. He also deduced that it was heat within the Earth that created new rocks and continents and thrust up mountain chains. It is not too much to say that geologists wouldn't grasp the full implications of this thought until two hundred years later, when finally they adopted the concept of plate tectonics. Above all, what Hutton's theories suggested was that the processes that shaped the Earth required huge amounts of time, far more than anyone had ever dreamed. There were enough insights here to transform utterly our understanding of the planet.

In 1785 Hutton worked his ideas up into a long paper, which was read at consecutive meetings of the Royal Society of Edinburgh. It attracted almost no notice at all. It's not hard to see why. Here, in part, is how he presented it to his audience:

> In the one case, the forming cause is in the body which is separated; for, after the body has been actuated by heat, it is by the reaction of the proper matter of the body, that the chasm which constitutes the vein is formed. In the other case, again, the cause is extrinsic in relation to the body in which the chasm is formed. There has been the most violent fracture and divulsion; but the cause is still to seek; and it appears not in the vein; for it is not every fracture and dislocation of the solid body of our earth, in which minerals, or the proper substances of mineral veins, are found.

Needless to say, almost no-one in the audience had the faintest idea what he was talking about. Encouraged by his friends to expand his theory, in the touching hope that he might somehow stumble onto clarity in a more expansive format, Hutton spent the next ten years preparing his magnum opus, which was published in two volumes in 1795.

Together the two books ran to nearly a thousand pages and were, remarkably, worse than even his most pessimistic friends had feared. Apart from anything else, nearly half the completed work now consisted of quotations from French sources, still in the original French. A third volume was so unenticing that it wasn't published until 1899, more than a century after Hutton's death, and the fourth and concluding volume was never published at all. Hutton's *Theory of the Earth* is a strong candidate for the least read important book in science (or at least, it would be if there weren't so many others). Even Charles Lyell, the greatest geologist of the following century and a man who read everything, admitted he couldn't get through it.

Luckily, Hutton had a Boswell in the form of John Playfair, a professor of mathematics at the University of Edinburgh and a close friend, who not only could write silken prose but – thanks to many years at Hutton's elbow – actually understood what Hutton was trying to say, most of the time. In 1802, five years after Hutton's death, Playfair produced a simplified exposition of the Huttonian principles, entitled *Illustrations of the Huttonian Theory of the Earth*. The book was gratefully received by those who took an active interest in geology, which in 1802 was not a large number. That, however, was about to change. And how.

In the winter of 1807, thirteen like-minded souls in London got together at the Freemasons Tavern at Long Acre, in Covent Garden, to form a dining club to be called the Geological Society. The idea was to meet once a month to swap geological notions over a glass or two of Madeira and a convivial dinner. The price of the meal was set at a deliberately hefty 15 shillings to discourage those whose qualifications were merely cerebral. It soon became apparent, however, that there was a demand for something more properly institutional, with a permanent headquarters, where people could gather to share and discuss new findings. In barely a decade membership grew to 400 – still all gentlemen, of course – and the Geological was threatening to eclipse the Royal as the premier scientific society in the country.

The members met twice a month from November until June, when virtually all of them went off to spend the summer doing fieldwork. These weren't people with a pecuniary interest in minerals, you understand, or even academics for the most part, but simply gentlemen with the wealth and time to indulge a hobby at a more or less professional level. By 1830 there were 745 of them, and the world would never see the like again.

It is hard to imagine now, but geology excited the nineteenth century – positively gripped it – in a way that no science ever had before or would again. In 1839, when Roderick Murchison published *The Silurian System*, a plump and ponderous study of a type of rock called greywacke, it was an instant bestseller, racing through four editions, even though it cost 8 guineas a copy and was, in true Huttonian style, unreadable. (As even a Murchison supporter conceded, it had 'a total want of literary attractiveness'.) And when, in 1841, the great Charles Lyell travelled to America to give a series of lectures in Boston, sellout audiences of three thousand at a time packed into the Lowell Institute to hear his tranquillizing descriptions of marine zeolites and seismic perturbations in Campania.

Throughout the modern, thinking world, but especially in Britain, men of learning ventured into the countryside to do a little 'stone-breaking', as they called it. It was a pursuit taken seriously and they tended to dress with appropriate gravity, in top hats and dark suits, except for the Reverend William Buckland of Oxford, whose habit it was to do his fieldwork in an academic gown.

The field attracted many extraordinary figures, not least the aforementioned Murchison, who spent the first thirty or so years of his life galloping after foxes, converting aeronautically challenged birds into puffs of drifting feathers with buckshot and showing no mental agility whatever beyond that needed to read *The Times* or play a hand of cards. Then he discovered an interest in rocks and became with rather astounding swiftness a titan of geological thinking.

Then there was Dr James Parkinson, who was also an early socialist and author of many provocative pamphlets with titles like 'Revolution without Bloodshed'. In 1794 he was implicated in a faintly lunatic-sounding conspiracy called 'the Pop-gun Plot', in which it was planned to shoot King George III in the neck with a poisoned dart as he sat in his box at the theatre. Parkinson was hauled before the Privy Council for questioning and came within an ace of being dispatched in irons to Australia before the charges against him were quietly dropped. Adopting a more conservative approach to life, he developed an interest in geology and became one of the founding members of the Geological Society and the author of an important geological text, *Organic Remains of a Former World*, which remained in print for half a century. He never caused trouble again. Today, however, we remember him for his landmark study of the affliction then called the 'shaking palsy', but known ever since as Parkinson's disease. (Parkinson had one other slight claim to fame. In 1785 he became possibly the only person in history to win a natural history museum in a raffle. The museum, in London's Leicester Square, had been founded by Sir Ashton Lever, who had driven himself bankrupt with his unrestrained collecting of natural wonders. Parkinson kept the museum until 1805, when he could no longer support it and the collection was broken up and sold.)

Not quite as remarkable in character but more influential than all the others combined was Charles Lyell. Lyell was born in the year that Hutton died and only 70 miles away, in the village of Kinnordy. Though Scottish by birth, he grew up in the far south of England, in the New Forest of Hampshire, because his mother was convinced that Scots were feckless drunks. As was generally the pattern with nineteenth-century gentlemen scientists, Lyell came from a background of comfortable wealth and intellectual vigour. His father, also named Charles, had the unusual distinction of being a leading authority on the poet Dante and on mosses. (*Orthotricium lyelli*, which most visitors to the English countryside will at some time have sat on, is named for him.) From his father Lyell gained an interest in natural history, but it was at Oxford, where he fell under the spell of the Reverend William Buckland – he of the flowing gowns – that the young Lyell began his lifelong devotion to geology.

Buckland was a bit of a charming oddity. He had some real achievements, but he is remembered at least as much for his eccentricities. He was particularly noted for a menagerie of wild animals, some large and dangerous, that were allowed to wander through his house and garden, and for his desire to eat his way through every animal in creation. Depending on whim and availability, guests to Buckland's house might be served baked guinea pig, mice in batter, roasted hedgehog or boiled southeast Asian sea slug. Buckland was able to find merit in them all, except the common garden mole, which he declared disgusting. Almost inevitably, he became the leading authority on coprolites – fossilized faeces – and had a table made entirely out of his collection of specimens.

Even when conducting serious science his manner was generally singular. Once Mrs Buckland found herself being shaken awake in the middle of the night, her husband crying in excitement: 'My dear, I believe that *Cheirotherium*'s footsteps are undoubtedly testudinal.' Together they hurried to the kitchen in their nightclothes. Mrs Buckland made a flour paste, which she spread across the table, while the Reverend Buckland fetched the family tortoise. Plunking it onto the paste, they goaded it forward and discovered to their delight that its footprints did indeed match those of the fossil Buckland had been studying. Charles Darwin thought Buckland a buffoon – that was the word he used – but Lyell appeared to find him inspiring and liked him well enough to go touring with him in Scotland in 1824. It was soon after this trip that Lyell decided to abandon a career in law and devote himself to geology full-time.

Lyell was extremely short-sighted and went through most of his life with a pained squint, which gave him a troubled air. (Eventually he would lose his sight altogether.) His other slight peculiarity was the habit, when distracted by thought, of taking up improbable positions on furniture – lying across two chairs at once or 'resting his head on the seat of a chair, while standing up' (to quote his friend Darwin). Often when lost in thought he would slink so low in a chair that his buttocks would all but touch the floor. Lyell's only real job in life was as professor of geology at King's College in London from 1831 to 1833. It was just at this time that he produced *The Principles of Geology*, published in three volumes between 1830 and 1833, which in many ways consolidated and elaborated upon the thoughts first voiced by Hutton a generation earlier. (Although Lyell never read Hutton in the original, he was a keen student of Playfair's reworked version.)

Between Hutton's day and Lyell's there arose a new geological controversy, which largely superseded, but is often confused with, the old Neptunian–Plutonian dispute. The new battle became an argument between catastrophism and uniformitarianism – unattractive terms for an important and very long-running dispute. Catastrophists, as you might expect from the name, believed that the Earth was shaped by abrupt cataclysmic events – floods, principally, which is why catastrophism and Neptunism are often wrongly bundled together. Catastrophism was particularly comforting to clerics like Buckland because it allowed them to incorporate the biblical flood of Noah into serious scientific discussions.

Uniformitarians, by contrast, believed that changes on Earth were gradual and that nearly all earth processes happened slowly, over immense spans of time. Hutton was much more the father of the notion than Lyell, but it was Lyell most people read, and so he became in most people's minds, then and now, the father of modern geological thought.

Lyell believed that the Earth's shifts were uniform and steady – that everything that had ever happened in the past could be explained by events still going on today. Lyell and his adherents didn't just disdain catastrophism, they detested it. Catastrophists believed that extinctions were part of a series in which animals were repeatedly wiped out and replaced with new sets – a belief that the naturalist T. H. Huxley mockingly likened to 'a succession of rubbers of whist, at the end of which the players upset the table and called for a new pack'. It was too convenient a way to explain the unknown. 'Never was there a dogma more calculated to foster indolence, and to blunt the keen edge of curiosity', sniffed Lyell.

Lyell's oversights were not inconsiderable. He failed to explain convincingly how mountain ranges were formed and overlooked glaciers as an agent of change. He refused to accept Agassiz's idea of ice ages – 'the refrigeration of the globe', as he dismissively termed it – and was confident that mammals 'would be found in the oldest fossiliferous beds'. He rejected the notion that animals and plants suffered sudden annihilations, and believed that all the principal animal groups – mammals, reptiles, fish and so on – had co-existed since the dawn of time. On all of these he would ultimately be proved wrong.

Yet it would be nearly impossible to overstate Lyell's influence. *The Principles of Geology* went through twelve editions in his lifetime and contained notions that shaped geological thinking far into the twentieth century. Darwin took a first edition with him on the *Beagle* voyage and wrote afterwards that 'the great merit of the *Principles* was that it altered the whole tone of one's mind, and therefore that, when seeing a thing never seen by Lyell, one yet saw it partially through his eyes.' In short, he thought him nearly a god, as did many of his generation. It is a testament to the strength of Lyell's sway that in the 1980s, when geologists had to abandon just a part of his theory to accommodate the impact theory of extinctions, it nearly killed them. But that is another chapter.

Meanwhile, geology had a great deal of sorting out to do, and not all of it went smoothly. From the outset geologists tried to categorize rocks by the periods in which they were laid down, but there were often bitter disagreements about where to put the dividing lines – none more so than a long-running debate that became known as the Great Devonian Controversy. The issue arose when the Reverend Adam Sedgwick of Cambridge claimed for the Cambrian period a layer of rock that Roderick Murchison believed belonged rightly to the Silurian. The dispute raged for years and grew extremely heated. 'De la Beche is a dirty dog,' Murchison wrote to a friend in a typical outburst.

Some sense of the strength of feeling can be gained by glancing through the chapter titles of Martin J. S. Rudwick's excellent and sombre account of the issue, *The Great Devonian Controversy*. These begin innocuously enough with headings such as 'Arenas of Gentlemanly Debate' and 'Unraveling the Greywacke', but then proceed on to 'The Greywacke Defended and Attacked', 'Reproofs and Recriminations', 'The Spread of Ugly Rumours', 'Weaver Recants his Heresy', 'Putting a Provincial in his Place' and (in case there was any doubt that this was war) 'Murchison Opens the Rhineland Campaign'. The fight was finally settled in 1879 with the simple expedient of coming up with a new period, the Ordovician, to be inserted between the Cambrian and Silurian.

Because the British were the most active in the early years of the discipline, British names are predominant in the geological lexicon. *Devonian* is of course from the English county of Devon. *Cambrian* comes from the Roman name for Wales, while *Ordovician* and *Silurian* recall ancient Welsh tribes, the Ordovices and Silures. But with the rise of geological prospecting elsewhere, names began to creep in from all over. *Jurassic* refers to the Jura Mountains on the border of France and Switzerland. *Permian* recalls the former Russian province of Perm in the Ural Mountains. For *Cretaceous* (from the Latin for chalk) we are indebted to a Belgian geologist with the perky name of J. J. d'Omalius d'Halloy.

Originally, geological history was divided into four spans of time: primary, secondary, tertiary and quaternary. The system was too neat to last, and soon geologists were contributing additional divisions while eliminating others. Primary and secondary fell out of use altogether, while quaternary was discarded by some but kept by others. Today only tertiary remains as a common designation everywhere, even though it no longer represents a third period of anything.

Lyell, in his *Principles*, introduced additional units known as epochs or series to cover the period since the age of the dinosaurs, among them Pleistocene ('most recent'), Pliocene ('more recent'), Miocene ('moderately recent') and the rather endearingly vague Oligocene ('but a little recent'). Lyell originally intended to employ '-synchronous' for his endings, giving us such crunchy designations as Meiosynchronous and Pleiosynchronous. The Reverend William Whewell, an influential man, objected on etymological grounds and suggested instead an '-eous' pattern, producing Meioneous, Pleioneous and so on. The '-cene' terminations were thus something of a compromise.

Nowadays, and speaking very generally, geological time is

divided first into four great chunks known as eras: Precambrian, Palaeozoic (from the Greek meaning 'old life'), Mesozoic ('middle life') and Cenozoic ('recent life'). These four eras are further divided into anywhere from a dozen to twenty subgroups, usually called periods though sometimes known as systems. Most of these are also reasonably well known: Cretaceous, Jurassic, Triassic, Silurian and so on.*

Then come Lyell's epochs – the Pleistocene, Miocene, and so on – which apply only to the most recent (but palaeontologically busy) 65 million years; and finally we have a mass of finer subdivisions known as stages or ages. Most of these are named, nearly always awkwardly, after places: *Illinoian*, *Desmoinesian*, *Croixian*, *Kimmeridgian* and so on in like vein. Altogether, according to John McPhee, these number in the 'tens of dozens'. Fortunately, unless you take up geology as a career, you are unlikely ever to hear any of them again.

Further confusing the matter is that the stages or ages in North America have different names from the stages in Europe and often only roughly intersect with them in time. Thus the North American Cincinnatian stage mostly corresponds with the Ashgillian stage in Europe, plus a tiny bit of the slightly earlier Caradocian stage.

Also, all this changes from textbook to textbook and from person to person, so that some authorities describe seven recent epochs, while others are content with four. In some books, too, you will find the tertiary and quaternary taken out and replaced by periods of different lengths called the Palaeogene and Neogene. Others divide the Precambrian

*There will be no testing here, but if you are ever required to memorize them you might wish to remember John Wilford's helpful advice to think of the eras (Precambrian, Palaeozoic, Mesozoic and Cenozoic) as seasons in a year and the periods (Permian, Triassic, Jurassic, etc.) as the months.

into two eras, the very ancient Archaean and the more recent Proterozoic. Sometimes, too, you will see the term Phanerozoic used to describe the span encompassing the Cenozoic, Mesozoic and Palaeozoic eras.

Moreover, all this applies only to units of *time*. Rocks are divided into quite separate units known as systems, series and stages. A distinction is also made between late and early (referring to time) and upper and lower (referring to layers of rock). It can all get terribly confusing to non-specialists, but to a geologist these can be matters of passion. 'I have seen grown men glow incandescent with rage over this metaphorical millisecond in life's history,' the British palaeontologist Richard Fortey has written with regard to a long-running twentieth-century dispute over where the boundary lies between the Cambrian and Ordovician.

At least today we can bring some sophisticated dating techniques to the table. For most of the nineteenth century, geologists could draw on nothing more than the most hopeful guesswork. The frustrating position then was that although they could place the various rocks and fossils in order by age, they had no idea how long any of those ages was. When Buckland speculated on the antiquity of an ichthyosaurus skeleton he could do no better than suggest that it had lived somewhere between 'ten thousand [and] more than ten thousand times ten thousand' years earlier.

Although there was no reliable way of dating periods, there was no shortage of people willing to try. The most well known early attempt was made in 1650, when Archbishop James Ussher of the Church of Ireland made a careful study of the Bible and other historical sources and concluded, in a hefty tome called *Annals of the Old Testament*, that the Earth had been created at midday on 23 October 4004 BC, an assertion that has amused historians and textbook writers ever since.*

There is a persistent myth, incidentally – and one propounded in many serious books – that Ussher's views dominated scientific beliefs well into the nineteenth century, and that it was Lyell who put everyone straight. Stephen Jay Gould in *Time's Arrow* cites as a typical example this sentence from a popular book of the 1980s: 'Until Lyell published his book, most thinking people accepted the idea that the earth was young.' In fact, no. As Martin J. S. Rudwick puts it, 'No geologist of any nationality whose work was taken seriously by other geologists advocated a timescale confined within the limits of a literalistic exegesis of Genesis.' Even the Reverend Buckland, as pious a soul as the nineteenth century produced, noted that nowhere did the Bible suggest that God made Heaven and Earth on the first day, but merely 'in the beginning'. That beginning, he reasoned, may have lasted 'millions upon millions of years'. Everyone agreed that the Earth was ancient. The question was simply: how ancient?

One of the better early ideas at dating the planet came from the ever-reliable Edmond Halley, who in 1715 suggested that if you divided the total amount of salt in the world's seas by the amount added each year, you would get the number of years that the oceans had been in existence, which would give you a rough idea of Earth's age. The logic was appealing, but unfortunately no-one knew how much salt was in the sea or by how much it increased each year, which rendered the experiment impracticable.

The first attempt at measurement that could be called

*Although virtually all books find a space for him, there is a striking variability in the details associated with Ussher. Some books say he made his pronouncement in 1650, others in 1654, still others in 1664. Many cite the date of the Earth's reputed beginning as 26 October. At least one book of note spells his name 'Usher'. The matter is interestingly surveyed in Stephen Jay Gould's *Eight Little Piggies*.

remotely scientific was made by the Frenchman Georges-Louis Leclerc, Comte de Buffon, in the 1770s. It had long been known that the Earth radiated appreciable amounts of heat – that was apparent to anyone who went down a coal mine – but there wasn't any way of estimating the rate of dissipation. Buffon's experiment consisted of heating spheres until they glowed white-hot and then estimating the rate of heat loss by touching them (presumably very lightly at first) as they cooled. From this he guessed the Earth's age to be somewhere between 75,000 and 168,000 years old. This was of course a wild underestimate; but it was a radical notion nonetheless, and Buffon found himself threatened with excommunication for expressing it. A practical man, he apologized at once for his thoughtless heresy, then cheerfully repeated the assertions throughout his subsequent writings.

By the middle of the nineteenth century most learned people thought the Earth was at least a few million years old, perhaps even some tens of millions years old, but probably not more than that. So it came as a surprise when in 1859, in *On the Origin of Species*, Charles Darwin announced that the geological processes that created the Weald, an area of southern England stretching across Kent, Surrey and Sussex, had taken, by his calculations, 306,662,400 years to complete. The assertion was remarkable partly for being so arrestingly specific but even more for flying in the face of accepted wisdom about the age of the Earth.* It proved so contentious that Darwin withdrew it from the third edition of the book. The problem at its heart remained, however. Darwin and his geological friends needed the Earth to be old, but no-one could come up with a way to make it so.

★ ★ ★

*Darwin loved an exact number. In a later work, he announced that the number of worms to be found in an average acre of English country soil was 53,767.

Unfortunately for Darwin, and for progress, the question came to the attention of the great Lord Kelvin (who, though indubitably great, was then still just plain William Thomson; he wouldn't be elevated to the peerage until 1892, when he was sixty-eight years old and nearing the end of his career, but I shall follow the convention here of using the name retroactively). Kelvin was one of the most extraordinary figures of the nineteenth century – indeed, of any century. The German scientist Hermann von Helmholtz, no intellectual slouch himself, wrote that Kelvin had by far the greatest 'intelligence and lucidity, and mobility of thought' of any man he had ever met. 'I felt quite wooden beside him sometimes,' he added, a bit dejectedly.

The sentiment is understandable, for Kelvin really was a kind of Victorian superman. He was born in 1824 in Belfast, the son of a professor of mathematics at the Royal Academical Institution who soon afterwards transferred to Glasgow. There Kelvin proved himself such a prodigy that he was admitted to Glasgow University at the exceedingly tender age of ten. By the time he had reached his early twenties, he had studied at institutions in London and Paris, graduated from Cambridge (where he won the university's top prizes for rowing and mathematics, and somehow found time to launch a musical society as well), been elected a fellow of Peterhouse, and written (in French and English) a dozen papers in pure and applied mathematics of such dazzling originality that he had to publish them anonymously for fear of embarrassing his superiors. At the age of twenty-two he returned to Glasgow to take up a professorship in natural philosophy, a position he would hold for the next fifty-three years.

In the course of a long career (he lived to 1907 and the age of eighty-three), he wrote 661 papers, accumulated 69 patents (from which he grew abundantly wealthy) and gained renown in nearly every branch of the physical sciences. Among much else, he suggested the method that led directly to the invention of refrigeration, devised the scale of absolute temperature that still bears his name, invented the boosting devices that allowed telegrams to be sent across oceans, and made innumerable improvements to shipping and navigation, from the invention of a popular marine compass to the creation of the first depth sounder. And those were merely his practical achievements.

His theoretical work, in electromagnetism, thermodynamics and the wave theory of light, was equally revolutionary.* He had really only one flaw and that was an inability to calculate the correct age of the Earth. The question occupied much of the second half of his career, but he never came anywhere near getting it right. His first effort, in 1862 for an article in a popular magazine called *Macmillan's*, suggested that the Earth was 98 million years old, but cautiously allowed that the figure could be as low as 20 million years or as high as 400 million. With remarkable prudence he acknowledged that his calculations could be wrong if 'sources now unknown to us are prepared in the great storehouse of creation' – but it was clear that he thought that unlikely.

With the passage of time Kelvin would become more forthright in his assertions and less correct. He continually

---

*In particular, he elaborated the Second Law of Thermodynamics. A discussion of these laws would be a book in itself, but I offer here this crisp summation by the chemist P. W. Atkins, just to provide a sense of them: 'There are four Laws. The third of them, the Second Law, was recognized first; the first, the Zeroth Law, was formulated last; the First Law was second; the Third Law might not even be a law in the same sense as the others.' In briefest terms, the second law states that a little energy is always wasted. You can't have a perpetual motion device because no matter how efficient, it will always lose energy and eventually run down. The first law says that you can't create energy and the third that you can't reduce temperatures to absolute zero; there will always be some residual warmth. As Dennis Overbye notes, the three principal laws are sometimes expressed jocularly as (1) you can't win, (2) you can't break even, and (3) you can't get out of the game.

revised his estimates downwards, from a maximum of 400 million years to 100 million years, then to 50 million years and finally, in 1897, to a mere 24 million years. Kelvin wasn't being wilful. It was simply that there was nothing in physics that could explain how a body the size of the Sun could burn continuously for more than a few tens of millions of years at most without exhausting its fuel. Therefore it followed that the Sun and its planets were relatively, but inescapably, youthful.

The problem was that nearly all the fossil evidence contradicted this, and suddenly in the nineteenth century there was a *lot* of fossil evidence.

6

SCIENCE RED IN TOOTH AND CLAW

In 1787, someone in New Jersey – exactly who now seems to be forgotten – found an enormous thigh bone sticking out of a stream bank at a place called Woodbury Creek. The bone clearly didn't belong to any species of creature still alive, certainly not in New Jersey. From what little is known now, it is thought to have belonged to a hadrosaur, a large duckbilled dinosaur. At the time, dinosaurs were unknown.

The bone was sent to Dr Caspar Wistar, the nation's leading anatomist, who described it at a meeting of the American Philosophical Society in Philadelphia that autumn. Unfortunately, Wistar failed completely to recognize the bone's significance and merely made a few cautious and uninspired remarks to the effect that it was indeed a whopper. He thus missed the chance, half a century ahead of anyone else, to be the discoverer of dinosaurs. Indeed, the bone excited so little interest that it was put in a storeroom and eventually disappeared altogether. So the first dinosaur bone ever found was also the first to be lost.

That the bone didn't attract greater interest is more than a little puzzling for its appearance came at a time when America was in a froth of excitement about the remains of large, ancient animals. The cause of this froth was a strange assertion by the great French naturalist the Comte de Buffon – he of the heated spheres from the previous chapter – that living things in the New World were inferior in nearly every way to those of the Old World. America, Buffon wrote in his vast and much-esteemed *Histoire naturelle*, was a land where the water was stagnant, the soil unproductive, and the animals without size or vigour, their constitutions weakened by the 'noxious vapours' that rose from its rotting swamps and sunless forests. In such an environment even the native Indians lacked virility. 'They have no beard or body hair,' Buffon sagely confided, 'and no ardour for the female.' Their reproductive organs were 'small and feeble'.

Buffon's observations found surprisingly eager support among other writers, especially those whose conclusions were not complicated by actual familiarity with the country. A Dutchman named Corneille de Pauw announced in a popular work called *Recherches philosophiques sur les américains* that native American males were not only reproductively unimposing, but 'so lacking in virility that they had milk in their breasts'. Such views enjoyed an improbable durability and could be found repeated or echoed in European texts until near the end of the nineteenth century.

Not surprisingly, such aspersions were indignantly met in America. Thomas Jefferson incorporated a furious (and, unless the context is understood, quite bewildering) rebuttal in his *Notes on the State of Virginia*, and induced his New Hampshire friend General John Sullivan to send twenty soldiers into the northern woods to find a bull moose to present to Buffon as proof of the stature and majesty of American quadrupeds. It took the men two weeks to track down a suitable subject. The moose, when shot, unfortunately lacked the imposing horns that Jefferson had specified, but Sullivan thoughtfully included a rack of antlers from an elk or stag with the suggestion that these be attached instead. Who in France, after all, would know?

Meanwhile, in Philadelphia – Wistar's city – naturalists had begun to assemble the bones of a giant elephant-like creature known at first as 'the great American incognitum' but later identified, not quite correctly, as a mammoth. The first of these bones had been discovered at a place called Big Bone Lick in Kentucky, but soon others were turning up all over. America, it appeared, had once been the home of a truly substantial creature – one that would surely disprove Buffon's foolish Gallic contentions.

In their keenness to demonstrate the incognitum's bulk and ferocity, the American naturalists appear to have become slightly carried away. They overestimated its size by a factor of six and gave it frightening claws, which in fact came from a *Megalonyx*, or giant ground sloth, found nearby. Rather remarkably, they persuaded themselves that the animal had enjoyed 'the agility and ferocity of the tiger', and portrayed it in illustrations as pouncing with feline grace onto prey from boulders. When tusks were discovered, they were forced into the animal's head in any number of inventive ways. One restorer screwed the tusks in upside down, like the fangs of a sabre-toothed cat, which gave it a satisfyingly aggressive aspect. Another arranged the tusks so that they curved backwards on the engaging theory that the creature had been aquatic and had used them to anchor itself to trees while dozing. The most pertinent consideration about the incognitum, however, was that it appeared to be extinct – a fact that Buffon cheerfully seized upon as proof of its incontestably degenerate nature.

Buffon died in 1788, but the controversy rolled on. In 1795 a selection of bones made their way to Paris, where they were examined by the rising star of palaeontology, the youthful and aristocratic Georges Cuvier. Cuvier was already dazzling people with his genius for taking heaps of disarticulated bones and whipping them into shapely forms. It was said that he could describe the look and nature of an animal from a single tooth or scrap of jaw, and often name the species and genus into the bargain. Realizing that no-one in America had thought to write a formal description of the lumbering beast, Cuvier did so, and thus became its official discoverer. He called it a *mastodon* (which means, a touch unexpectedly, 'nipple-teeth').

Inspired by the controversy, in 1796 Cuvier wrote a landmark paper, *Note on the Species of Living and Fossil Elephants*, in which he put forward for the first time a formal theory of extinctions. His belief was that from time to time the Earth experienced global catastrophes in which groups of creatures were wiped out. For religious people, including Cuvier himself, the idea raised uncomfortable implications since it suggested an unaccountable casualness on the part of Providence. To what end would God create species only to wipe them out later? The notion was contrary to the belief in the Great Chain of Being, which held that the world was carefully ordered and that every living thing within it had a place and purpose, and always had and always would. Jefferson for one couldn't abide the thought that whole species would ever be permitted to vanish (or, come to that, to evolve). So when it was put to him that there might be scientific and political value in sending a party to explore the interior of America beyond the Mississippi he leaped at the idea, hoping the intrepid adventurers would find herds of healthy mastodons and other outsized creatures grazing on the bounteous plains. Jefferson's personal secretary and trusted friend Meriwether Lewis was chosen co-leader, with William Clark, and chief naturalist for the expedition. The person selected to advise him on what to look out for with regard to animals living and deceased was none other than Caspar Wistar.

In the same year – in fact, the same month – that the aristocratic and celebrated Cuvier was propounding his extinction theories in Paris, on the other side of the English

Channel a rather more obscure Englishman was having an insight into the value of fossils that would also have lasting ramifications. William Smith was a young supervisor of construction on the Somerset Coal Canal. On the evening of 5 January 1796, he was sitting in a coaching inn in Somerset when he jotted down the notion that would eventually make his reputation. To interpret rocks, there needs to be some means of correlation, a basis on which you can tell that those carboniferous rocks from Devon are younger than these Cambrian rocks from Wales. Smith's insight was to realize that the answer lay with fossils. At every change in rock strata certain species of fossils disappeared while others carried on into subsequent levels. By noting which species appeared in which strata, you could work out the relative ages of rocks wherever they appeared. Drawing on his knowledge as a surveyor, Smith began at once to make a map of Britain's rock strata, which would be published after many trials in 1815 and would become a cornerstone of modern geology. (The story is comprehensively covered in Simon Winchester's popular book *The Map that Changed the World*.)

Unfortunately, having had his insight, Smith was curiously uninterested in understanding why rocks were laid down in the way they were. 'I have left off puzzling about the origin of Strata and content myself with knowing that it is so,' he recorded. 'The whys and wherefores cannot come within the Province of a Mineral Surveyor.'

Smith's revelation regarding strata heightened the moral awkwardness concerning extinctions. To begin with, it confirmed that God had wiped out creatures not occasionally but repeatedly. This made Him seem not so much careless as peculiarly hostile. It also made it inconveniently necessary to explain how some species were wiped out while others continued unimpeded into succeeding eons. Clearly there was more to extinctions than could be accounted for by a single

Noachian deluge, as the biblical flood was known. Cuvier resolved the matter to his own satisfaction by suggesting that Genesis applied only to the most recent inundation. God, it appeared, hadn't wished to distract or alarm Moses with news of earlier, irrelevant extinctions.

So, by the early years of the nineteenth century, fossils had taken on a certain inescapable importance, which makes Wistar's failure to see the significance of his dinosaur bone all the more unfortunate. Suddenly, in any case, bones were turning up all over. Several other opportunities arose for Americans to claim the discovery of dinosaurs, but all were wasted. In 1806 the Lewis and Clark expedition passed through the Hell Creek Formation in Montana, an area where fossil hunters would later literally trip over dinosaur bones, and even examined what was clearly a dinosaur bone embedded in rock, but failed to make anything of it. Other bones and fossilized footprints were found in the Connecticut river valley of New England after a farm boy named Plinus Moody spied ancient tracks on a rock ledge at South Hadley, Massachusetts. Some of these at least survive – notably the bones of an anchisaurus, which are in the collection of the Peabody Museum at Yale. Found in 1818, they were the first dinosaur bones to be examined and saved, but unfortunately weren't recognized for what they were until 1855. In that same year, 1818, Caspar Wistar died, but he did gain a certain unexpected immortality when a botanist named Thomas Nuttall named a delightful climbing shrub after him. Some botanical purists still insist on spelling it *wistaria*.

By this time, however, palaeontological momentum had moved to England. In 1812, at Lyme Regis on the Dorset coast, an extraordinary child named Mary Anning – aged eleven, twelve or thirteen, depending on whose account you read – found a strange fossilized sea monster, 17 feet long and now known as the ichthyosaurus, embedded in the steep and dangerous cliffs along the English Channel.

It was the start of a remarkable career. Anning would spend the next thirty-five years gathering fossils, which she sold to visitors. (She is commonly held to be the source for the famous tongue-twister 'She sells sea-shells on the sea-shore.') She would also find the first plesiosaurus, another marine monster, and one of the first and best pterodactyls. Though none of these was technically a dinosaur, that wasn't terribly relevant at the time since nobody then knew what a dinosaur was. It was enough to realize that the world had once held creatures strikingly unlike anything we might now find.

It wasn't simply that Anning was good at spotting fossils – though she was unrivalled at that – but that she could extract them with the greatest delicacy and without damage. If you ever have the chance to visit the hall of ancient marine reptiles at the Natural History Museum in London, I urge you to take it, for there is no other way to appreciate the scale and beauty of what this young woman achieved working virtually unaided with the most basic tools in nearly impossible conditions. The plesiosaur alone took her ten years of patient excavation. Although untrained, Anning was also able to provide competent drawings and descriptions for scholars. But even with the advantage of her skills, significant finds were rare and she passed most of her life in considerable poverty.

It would be hard to think of a more overlooked person in the history of palaeontology than Mary Anning, but in fact there was one who came painfully close. His name was Gideon Algernon Mantell and he was a country doctor in Sussex.

Mantell was a lanky assemblage of shortcomings – he was vain, self-absorbed, priggish, neglectful of his family – but never was there a more committed amateur palaeontologist.

He was also lucky to have a devoted and observant wife. In 1822, while he was making a house call on a patient in rural Sussex, Mrs Mantell went for a stroll down a nearby lane and in a pile of rubble that had been left to fill potholes she found a curious object – a curved brown stone, about the size of a small walnut. Knowing her husband's interest in fossils, and thinking it might be one, she took it to him. Mantell could see at once it was a fossilized tooth, and after a little study became certain that it was from an animal that was herbivorous, reptilian, extremely large – tens of feet long – and from the Cretaceous period. He was right on all counts; but these were bold conclusions, since nothing like it had been seen before or even imagined.

Aware that his finding would entirely upend what was understood about the past, and urged by his friend the Reverend William Buckland – he of the gowns and experimental appetite – to proceed with caution, Mantell devoted three painstaking years to seeking evidence to support his conclusions. He sent the tooth to Cuvier in Paris for an opinion, but the great Frenchman dismissed it as being from a hippopotamus. (Cuvier later apologized handsomely for this uncharacteristic error.) One day, while doing research at the Hunterian Museum in London, Mantell fell into conversation with a fellow researcher who told him the tooth looked very like those of animals he had been studying, South American iguanas. A hasty comparison confirmed the resemblance. And so Mantell's creature became the iguanodon, after a basking tropical lizard to which it was not in any manner related.

Mantell prepared a paper for delivery to the Royal Society. Unfortunately, it emerged that another dinosaur had been found at a quarry in Oxfordshire and had just been formally described – by the Reverend Buckland, the very man who had urged him not to work in haste. It was the megalosaurus, and the name was actually suggested to

Buckland by his friend Dr James Parkinson, the would-be radical and eponym for Parkinson's disease. Buckland, it may be recalled, was foremost a geologist, and he showed it with his work on megalosaurus. In his report, for the *Transactions of the Geological Society of London*, he noted that the creature's teeth were not attached directly to the jawbone, as in lizards, but placed in sockets, in the manner of crocodiles. But, having noticed this much, Buckland failed to realize what it meant: namely, that megalosaurus was an entirely new type of creature. Still, although his report demonstrated little acuity or insight, it was the first published description of a dinosaur – and so it is to Buckland, rather than the far more deserving Mantell, that the credit goes for the discovery of this ancient line of beings.

Unaware that disappointment was going to be a continuing feature of his life, Mantell continued hunting for fossils – he found another giant, the hylaeosaurus, in 1833 – and purchasing others from quarrymen and farmers until he had probably the largest fossil collection in Britain. Mantell was an excellent doctor and equally gifted bone hunter, but he was unable to support both his talents. As his collecting mania grew, he neglected his medical practice. Soon, fossils filled nearly the whole of his house in Brighton and consumed much of his income. A good deal of the rest went to underwriting the publication of books that few cared to own. *Illustrations of the Geology of Sussex*, published in 1827, sold only fifty copies and left him £300 out of pocket – an uncomfortably substantial sum for the times.

In some desperation Mantell hit on the idea of turning his house into a museum and charging admission, then belatedly realized that such a mercenary act would ruin his standing as a gentleman, not to mention as a scientist – so he allowed people to visit the house for free. They came in their hundreds, week after week, disrupting both his practice and his home life. Eventually he was forced to sell most of his collection to pay off his debts. Soon after, his wife left him, taking their four children with her.

Remarkably, his troubles were only just beginning.

In the district of Sydenham in south London, at a place called Crystal Palace Park, there stands a strange and forgotten sight: the world's first life-sized models of dinosaurs. Not many people travel there these days, but once this was one of the most popular attractions in London – in effect, as Richard Fortey has noted, the world's first theme park. Quite a lot about the models is not strictly correct. The iguanodon's thumb has been placed on its nose, as a kind of spike, and it stands on four sturdy legs, making it look like a rather stout and awkwardly overgrown dog. (In life, the iguanodon did not crouch on all fours, but was bipedal.) Looking at them now you would scarcely guess that these odd and lumbering beasts could cause great rancour and bitterness, but they did. Perhaps nothing in natural history has been at the centre of fiercer and more enduring hatreds than the line of ancient beasts known as dinosaurs.

At the time of the dinosaurs' construction, Sydenham was on the edge of London and its spacious park was considered an ideal place to re-erect the famous Crystal Palace, the glass and cast-iron structure that had been the centrepiece of the Great Exhibition of 1851, and from which the new park naturally took its name. The dinosaurs, built of concrete, were a kind of bonus attraction. On New Year's Eve 1853 a famous dinner for twenty-one prominent scientists was held inside the unfinished iguanodon. Gideon Mantell, the man who had found and identified the iguanodon, was not among them. The person at the head of the table was the greatest star of the young science of palaeontology. His name was Richard Owen and by this time he had already devoted several productive years to making Gideon Mantell's life hell.

Owen had grown up in Lancaster, in the north of England, where he had trained as a doctor. He was a born anatomist and so devoted to his studies that he sometimes illicitly borrowed limbs, organs and other parts from corpses and took them home for leisurely dissection. Once, while carrying a sack containing the head of a black African sailor that he had just removed, Owen slipped on a wet cobble and watched in horror as the head bounced away from him down the lane and through the open doorway of a cottage, where it came to rest in the front parlour. What the occupants had to say upon finding an unattached head rolling to a halt at their feet can only be imagined. One assumes that they had not formed any terribly advanced conclusions when, an instant later, a fraught-looking young man rushed in, wordlessly retrieved the head and rushed out again.

In 1825, aged just twenty-one, Owen moved to London and soon after was engaged by the Royal College of Surgeons to help organize their extensive, but disordered, collections of medical and anatomical specimens. Most of these had been left to the institution by John Hunter, a distinguished surgeon and tireless collector of medical curiosities, but had never been catalogued or organized, largely because the paperwork explaining the significance of each had gone missing soon after Hunter's death.

Owen swiftly distinguished himself with his powers of organization and deduction. At the same time he showed himself to be a peerless anatomist with instincts for reconstruction almost on a par with the great Cuvier in Paris. He became such an expert on the anatomy of animals that he was granted first refusal on any animal that died at the London Zoological Gardens, and these he would invariably have delivered to his house for examination. Once his wife returned home to find a freshly deceased rhinoceros filling the front hallway. He quickly became a leading expert on all kinds of animals living and extinct – from platypuses, echidnas and other newly discovered marsupials to the hapless dodo and the extinct giant birds called moas that had roamed New Zealand until eaten out of existence by the Maoris. He was the first to describe the archaeopteryx after its discovery in Bavaria in 1861 and the first to write a formal epitaph for the dodo. Altogether he produced some six hundred anatomical papers, a prodigious output.

But it was for his work with dinosaurs that Owen is remembered. He coined the term *dinosauria* in 1841. It means 'terrible lizard' and was a curiously inapt name. Dinosaurs, as we now know, weren't all terrible – some were no bigger than rabbits and probably extremely retiring – and the one thing they most emphatically were not was lizards, which are actually of a much older (by 30 million years) lineage. Owen was well aware that the creatures were reptilian and had at his disposal a perfectly good Greek word, *herpeton*, but for some reason chose not to use it. Another, more excusable error (given the paucity of specimens at the time) was his failure to note that dinosaurs constitute not one but two orders of reptiles: the bird-hipped ornithischians and the lizard-hipped saurishchians.

Owen was not an attractive person, in appearance or in temperament. A photograph from his late middle years shows him as gaunt and sinister, like the villain in a Victorian melodrama, with long, lank hair and bulging eyes – a face to frighten babies. In manner he was cold and imperious, and he was without scruple in the furtherance of his ambitions. He was the only person Charles Darwin was ever known to hate. Even Owen's son (who soon after killed himself) referred to his father's 'lamentable coldness of heart'.

His undoubted gifts as an anatomist allowed him to get away with the most barefaced dishonesties. In 1857, the naturalist T. H. Huxley was leafing through a new edition of *Churchill's Medical Directory* when he noticed that Owen was listed as Professor of Comparative Anatomy and Physiology at the Government School of Mines, which rather surprised Huxley as that was the position he held. Upon enquiring how Churchill's had made such an elemental error, he was told that the information had been provided to them by Dr Owen himself. A fellow naturalist named Hugh Falconer, meanwhile, caught Owen taking credit for one of his discoveries. Others accused him of borrowing specimens, then denying he had done so. Owen even fell into a bitter dispute with the Queen's dentist over the credit for a theory concerning the physiology of teeth.

He did not hesitate to persecute those whom he disliked. Early in his career Owen used his influence at the Zoological Society to blackball a young man named Robert Grant, whose only crime was to have shown promise as a fellow anatomist. Grant was astonished to discover that he was suddenly denied access to the anatomical specimens he needed to conduct his research. Unable to pursue his work, he sank into an understandably dispirited obscurity.

But no-one suffered more from Owen's unkindly attentions than the hapless and increasingly tragic Gideon Mantell. After losing his wife, his children, his medical practice and most of his fossil collection, Mantell moved to London. There, in 1841 – the fateful year in which Owen would achieve his greatest glory for naming and identifying the dinosaurs – Mantell was involved in a terrible accident. While crossing Clapham Common in a carriage, he somehow fell from his seat, grew entangled in the reins and was dragged at a gallop over rough ground by the panicked horses. The accident left him bent, crippled and in chronic pain, with a spine damaged beyond repair.

Capitalizing on Mantell's enfeebled state, Owen set about systematically expunging his contributions from the record, renaming species that Mantell had named years before and claiming credit for their discovery for himself. Mantell continued to try to do original research, but Owen used his influence at the Royal Society to ensure that most of his papers were rejected. In 1852, unable to bear any more pain or persecution, Mantell took his own life. His deformed spine was removed and sent to the Royal College of Surgeons where – now here's an irony for you – it was placed in the care of Richard Owen, director of the college's Hunterian Museum.

But the insults had not quite finished. Soon after Mantell's death, an arrestingly uncharitable obituary appeared in the *Literary Gazette*. In it Mantell was characterized as a mediocre anatomist whose modest contributions to palaeontology were limited by a 'want of exact knowledge'. The obituary even removed the discovery of the iguanodon from him and credited it instead to Cuvier and Owen, among others. Though the piece carried no byline, the style was Owen's and no-one in the world of the natural sciences doubted the authorship.

By this stage, however, Owen's transgressions were beginning to catch up with him. His undoing began when a committee of the Royal Society – a committee of which he happened to be chairman – decided to award him its highest honour, the Royal Medal, for a paper he had written on an extinct mollusc called the belemnite. 'However,' as Deborah Cadbury notes in her excellent history of the period, *Terrible Lizard*, 'this piece of work was not quite as original as it appeared.' The belemnite, it turned out, had been discovered four years earlier by an amateur naturalist named Chaning Pearce, and the discovery had been fully reported at a meeting of the Geological Society. Owen had been at that meeting, but failed to mention this when he presented a report of his own to the Royal Society – at which, not incidentally, he rechristened the creature *Belemnites owenii* in his own honour. Although Owen was allowed to keep the Royal Medal, the episode left a permanent tarnish on his reputation, even among his few remaining supporters.

Eventually Huxley managed to do to Owen what Owen had done to so many others: he had him voted off the councils of the Zoological and Royal Societies. To round off the retribution, Huxley became the new Hunterian Professor at the Royal College of Surgeons.

Owen would never again do important research, but the latter half of his career was devoted to one unexceptionable pursuit for which we can all be grateful. In 1856 he became head of the natural history section of the British Museum, in which capacity he became the driving force behind the creation of London's Natural History Museum. The grand and beloved gothic heap in South Kensington, opened in 1880, is almost entirely a testament to his vision.

Before Owen, museums were designed primarily for the use and edification of the elite, and even they found it difficult to gain access. In the early days of the British Museum, prospective visitors had to make a written application and undergo a brief interview to determine if they were fit to be admitted at all. They then had to return a second time to pick up a ticket – that is, assuming they had passed the interview – and finally come back a third time to view the museum's treasures. Even then they were whisked through in groups and not allowed to linger. Owen's plan was to welcome everyone, even to the point of encouraging working men to visit in the evening, and to devote most of the museum's space to public displays. He even proposed, very radically, to put informative labels on each display so that people could appreciate what they were viewing. In this, somewhat unexpectedly, he was opposed by T. H. Huxley, who believed that museums should be primarily research institutes. By making the Natural History Museum an institution for everyone, Owen transformed our expectations of what museums are for.

Still, his altruism towards his fellow man generally did not deflect him from more personal rivalries. One of his last official acts was to lobby against a proposal to erect a statue in memory of Charles Darwin. In this he failed – though he did achieve a certain belated, inadvertent triumph. Today his own statue commands a masterful view from the staircase of the main hall in the Natural History Museum, while Darwin and T. H. Huxley are consigned somewhat obscurely to the museum coffee shop, where they stare gravely over people snacking on cups of tea and jam doughnuts.

It would be reasonable to suppose that Richard Owen's petty rivalries marked the low point of nineteenth-century palaeontology, but in fact worse was to come, this time from overseas. In America in the closing decades of the century there arose a rivalry even more spectacularly venomous, if not quite as destructive. It was between two strange and ruthless men, Edward Drinker Cope and Othniel Charles Marsh.

They had much in common. Both were spoiled, driven, self-centred, quarrelsome, jealous, mistrustful and ever unhappy. Between them they changed the world of palaeontology.

They began as friends and admirers, even naming fossil species after each other, and spent a pleasant week together in 1868. However, something then went wrong between them – nobody is quite sure what – and by the following year they had developed an enmity that would grow into consuming hatred over the next three decades. It is probably safe to say that no two people in the natural sciences have ever despised each other more.

Marsh, the elder of the two by eight years, was a retiring and bookish fellow, with a trim beard and dapper manner, who spent little time in the field and was seldom very good at finding things when he was there. On a visit to the famous dinosaur fields of Como Bluff, Wyoming, he failed to notice the bones that were, in the words of one historian, 'lying everywhere like logs'. But he had the means to buy almost anything he wanted. Although he came from a modest background – his father was a farmer in upstate New York – his uncle was the supremely rich and extraordinarily indulgent financier George Peabody. When Marsh showed an interest in natural history, Peabody had a museum built for him at Yale and provided funds sufficient for him to fill it with almost whatever took his fancy.

Cope was born more directly into privilege – his father was a rich Philadelphia businessman – and was by far the more adventurous of the two. In the summer of 1876 in Montana, while George Armstrong Custer and his troops were being cut down at Little Big Horn, Cope was out hunting for bones nearby. When it was pointed out to him that this was probably not the most prudent time to be taking treasures from Indian lands, Cope thought for a minute and decided to press on anyway. He was having too good a season. At one point he ran into a party of suspicious Crow Indians, but he managed to win them over by repeatedly taking out and replacing his false teeth.

For a decade or so, Marsh and Cope's mutual dislike primarily took the form of quiet sniping, but in 1877 it erupted into grandiose dimensions. In that year a Colorado schoolteacher named Arthur Lakes found bones near Morrison while out hiking with a friend. Recognizing the bones as coming from a 'gigantic saurian', Lakes thoughtfully dispatched some samples to both Marsh and Cope. A delighted Cope sent Lakes $100 for his trouble and asked him not to tell anyone of his discovery, especially Marsh. Confused, Lakes now asked Marsh to pass the bones on to Cope. Marsh did so, but it was an affront that he would never forget.

It also marked the start of a war between the two that became increasingly bitter, underhand and often ridiculous.

It sometimes stooped to one team's diggers throwing rocks at the other team's. Cope was caught at one point prising open crates that belonged to Marsh. They insulted each other in print and poured scorn on each other's results. Seldom – perhaps never – has science been driven forward more swiftly and successfully by animosity. Over the next several years the two men between them increased the number of known dinosaur species in America from nine to almost one hundred and fifty. Nearly every dinosaur that the average person can name – stegosaurus, brontosaurus, diplodocus, triceratops – was found by one or the other of them.* Unfortunately, they worked in such reckless haste that they often failed to note that a new discovery was something already known. Between them they managed to 'discover' a species called *Uintatheres anceps* no fewer than twenty-two times. It took years to sort out some of the classification messes they made. Some are not sorted out yet.

Of the two, Cope's scientific legacy was much the more substantial. In a breathtakingly industrious career, he wrote some fourteen hundred learned papers and described almost thirteen hundred new species of fossil (of all types, not just dinosaurs) – more than double Marsh's output in both cases. Cope might have done even more, but unfortunately he went into a rather precipitous descent in his later years. Having inherited a fortune in 1875, he invested unwisely in silver and lost everything. He ended up living in a single room in a Philadelphia boarding house, surrounded by books, papers and bones. Marsh, by contrast, finished his days in a splendid mansion in New Haven. Cope died in 1897, Marsh two years later.

In his final years, Cope developed one other interesting obsession. It became his earnest wish to be declared the type

*The notable exception being the *Tyrannosaurus rex*, which was found by Barnum Brown in 1902.

specimen for *Homo sapiens* – that is, to have his bones be the official set for the human race. Normally, the type specimen of a species is the first set of bones found, but since no first set of *Homo sapiens* bones exists, there was a vacancy, which Cope desired to fill. It was an odd and vain wish, but no-one could think of any grounds to oppose it. To that end, Cope willed his bones to the Wistar Institute, a learned society in Philadelphia endowed by the descendants of the seemingly inescapable Caspar Wistar. Unfortunately, after his bones were prepared and assembled, it was found that they showed signs of incipient syphilis, hardly a feature one would wish to preserve in the type specimen for one's own race. So Cope's petition and his bones were quietly shelved. There is still no type specimen for modern humans.

As for the other players in this drama, Owen died in 1892, a few years before Cope or Marsh. Buckland ended up by losing his mind and finished his days a gibbering wreck in a lunatic asylum in Clapham, not far from where Mantell had suffered his crippling accident. Mantell's twisted spine remained on display at the Hunterian Museum for nearly a century before being mercifully obliterated by a German bomb in the Blitz. What remained of Mantell's collection after his death passed on to his children and much of it was taken to New Zealand by his son Walter, who emigrated there in 1840. Walter became a distinguished Kiwi, eventually attaining the office of Minister of Native Affairs. In 1865 he donated the prime specimens from his father's collection, including the famous iguanodon tooth, to the Colonial Museum (now the Museum of New Zealand) in Wellington, where they have remained ever since. The iguanodon tooth that started it all – arguably the most important tooth in palaeontology – is no longer on display.

Of course, dinosaur hunting didn't end with the deaths of the great nineteenth-century fossil hunters. Indeed, to a surprising extent it had only just begun. In 1898, the year that fell between the deaths of Cope and Marsh, a trove greater by far than anything found before was discovered – noticed, really – at a place called Bone Cabin Quarry, only a few miles from Marsh's prime hunting ground at Como Bluff, Wyoming. There, hundreds and hundreds of fossil bones were to be found weathering out of the hills. They were so numerous, in fact, that someone had built a cabin out of them – hence the name. In just the first two seasons, one hundred thousand pounds of ancient bones were excavated from the site, and tens of thousands of pounds more came in each of the half dozen years that followed.

The upshot is that by the turn of the twentieth century, palaeontologists had literally tons of old bones to pick over. The problem was that they still didn't have any idea how old any of these bones were. Worse, the agreed ages for the Earth couldn't comfortably support the numbers of aeons and ages and epochs that the past obviously contained. If Earth were really only twenty million years old or so, as the great Lord Kelvin insisted, then whole orders of ancient creatures must have come into being and gone out again practically in the same geological instant. It just made no sense.

Other scientists besides Kelvin turned their minds to the problem and came up with results that only deepened the uncertainty. Samuel Haughton, a respected geologist at Trinity College in Dublin, announced an estimated age for the Earth of 2,300 million years – way beyond anything anybody else was suggesting. When this was drawn to his attention, he recalculated using the same data and put the figure at 153 million years. John Joly, also of Trinity, decided to give Edmond Halley's ocean salts idea a whirl, but his method was based on so many faulty assumptions that he was hopelessly adrift. He calculated that the Earth was 89 million years old – an age that fitted neatly enough with

Kelvin's assumptions but unfortunately not with reality.

Such was the confusion that by the close of the nineteenth century, depending on which text you consulted, you could learn that the number of years that stood between us and the dawn of complex life in the Cambrian period was 3 million, 18 million, 600 million, 794 million, or 2.4 billion – or some other number within that range. As late as 1910, one of the most respected estimates, by the American George Becker, put the Earth's age at perhaps as little as 55 million years.

Just when matters seemed most intractably confused, along came another extraordinary figure with a novel approach. He was a bluff and brilliant New Zealand farm boy named Ernest Rutherford, and he produced pretty well irrefutable evidence that the Earth was at least many hundreds of millions of years old, probably rather more.

Remarkably, his evidence was based on alchemy – natural, spontaneous, scientifically credible and wholly non-occult, but alchemy nonetheless. Newton, it turned out, had not been so wrong after all. And exactly how *that* became evident is, of course, another story.

7
ELEMENTAL MATTERS

Chemistry as an earnest and respectable science is often said to date from 1661, when Robert Boyle of Oxford published *The Sceptical Chymist* – the first work to distinguish between chemists and alchemists – but it was a slow and often erratic transition. Into the eighteenth century scholars could feel oddly comfortable in both camps – like the German Johann Becher, who produced a sober and unexceptionable work on mineralogy called *Physica Subterranea*, but who also was certain that, given the right materials, he could make himself invisible.

Perhaps nothing better typifies the strange and often accidental nature of chemical science in its early days than a discovery made by a German named Hennig Brand in 1675. Brand became convinced that gold could somehow be distilled from human urine. (The similarity of colour seems to have been a factor in his conclusion.) He assembled fifty buckets of human urine, which he kept for months in his cellar. By various recondite processes, he converted the urine first into a noxious paste and then into a translucent waxy substance. None of it yielded gold, of course, but a strange and interesting thing did happen. After a time, the substance began to glow. Moreover, when exposed to air, it often spontaneously burst into flame.

The commercial potential for the stuff – which soon became known as phosphorus, from Greek and Latin roots meaning 'light-bearing' – was not lost on eager business people, but the difficulties of manufacture made it too costly to exploit. An ounce of phosphorus retailed for 6 guineas – perhaps £300 in today's money – or more than gold.

At first, soldiers were called on to provide the raw material, but such an arrangement was hardly conducive to industrial-scale production. In the 1750s a Swedish chemist named Karl (or Carl) Scheele devised a way to manufacture phosphorus in bulk without the slop or smell of urine. It was largely because of this mastery of phosphorus that Sweden became, and remains, a leading producer of matches.

Scheele was both an extraordinary and an extraordinarily luckless fellow. A humble pharmacist with little in the way of advanced apparatus, he discovered eight elements – chlorine, fluorine, manganese, barium, molybdenum, tungsten, nitrogen and oxygen – and got credit for none of them. In every case, his finds either were overlooked or made it into publication after someone else had made the same discovery independently. He also discovered many useful compounds, among them ammonia, glycerin and tannic acid, and was the first to see the commercial potential of chlorine as a bleach – all breakthroughs that made other people extremely wealthy.

Scheele's one notable shortcoming was a curious insistence on tasting a little of everything he worked with, including such notoriously disagreeable substances as mercury and hydrocyanic acid (another of his discoveries) – a compound so famously poisonous that 150 years later Erwin Schrödinger chose it as his toxin of choice in a famous thought experiment (see page 190). Scheele's rashness eventually caught up with him. In 1786, aged just forty-three, he was found dead at his workbench surrounded by an array of toxic chemicals, any one of which could have accounted for the stunned and terminal look on his face.

Were the world just and Swedish-speaking, Scheele would have enjoyed universal acclaim. As it is, the plaudits have tended to go to more celebrated chemists, mostly from the English-speaking world. Scheele discovered oxygen in 1772, but for various heartbreakingly complicated reasons could not get his paper published in a timely manner. Credit went instead to Joseph Priestley, who discovered the same element independently, but latterly, in the summer of 1774. Even more remarkable was Scheele's failure to receive credit for the discovery of chlorine. Nearly all textbooks still attribute chlorine's discovery to Humphry Davy, who did indeed find it, but *thirty-six years* after Scheele.

Although chemistry had come a long way in the century that separated Newton and Boyle from Scheele and Priestley and Henry Cavendish, it still had a long way to go. Right up to the closing years of the eighteenth century (and in Priestley's case a little beyond) scientists everywhere searched for, and sometimes believed they had actually found, things that just weren't there: vitiated airs, dephlogisticated marine acids, phloxes, calxes, terraqueous exhalations and, above all, phlogiston, the substance that was thought to be the active agent in combustion. Somewhere in all this, it was thought, there also resided a mysterious *élan vital*, the force that brought inanimate objects to life. No-one knew where this ethereal essence lay, but two things seemed probable: that you could enliven it with a jolt of electricity (a notion Mary Shelley exploited to full effect in her novel *Frankenstein*); and that it existed in some substances but not others, which is why we ended up with two branches of chemistry: organic (for those substances that were thought to have it) and inorganic (for those that did not).

Someone of insight was needed to thrust chemistry into the modern age, and it was the French who provided him.

His name was Antoine-Laurent Lavoisier. Born in 1743, Lavoisier was a member of the minor nobility (his father had purchased a title for the family). In 1768 he bought a practising share in a deeply despised institution called the Ferme Générale (or General Farm), which collected taxes and fees on behalf of the government. Although Lavoisier himself was by all accounts mild and fair-minded, the company he worked for was neither. For one thing, it did not tax the rich but only the poor, and then often arbitrarily. For Lavoisier, the appeal of the institution was that it provided him with the wealth to follow his principal devotion, science. At his peak, his personal earnings reached 150,000 livres a year – perhaps £12 million in today's money.

Three years after embarking on this lucrative career path, he married the fourteen-year-old daughter of one of his bosses. The marriage was a meeting of hearts and minds. Mme Lavoisier had an incisive intellect and soon was working productively alongside her husband. Despite the demands of his job and busy social life, they managed on most days to put in five hours of science – two in the early morning and three in the evening – as well as the whole of Sunday, which they called their *jour de bonheur* (day of happiness). Somehow Lavoisier also found the time to be commissioner of gunpowder, supervise the building of a wall around Paris to deter smugglers, help found the metric system and co-author the handbook *Méthode de Nomenclature Chimique*, which became the bible for agreeing the names of the elements.

As a leading member of the Académie Royale des Sciences, he was also required to take an informed and active interest in whatever was topical – hypnotism, prison reform, the respiration of insects, the water supply of Paris. It was in such a capacity in 1780 that Lavoisier made some dismissive remarks about a new theory of combustion that had been submitted to the academy by a hopeful young scientist. The theory was indeed wrong, but the scientist never forgave him. His name was Jean-Paul Marat.

The one thing Lavoisier never did was discover an element. At a time when it seemed as if almost anybody with a beaker, a flame and some interesting powders could discover something new – and when, not incidentally, some two-thirds of the elements were yet to be found – Lavoisier failed to uncover a single one. It certainly wasn't for want of beakers. Lavoisier had thirteen thousand of them in what was, to an almost preposterous degree, the finest private laboratory in existence.

Instead, he took the discoveries of others and made sense of them. He threw out phlogiston and mephitic airs. He identified oxygen and hydrogen for what they were and gave them both their modern names. In short, he helped to bring rigour, clarity and method to chemistry.

And his fancy equipment did in fact come in very handy. For years, he and Mme Lavoisier occupied themselves with extremely exacting studies requiring the finest measurements. They determined, for instance, that a rusting object doesn't lose weight, as everyone had long assumed, but gains weight – an extraordinary discovery. Somehow, as it rusted the object was attracting elemental particles from the air. It was the first realization that matter can be transformed but not eliminated. If you burned this book now, its matter would be changed to ash and smoke, but the net amount of stuff in the universe would be the same. This became known as the conservation of mass, and it was a revolutionary concept. Unfortunately, it coincided with another type of revolution – the French one – and in this one Lavoisier was entirely on the wrong side.

Not only was he a member of the hated Ferme Générale, but he had enthusiastically built the wall that enclosed Paris – an edifice so loathed that it was the first thing attacked by the rebellious citizens. Capitalizing on this, in 1791 Marat, now a leading voice in the National Assembly, denounced Lavoisier and suggested that it was well past time for his hanging. Soon afterwards the Ferme Générale was shut down. Not long after this Marat was murdered in his bath by an aggrieved young woman named Charlotte Corday, but by this time it was too late for Lavoisier.

In 1793 the Reign of Terror, already intense, ratcheted up to a higher gear. In October Marie Antoinette was sent to the guillotine. The following month, as he and his wife were making tardy plans to slip away to Scotland, Lavoisier was arrested. In May he and thirty-one fellow farmers-general were brought before the Revolutionary Tribunal (in a courtroom presided over by a bust of Marat). Eight were granted acquittals, but Lavoisier and the others were taken directly to the Place de la Révolution (now the Place de la Concorde), site of the busiest of French guillotines. Lavoisier watched his father-in-law beheaded, then stepped up and accepted his fate. Less than three months later, on 27 July, Robespierre himself was dispatched in the same way and in the same place and the Reign of Terror swiftly ended.

A hundred years after his death, a statue of Lavoisier was erected in Paris and much admired until someone pointed out that it looked nothing like him. Under questioning, the sculptor admitted that he had used the head of the mathematician and philosopher the Marquis de Condorcet – apparently he had a spare – in the hope that no-one would notice or, having noticed, would care. In the second regard he was correct. The statue of Lavoisier-cum-Condorcet was allowed to remain in place for another half-century until the Second World War, when, one morning, it was taken away and melted down for scrap.

In the early 1800s there arose in England a fashion for inhaling nitrous oxide, or laughing gas, after it was discovered that its use 'was attended by a highly pleasurable thrilling'.

For the next half-century it would be the drug of choice for young people. One learned body, the Askesian Society, was for a time devoted to little else. Theatres put on 'laughing gas evenings' where volunteers could refresh themselves with a robust inhalation and then entertain the audience with their comical staggerings.

It wasn't until 1846 that anyone got around to finding a practical use for nitrous oxide, as an anaesthetic. Goodness knows how many tens of thousands of people suffered unnecessary agonies under the surgeon's knife because no-one had thought of the gas's most obvious practical application.

I mention this to make the point that chemistry, having come so far in the eighteenth century, rather lost its bearings in the first decades of the nineteenth, in much the way that geology would in the early years of the twentieth. Partly it was to do with the limitations of equipment – there were, for instance, no centrifuges until the second half of the century, severely restricting many kinds of experiments – and partly it was social. Chemistry was, generally speaking, a science for business people, for those who worked with coal and potash and dyes, and not for gentlemen, who tended to be drawn to geology, natural history and physics. (This was slightly less true in continental Europe than in Britain, but only slightly.) It is perhaps telling that one of the most important observations of the century, Brownian motion, which established the active nature of molecules, was made not by a chemist but by a Scottish botanist, Robert Brown. (What Brown noticed, in 1827, was that tiny grains of pollen suspended in water remained indefinitely in motion no matter how long he gave them to settle. The cause of this perpetual motion – namely, the actions of invisible molecules – was long a mystery.)

Things might have been worse had it not been for a splendidly improbable character named Count von

Rumford, who, despite the grandeur of his title, began life in Woburn, Massachusetts, in 1753 as plain Benjamin Thompson. Thompson was dashing and ambitious, 'handsome in feature and figure', occasionally courageous and exceedingly bright, but untroubled by anything so inconveniencing as a scruple. At nineteen he married a rich widow fourteen years his senior, but at the outbreak of revolution in the colonies he unwisely sided with the loyalists, for a time spying on their behalf. In the fateful year of 1776, facing arrest 'for lukewarmness in the cause of liberty', he abandoned his wife and child and fled just ahead of a mob of anti-royalists armed with buckets of hot tar, bags of feathers and an earnest desire to adorn him with both.

He decamped first to England and then to Germany, where he served as a military adviser to the government of Bavaria, so impressing the authorities that in 1791 he was named Count von Rumford of the Holy Roman Empire. While in Munich, he also designed and laid out the famous park known as the English Garden.

In between these undertakings, he somehow found time to conduct a good deal of solid science. He became the world's foremost authority on thermodynamics and the first to elucidate the principles of the convection of fluids and the circulation of ocean currents. He also invented several useful objects, including a drip coffee-maker, thermal underwear, and a type of range still known as the Rumford fireplace. In 1805, during a sojourn in France, he wooed and married Mme Lavoisier, widow of Antoine-Laurent. The marriage was not a success and they soon parted. Rumford stayed on in France where he died, universally esteemed by all but his former wives, in 1814.

Our purpose in mentioning him here is that in 1799, during a comparatively brief interlude in London, he founded the Royal Institution, yet another of the many learned societies that popped into being all over Britain in the late eighteenth and early nineteenth centuries. For a time it was almost the only institution of standing to actively promote the young science of chemistry, and that was thanks almost entirely to a brilliant young man named Humphry Davy, who was appointed the institution's professor of chemistry shortly after its inception and rapidly gained fame as an outstanding lecturer and productive experimentalist.

Soon after taking up his position, Davy began to bang out new elements one after another – potassium, sodium, magnesium, calcium, strontium, and aluminum or aluminium (depending on which branch of English you favour).* He discovered so many elements not so much because he was serially astute as because he developed an ingenious technique of applying electricity to a molten substance – electrolysis, as it is known. Altogether he discovered a dozen elements, a fifth of the known total of his day. Davy might have done far more, but unfortunately as a young man he developed an abiding attachment to the buoyant pleasures of nitrous oxide. He grew so attached to the gas that he drew on it (literally) three or four times a day. Eventually, in 1829, it is thought to have killed him.

Fortunately, more sober types were at work elsewhere. In 1808, a dour Quaker named John Dalton became the first person to intimate the nature of an atom (progress that will be discussed more completely a little further on) and in 1811 an Italian with the splendidly operatic name of Lorenzo Romano Amadeo Carlo Avogadro, Count of Quarequa and

*The confusion over the aluminum/aluminium spelling arose because of some uncharacteristic indecisiveness on Davy's part. When he first isolated the element in 1808, he called it *alumium*. For some reason he thought better of that and changed it to *aluminum* four years later. Americans dutifully adopted the new term, but many British users disliked *aluminum*, pointing out that it disrupted the *-ium* pattern established by sodium, calcium and strontium, so they added a vowel and syllable. Among his other achievements, Davy also invented the miner's safety lamp.

Cerreto, made a discovery that would prove highly significant in the long term – namely, that two equal volumes of gases of any type, if kept at the same pressure and temperature, will contain identical numbers of molecules.

Two things were notable about the appealingly simple Avogadro's Principle, as it became known. First, it provided a basis for more accurately measuring the size and weight of atoms. Using Avogadro's mathematics, chemists were eventually able to work out, for instance, that a typical atom had a diameter of 0.00000008 centimetres, which is very little indeed. And second, almost no one knew about it for almost fifty years.*

Partly this was because Avogadro himself was a retiring fellow – he worked alone, corresponded very little with fellow scientists, published few papers and attended no meetings – but also it was because there were no meetings to attend and few chemical journals in which to publish. This is a fairly extraordinary fact. The Industrial Revolution was driven in large part by developments in chemistry and yet as an organized science chemistry barely existed for decades.

The Chemical Society of London was not founded until 1841 and didn't begin to produce a regular journal until 1848, by which time most learned societies in Britain – Geological, Geographical, Zoological, Horticultural and

*The principle led to the much later adoption of Avogadro's Number, a basic unit of measure in chemistry, which was named for Avogadro long after his death. It is the number of molecules found in 2.016 grams of hydrogen gas (or an equal volume of any other gas). Its value is placed at 6.0221367×10^{23}, which is an enormously large number. Chemistry students have long amused themselves by computing just how large a number it is, so I can report that it is equivalent to the number of popcorn kernels needed to cover the United States to a depth of nine miles, or cupfuls of water in the Pacific Ocean, or soft-drink cans that would, evenly stacked, cover the Earth to a depth of two hundred miles. An equivalent number of American pennies would be enough to make every person on Earth a dollar trillionaire. It is a big number.

Linnaean (for naturalists and botanists) – were at least twenty years old and in several cases much more. The rival Institute of Chemistry didn't come into being until 1877, a year after the founding of the American Chemical Society. Because chemistry was so slow to get organized, news of Avogadro's important breakthrough of 1811 didn't begin to become general until the first international chemistry congress, in Karlsruhe, in 1860.

Because chemists worked for so long in isolation, conventions were slow to emerge. Until well into the second half of the century, the formula H_2O_2 might mean water to one chemist but hydrogen peroxide to another. C_2H_4 could signify ethylene or marsh gas. There was hardly a molecule that was uniformly represented everywhere.

Chemists also used a bewildering variety of symbols and abbreviations, often self-invented. Sweden's J. J. Berzelius brought a much-needed measure of order to matters by decreeing that the elements be abbreviated on the basis of their Greek or Latin names, which is why the abbreviation for iron is Fe (from the Latin *ferrum*) and for silver is Ag (from the Latin *argentum*). That so many of the other abbreviations accord with their English names (N for nitrogen, O for oxygen, H for hydrogen and so on) reflects English's latinate nature, not its exalted status. To indicate the number of atoms in a molecule, Berzelius employed a superscript notation, as in H^2O. Later, for no special reason, the fashion became to render the number as subscript: H_2O.

Despite the occasional tidyings-up, chemistry by the second half of the nineteenth century was in something of a mess, which is why everybody was so pleased by the rise to prominence in 1869 of an odd and crazed-looking professor at the University of St Petersburg named Dmitri Ivanovich Mendeleyev.

Mendeleyev (also sometimes spelled Mendeleev or Mendeléef) was born in 1834 at Tobolsk, in the far west of

Siberia, into a well-educated, reasonably prosperous and very large family – so large, in fact, that history has lost track of exactly how many Mendeleyevs there were: some sources say there were fourteen children, some say seventeen. All agree, at any rate, that Dmitri was the youngest. Luck was not always with the Mendeleyevs. When Dmitri was small his father, the headmaster of a local school, went blind and his mother had to go out to work. Clearly an extraordinary woman, she eventually became the manager of a successful glass factory. All went well until 1848, when the factory burned down and the family was reduced to penury. Determined to get her youngest child an education, the indomitable Mrs Mendeleyev hitchhiked with young Dmitri four thousand miles to St Petersburg – that's equivalent to travelling from London to Equatorial Guinea – and deposited him at the Institute of Pedagogy. Worn out by her efforts, she died soon after.

Mendeleyev dutifully completed his studies and eventually landed a position at the local university. There he was a competent but not terribly outstanding chemist, known more for his wild hair and beard, which he had trimmed just once a year, than for his gifts in the laboratory.

However, in 1869, at the age of thirty-five, he began to toy with a way to arrange the elements. At the time, elements were normally grouped in two ways – either by atomic weight (using Avogadro's Principle) or by common properties (whether they were metals or gases, for instance). Mendeleyev's breakthrough was to see that the two could be combined in a single table.

As is often the way in science, the principle had actually been anticipated three years previously by an amateur chemist in England named John Newlands. He suggested that when elements were arranged by weight they appeared to repeat certain properties – in a sense to harmonize – at every eighth place along the scale. Slightly unwisely, for this was an idea whose time had not quite yet come, Newlands called it the Law of Octaves and likened the arrangement to the octaves on a piano keyboard. Perhaps there was something in Newlands' manner of presentation, but the idea was considered fundamentally preposterous and widely mocked. At gatherings, droller members of the audience would sometimes ask him if he could get his elements to play them a little tune. Discouraged, Newlands gave up pushing the idea and soon dropped out of sight altogether.

Mendeleyev used a slightly different approach, placing his elements into groups of seven, but employed fundamentally the same premise. Suddenly the idea seemed brilliant and wondrously perceptive. Because the properties repeated themselves periodically, the invention became known as the Periodic Table.

Mendeleyev was said to have been inspired by the card game known as solitaire in North America and patience elsewhere, wherein cards are arranged by suit horizontally and by number vertically. Using a broadly similar concept, he arranged the elements in horizontal rows called periods and vertical columns called groups. This instantly showed one set of relationships when read up and down and another when read from side to side. Specifically, the vertical columns put together chemicals that have similar properties. Thus copper sits on top of silver and silver sits on top of gold because of their chemical affinities as metals, while helium, neon and argon are in a column made up of gases. (The actual, formal determinant in the ordering is something called their electron valences, and if you want to understand them you will have to enrol in evening classes.) The horizontal rows, meanwhile, arrange the chemicals in ascending order by the number of protons in their nuclei – what is known as their atomic number.

The structure of atoms and the significance of protons will come in a following chapter; for the moment, all that is necessary is to appreciate the organizing principle: hydrogen has just one proton and so it has an atomic number of 1 and comes first on the chart; uranium has 92 protons and so it comes near the end and has an atomic number of 92. In this sense, as Philip Ball has pointed out, chemistry really is just a matter of counting. (Atomic number, incidentally, is not to be confused with atomic weight, which is the number of protons plus the number of neutrons in a given element.)

There was still a great deal that wasn't known or understood. Hydrogen is the most common element in the universe, and yet no-one would guess as much for another thirty years. Helium, the second most abundant element, had only been found the year before – its existence hadn't even been suspected before that – and then not on the Earth, but in the Sun, where it was found with a spectroscope during a solar eclipse, which is why it honours the Greek sun god Helios. It wouldn't be isolated until 1895. Even so, thanks to Mendeleyev's invention, chemistry was now on a firm footing.

For most of us, the Periodic Table is a thing of beauty in the abstract, but for chemists it established an immediate orderliness and clarity that can hardly be overstated. 'Without a doubt, the Periodic Table of the Chemical Elements is the most elegant organizational chart ever devised,' wrote Robert E. Krebs in *The History and Use of Our Earth's Chemical Elements* – and you can find similar sentiments in virtually every history of chemistry in print.

Today we have '120 or so' known elements – 92 naturally occurring ones plus a couple of dozen that have been created in labs. The actual number is slightly contentious because the heavy, synthesized elements exist for only millionths of seconds and chemists sometimes argue over whether they have really been detected or not. In Mendeleyev's day just sixty-three elements were known, but part of his cleverness was to realize that the elements as then known didn't make a complete picture, that many pieces

The PERIODIC TABLE *of* CHEMICAL ELEMENTS

1 H 1																	2 He 4
3 Li 6,9	4 Be 9											5 B 10,8	6 C 12	7 N 14	8 O 16	9 F 19	10 Ne 20,2
11 Na 23	12 Mg 24,3											13 Al 27	14 Si 28,1	15 P 31	16 S 32,1	17 Cl 35,5	18 Ar 39,9
19 K 39,1	20 Ca 40,1	21 Sc 45	22 Ti 47,9	23 V 51	24 Cr 52	25 Mn 54,9	26 Fe 55,8	27 Co 58,9	28 Ni 58,7	29 Cu 63,5	30 Zn 65,4	31 Ga 69,7	32 Ge 72,6	33 As 74,9	34 Se 79	35 Br 79,9	36 Kr 83,8
37 Rb 85,5	38 Sr 87,6	39 Y 88,9	40 Zr 91,2	41 Nb 92,9	42 Mo 96	43 Tc 96	44 Ru 101,7	45 Rh 102,9	46 Pd 106,7	47 Ag 107,9	48 Cd 112,4	49 In 114,8	50 Sn 118,7	51 Sb 121,8	52 Te 127,6	53 I 126,9	54 Xe 131,3
55 Cs 132,9	56 Ba 137,4		72 Hf 178,6	73 Ta 180,9	74 W 183,9	75 Re 186,3	76 Os 190,2	77 Ir 193,1	78 Pt 195,2	79 Au 197,2	80 Hg 200,6	81 Tl 204,4	82 Pb 207,2	83 Bi 209	84 Po 210	85 At 210	86 Rn 222
87 Fr 221	88 Ra 226		104 Rf 261	105 Db 262	106 Sg 263	107 Bh 262	108 Hs 265	109 Mt 266	110 Uun 269	111 Uuu 272	112 Uub 277						
LANTHANOIDS			57 La 138,9	58 Ce 140,1	59 Pr 140,9	60 Nd 144,2	61 Pm 145	62 Sm 150,4	63 Eu 152	64 Gd 156,9	65 Tb 159,2	66 Dy 162,5	67 Ho 164,9	68 Er 167,2	69 Tm 169,4	70 Yb 173	71 Lu 175
ACTINOIDS			89 Ac 227	90 Th 232,1	91 Pa 231	92 U 238,1	93 Np 237	94 Pu 244	95 Am 243	96 Cm 247	97 Bk 247	98 Cf 251	99 Es 252	100 Fm 257	101 Md 258	102 No 259	103 Lr 262

were missing. His table predicted, with pleasing accuracy, where new elements would slot in when they were found.

No-one knows, incidentally, how high the number of elements might go, though anything beyond 168 as an atomic weight is considered 'purely speculative'; but what is certain is that anything that is found will fit neatly into Mendeleyev's great scheme.

The nineteenth century held one last important surprise for chemists. It began in 1896 when Henri Becquerel in Paris carelessly left a packet of uranium salts on a wrapped photographic plate in a drawer. When he took the plate out some time later, he was surprised to discover that the salts had burned an impression in it, just as if the plate had been exposed to light. The salts were emitting rays of some sort.

Considering the importance of what he had found, Becquerel did a very strange thing: he turned the matter over to a graduate student for investigation. Fortunately the student was a recent émigré from Poland named Marie Curie. Working with her new husband, Pierre, Curie found that certain kinds of rocks poured out constant and extraordinary amounts of energy, yet without diminishing in size or changing in any detectable way. What she and her husband couldn't know – what no-one could know until Einstein explained things the following decade – was that the rocks were converting mass into energy in an exceedingly efficient way. Marie Curie dubbed the effect 'radioactivity'. In the process of their work, the Curies also found two new elements – polonium, which they named after her native country, and radium. In 1903 the Curies and Becquerel were jointly awarded the Nobel Prize in physics. (Marie Curie would win a second prize, in chemistry, in 1911; the only person to win in both chemistry and physics.)

At McGill University in Montreal the young New Zealand-born Ernest Rutherford became interested in the new radioactive materials. With a colleague named Frederick Soddy he discovered that immense reserves of energy were bound up in these small amounts of matter, and that the radioactive decay of these reserves could account for most of the Earth's warmth. They also discovered that radioactive elements decayed into other elements – that one day you had an atom of uranium, say, and the next you had an atom of lead. This was truly extraordinary. It was alchemy pure and simple; no-one had ever imagined that such a thing could happen naturally and spontaneously.

Ever the pragmatist, Rutherford was the first to see that there could be a valuable practical application in this. He noticed that in any sample of radioactive material, it always took the same amount of time for half the sample to decay – the celebrated half-life* – and that this steady, reliable rate of decay could be used as a kind of clock. By calculating backwards from how much radiation a material had now and how swiftly it was decaying, you could work out its age. He tested a piece of pitchblende, the principal ore of uranium, and found it to be 700 million years old – very much older than the age most people were prepared to grant the Earth.

In the spring of 1904, Rutherford travelled to London to

*If you have ever wondered how the atoms determine which 50 per cent will die and which 50 per cent will survive for the next session, the answer is that the half-life is really just a statistical convenience – a kind of actuarial table for elemental things. Imagine you had a sample of material with a half-life of 30 seconds. It isn't that every atom in the sample will exist for exactly 30 seconds or 60 seconds or 90 seconds or some other tidily ordained period. Each atom will in fact survive for an entirely random length of time that has nothing to do with multiples of 30; it might last until two seconds from now or it might oscillate away for years or decades or centuries to come. No-one can say. But what we can say is that for the sample as a whole the rate of disappearance will be such that half the atoms will disappear every 30 seconds. It's an average rate, in other words, and you can apply it to any large sampling. Someone once worked out, for instance, that American dimes have a half-life of about thirty years.

give a lecture at the Royal Institution – the august organization founded by Count von Rumford only 105 years before, though that powdery and periwigged age now seemed a distant aeon compared with the roll-your-sleeves-up robustness of the late Victorians. Rutherford was there to talk about his new disintegration theory of radioactivity, as part of which he brought out his piece of pitchblende. Tactfully – for the ageing Kelvin was present, if not always fully awake – Rutherford noted that Kelvin himself had suggested that the discovery of some other source of heat would throw his calculations out. Rutherford had found that other source. Thanks to radioactivity the Earth could be – and self-evidently was – much older than the 24 million years Kelvin's final calculations allowed.

Kelvin beamed at Rutherford's respectful presentation, but was in fact unmoved. He never accepted the revised figures and to his dying day believed his work on the age of the Earth his most astute and important contribution to science – far greater than his work on thermodynamics.

As with most scientific revolutions, Rutherford's new findings were not universally welcomed. John Joly of Dublin strenuously insisted well into the 1930s that the Earth was no more than 89 million years old, and was stopped only then by his own death. Others began to worry that Rutherford had now given them too much time. But even with radiometric dating, as decay measurements became known, it would be decades before we got within a billion years or so of the Earth's actual age. Science was on the right track, but still way out.

Kelvin died in 1907. That year also saw the death of Dmitri Mendeleyev. Like Kelvin, his productive work was far behind him, but his declining years were notably less serene. As he aged, Mendeleyev became increasingly eccentric – he refused to acknowledge the existence of radiation or the electron or anything else much that was new

– and difficult. His final decades were spent mostly storming out of labs and lecture halls all across Europe. In 1955, element 101 was named mendelevium in his honour. 'Appropriately,' notes Paul Strathern, 'it is an unstable element.'

Radiation, of course, went on and on, literally and in ways nobody expected. In the early 1900s Pierre Curie began to experience clear signs of radiation sickness – notably dull aches in his bones and chronic feelings of malaise – which doubtless would have progressed unpleasantly. We shall never know for certain because in 1906 he was fatally run over by a carriage while crossing a Paris street.

Marie Curie spent the rest of her life working with distinction in the field, helping to found the celebrated Radium Institute of the University of Paris in 1914. Despite her two Nobel Prizes, she was never elected to the Academy of Sciences, in large part because after the death of Pierre she conducted an affair with a married physicist sufficiently indiscreet to scandalize even the French – or at least the old men who ran the academy, which is perhaps another matter.

For a long time it was assumed that anything so miraculously energetic as radioactivity must be beneficial. For years, manufacturers of toothpaste and laxatives put radioactive thorium in their products, and at least until the late 1920s the Glen Springs Hotel in the Finger Lakes region of New York (and doubtless others as well) featured with pride the therapeutic effects of its 'Radio-active mineral springs'. It wasn't banned in consumer products until 1938. By this time it was much too late for Mme Curie, who died of leukaemia in 1934. Radiation, in fact, is so pernicious and long-lasting that even now her papers from the 1890s – even her cookbooks – are too dangerous to handle. Her lab books are kept in lead-lined boxes and those who wish to see them must don protective clothing.

Thanks to the devoted and unwittingly high-risk work of the first atomic scientists, by the early years of the twentieth century, it was becoming clear that the Earth was unquestionably venerable, though another half-century of science would have to be done before anyone could confidently say quite how venerable. Science, meanwhile, was about to get a new age of its own – the atomic one.

III

A NEW AGE DAWNS

A physicist is the atoms' way of thinking about atoms.

Anonymous

8

EINSTEIN'S UNIVERSE

As the nineteenth century drew to a close, scientists could reflect with satisfaction that they had pinned down most of the mysteries of the physical world: electricity, magnetism, gases, optics, acoustics, kinetics and statistical mechanics, to name just a few, had all fallen into order before them. They had discovered the X-ray, the cathode ray, the electron and radioactivity, invented the ohm, the watt, the kelvin, the joule, the amp and the little erg.

If a thing could be oscillated, accelerated, perturbed, distilled, combined, weighed or made gaseous they had done it, and in the process produced a body of universal laws so weighty and majestic that we still tend to write them out in capitals: the Electromagnetic Field Theory of Light, Richter's Law of Reciprocal Proportions, Charles's Law of Gases, the Law of Combining Volumes, the Zeroth Law, the Valence Concept, the Laws of Mass Actions, and others beyond counting. The whole world clanged and chuffed with the machinery and instruments that their ingenuity had produced. Many wise people believed that there was nothing much left for science to do.

In 1875, when a young German in Kiel named Max Planck was deciding whether to devote his life to mathematics or to physics, he was urged most heartily not to choose physics because the breakthroughs had all been made there. The coming century, he was assured, would be one of consolidation and refinement, not revolution. Planck didn't listen. He studied theoretical physics and threw himself body and soul into work on entropy, a process at the heart of thermodynamics, which seemed to hold much promise for an ambitious young man.* In 1891 he produced his results and learned to his dismay that the important work on entropy *had* in fact been done already, in this instance by a retiring scholar at Yale University named J. Willard Gibbs.

Gibbs is perhaps the most brilliant person most people have never heard of. Modest to the point of near-invisibility, he passed virtually the whole of his life, apart from three years spent studying in Europe, within a three-block area bounded by his house and the Yale campus in New Haven, Connecticut. For his first ten years at Yale he didn't even bother to draw a salary. (He had independent means.) From 1871, when he joined the university as a professor, to his death in 1903, his courses attracted an average of slightly over one student a semester. His written work was difficult to follow and employed a private form of notation that many found incomprehensible. But buried among his arcane formulations were insights of the loftiest brilliance.

In 1875–8, Gibbs produced a series of papers, collectively titled *On the Equilibrium of Heterogeneous Substances*, which

*Specifically, it is a measure of randomness or disorder in a system. Darrell Ebbing, in the textbook *General Chemistry*, very usefully suggests thinking of a deck of cards. A new pack fresh out of the box, arranged by suit and in sequence from ace to king, can be said to be in its ordered state. Shuffle the cards and you put them in a disordered state. Entropy is a way of measuring just how disordered that state is and of determining the likelihood of particular outcomes with further shuffles. To grasp entropy fully, it is also necessary to understand concepts such as thermal non-uniformities, lattice distances and stoichiometric relationships, but that's the general idea.

dazzlingly elucidated the thermodynamic principles of, well, nearly everything – 'gases, mixtures, surfaces, solids, phase changes . . . chemical reactions, electrochemical cells, sedimentation, and osmosis', to quote William H. Cropper. In essence, what Gibbs did was show that thermodynamics didn't apply simply to heat and energy at the sort of large and noisy scale of the steam engine, but was also present and influential at the atomic level of chemical reactions. Gibbs's *Equilibrium* has been called 'the *Principia* of thermodynamics', but for reasons that defy speculation Gibbs chose to publish these landmark observations in the *Transactions of the Connecticut Academy of Arts and Sciences*, a journal that managed to be obscure even in Connecticut, which is why Planck did not hear of him until too late.

Undaunted – well, perhaps mildly daunted – Planck turned to other matters.* We shall turn to these ourselves in a moment, but first we must make a slight (but relevant!) detour to Cleveland, Ohio, and an institution then known as the Case School of Applied Science. There, in the 1880s, a physicist of early middle years named Albert Michelson, assisted by his friend the chemist Edward Morley, embarked on a series of experiments that produced curious and disturbing results that would have great ramifications for much of what followed.

What Michelson and Morley did, without actually intending to, was undermine a longstanding belief in something called the luminiferous ether, a stable, invisible,

*Planck was often unlucky in life. His beloved first wife died early, in 1909, and the younger of his two sons was killed in the First World War. He also had twin daughters whom he adored. One died giving birth. The surviving twin went to look after the baby and fell in love with her sister's husband. They married and two years later *she* died in childbirth. In 1944, when Planck was eighty-five, an Allied bomb fell on his house and he lost everything – papers, diaries, a lifetime of accumulations. The following year his surviving son was caught in a conspiracy to assassinate Hitler and executed.

weightless, frictionless and unfortunately wholly imaginary medium that was thought to permeate the universe. Conceived by Descartes, embraced by Newton, and venerated by nearly everyone ever since, the ether held a position of absolute centrality in nineteenth-century physics as a way of explaining how light travelled across the emptiness of space. It was especially needed in the 1800s because light and electromagnetism were now seen as waves, which is to say types of vibrations. Vibrations must occur *in* something; hence the need for, and lasting devotion to, an ether. As late as 1909, the great British physicist J. J. Thomson was insisting: 'The ether is not a fantastic creation of the speculative philosopher; it is as essential to us as the air we breathe' – this more than four years after it was pretty incontestably established that it didn't exist. People, in short, were really attached to the ether.

If you needed to illustrate the idea of nineteenth-century America as a land of opportunity, you could hardly improve on the life of Albert Michelson. Born in 1852 on the German–Polish border to a family of poor Jewish merchants, he came to the United States with his family as an infant and grew up in a mining camp in California's gold rush country where his father ran a dry goods business. Too poor to pay for college, he travelled to Washington, DC, and took to loitering by the front door of the White House so that he could fall in beside Ulysses S. Grant when the President emerged for his daily constitutional. (It was clearly a more innocent age.) In the course of these walks, Michelson so ingratiated himself with the President that Grant agreed to secure for him a free place at the US Naval Academy. It was there that Michelson learned his physics.

Ten years later, by now a professor at the Case School in Cleveland, Michelson became interested in trying to measure something called the ether drift – a kind of headwind produced by moving objects as they ploughed through space. One of the predictions of Newtonian physics was that the speed of light as it pushed through the ether should vary with respect to an observer depending on whether the observer was moving towards the source of light or away from it, but no-one had figured out a way to measure this. It occurred to Michelson that for half the year the Earth is travelling towards the Sun and for half the year it is moving away from it, and he reasoned that if you took careful enough measurements at opposite seasons, and compared light's travel time between the two, you would have your answer.

Michelson talked Alexander Graham Bell, newly enriched inventor of the telephone, into providing the funds to build an ingenious and sensitive instrument of Michelson's own devising called an interferometer, which could measure the velocity of light with great precision. Then, assisted by the genial but shadowy Morley, Michelson embarked on years of fastidious measurements. The work was delicate and exhausting, and had to be suspended for a time to permit Michelson a brief but comprehensive nervous breakdown, but by 1887 they had their results. They were not at all what the two scientists had expected to find.

As Caltech astrophysicist Kip S. Thorne has written: 'The speed of light turned out to be the same in *all* directions and at *all* seasons.' It was the first hint in two hundred years – in exactly two hundred years, in fact – that Newton's laws might not apply all the time everywhere. The Michelson–Morley outcome became, in the words of William H. Cropper, 'probably the most famous negative result in the history of physics'. Michelson was awarded a Nobel Prize in physics for the work – the first American so honoured – but not for twenty years. Meanwhile, the Michelson–Morley experiments would hover unpleasantly, like a musty odour, in the background of scientific thought.

Remarkably, and despite his findings, when the twentieth

century dawned Michelson counted himself among those who believed that the work of science was nearly at an end, with 'only a few turrets and pinnacles to be added, a few roof bosses to be carved', in the words of a writer in *Nature*.

In fact, of course, the world was about to enter a century of science where many people wouldn't understand anything and none would understand everything. Scientists would soon find themselves adrift in a bewildering realm of particles and antiparticles, where things pop in and out of existence in spans of time that make nanoseconds look plodding and uneventful, where everything is strange. Science was moving from a world of macrophysics, where objects could be seen and held and measured, to one of microphysics, where events transpire with inconceivable swiftness on scales of magnitude far below the limits of imagining. We were about to enter the quantum age, and the first person to push on the door was the so-far unfortunate Max Planck.

In 1900, now a theoretical physicist at the University of Berlin, and at the somewhat advanced age of forty-two, Planck unveiled a new 'quantum theory', which posited that energy is not a continuous thing like flowing water but comes in individualized packets, which he called quanta. This *was* a novel concept, and a good one. In the short term it would help to provide a solution to the puzzle of the Michelson–Morley experiments in that it demonstrated that light needn't be a wave after all. In the longer term it would lay the foundation for the whole of modern physics. It was, at all events, the first clue that the world was about to change.

But the landmark event – the dawn of a new age – came in 1905 when there appeared in the German physics journal *Annalen der Physik* a series of papers by a young Swiss bureaucrat who had no university affiliation, no access to a laboratory and the regular use of no library greater than that of the national patent office in Bern, where he was employed as a technical examiner third class. (An application to be promoted to technical examiner second class had recently been rejected.)

His name was Albert Einstein, and in that one eventful year he submitted to *Annalen der Physik* five papers, of which three, according to C. P. Snow, 'were among the greatest in the history of physics' – one examining the photoelectric effect by means of Planck's new quantum theory, one on the behaviour of small particles in suspension (what is known as Brownian motion), and one outlining a Special Theory of Relativity.

The first won its author a Nobel Prize and explained the nature of light (and also helped to make television possible, among other things).* The second provided proof that atoms do indeed exist – a fact that had, surprisingly, been in some dispute. The third merely changed the world.

Einstein was born in Ulm, in southern Germany, in 1879, but grew up in Munich. Little in his early life suggested the greatness to come. Famously, he didn't learn to speak until he was three. In the 1890s, his father's electrical business failing, the family moved to Milan, but Albert, by now a teenager, went to Switzerland to continue his education – though he failed his college entrance exams on the first try. In 1896 he gave up his German citizenship to avoid military

*Einstein was honoured, somewhat vaguely, 'for services to theoretical physics'. He had to wait sixteen years, until 1921, to receive the award – quite a long time, all things considered, but nothing at all compared with Frederick Reines, who detected the neutrino in 1957 but wasn't honoured with a Nobel until 1995, thirty-eight years later, or the German Ernst Ruska, who invented the electron microscope in 1932 and received his Nobel Prize in 1986, more than half a century after the fact. Since Nobel Prizes are never awarded posthumously, longevity can be as important a factor as ingenuity in securing one.

conscription and entered the Zurich Polytechnic Institute on a four-year course designed to churn out high-school science teachers. He was a bright but not outstanding student.

In 1900 he graduated and within a few months was beginning to contribute papers to *Annalen der Physik*. His very first paper, on the physics of fluids in drinking straws (of all things), appeared in the same issue as Planck's quantum theory. From 1902 to 1904 he produced a series of papers on statistical mechanics, only to discover that the quietly productive J. Willard Gibbs in Connecticut had done that work as well, in his *Elementary Principles of Statistical Mechanics* of 1901.

Albert had fallen in love with a fellow student, a Hungarian named Mileva Maric. In 1901 they had a child out of wedlock, a daughter, who was discreetly put up for adoption. Einstein never saw his child. Two years later, he and Maric were married. In between these events, in 1902, Einstein took a job with the Swiss patent office, where he stayed for the next seven years. He enjoyed the work: it was challenging enough to engage his mind, but not so challenging as to distract him from his physics. This was the background against which he produced the Special Theory of Relativity in 1905.

'On the Electrodynamics of Moving Bodies' is one of the most extraordinary scientific papers ever published, as much for how it was presented as for what it said. It had no footnotes or citations, contained almost no mathematics, made no mention of any work that had influenced or preceded it, and acknowledged the help of just one individual, a colleague at the patent office named Michele Besso. It was, wrote C. P. Snow, as if Einstein 'had reached the conclusions by pure thought, unaided, without listening to the opinions of others. To a surprisingly large extent, that is precisely what he had done.'

His famous equation, $E = mc^2$, did not appear with the paper, but came in a brief supplement that followed a few months later. As you will recall from schooldays, E in the equation stands for energy, m for mass and c^2 for the speed of light squared.

In simplest terms, what the equation says is that mass and energy have an equivalence. They are two forms of the same thing: energy is liberated matter; matter is energy waiting to happen. Since c^2 (the speed of light times itself) is a truly enormous number, what the equation is saying is that there is a huge amount – a really huge amount – of energy bound up in every material thing.*

You may not feel outstandingly robust, but if you are an average-sized adult you will contain within your modest frame no less than 7×10^{18} joules of potential energy – enough to explode with the force of thirty very large hydrogen bombs, assuming you knew how to liberate it and really wished to make a point. Everything has this kind of energy trapped within it. We're just not very good at getting it out. Even a uranium bomb – the most energetic thing we have produced yet – releases less than 1 per cent of the energy it could release if only we were more cunning.

Among much else, Einstein's theory explained how radiation worked: how a lump of uranium could throw out constant streams of high-level energy without melting away like an ice cube. (It could do it by converting mass to energy extremely efficiently à la $E = mc^2$.) It explained how stars could burn for billions of years without racing through their fuel. (Ditto.) At a stroke, in a simple formula, Einstein

*How *c* came to be the symbol for the speed of light is something of a mystery, but David Bodanis suggests it probably came from the Latin *celeritas*, meaning swiftness. The relevant volume of the *Oxford English Dictionary*, compiled a decade before Einstein's theory, recognizes *c* as a symbol for many things, from carbon to cricket, but makes no mention of it as a symbol for light or swiftness.

endowed geologists and astronomers with the luxury of billions of years. Above all, the special theory showed that the speed of light was constant and supreme. Nothing could overtake it. It brought light (no pun intended exactly) to the very heart of our understanding of the nature of the universe. Not incidentally, it also solved the problem of the luminiferous ether by making it clear that it didn't exist. Einstein gave us a universe that didn't need it.

Physicists as a rule are not over-attentive to the pronouncements of Swiss patent-office clerks and so, despite the abundance of useful tidings they offered, Einstein's papers attracted little notice. Having just solved several of the deepest mysteries of the universe, Einstein applied for a job as a university lecturer and was rejected, and then for one as a high-school teacher and was rejected there as well. So he went back to his job as an examiner third class – but of course he kept thinking. He hadn't even come close to finishing yet.

When the poet Paul Valéry once asked Einstein if he kept a notebook to record his ideas, Einstein looked at him with mild but genuine surprise. 'Oh, that's not necessary,' he replied. 'It's so seldom I have one.' I need hardly point out that when he did get one it tended to be good. Einstein's next idea was one of the greatest that anyone has ever had – indeed, the very greatest, according to Boorse, Motz and Weaver in their thoughtful history of atomic science. 'As the creation of a single mind,' they write, 'it is undoubtedly the highest intellectual achievement of humanity,' which is of course as good as a compliment can get.

In 1907, or so it has sometimes been written, Albert Einstein saw a workman fall off a roof and began to think about gravity. Alas, like many good stories this one appears to be apocryphal. According to Einstein himself, he was simply sitting in a chair when the problem of gravity occurred to him.

Actually, what occurred to Einstein was something more like the beginning of a solution to the problem of gravity, since it had been evident to him from the outset that one thing missing from the special theory was gravity. What was 'special' about the special theory was that it dealt with things moving in an essentially unimpeded state. But what happened when a thing in motion – light, above all – encountered an obstacle such as gravity? It was a question that would occupy his thoughts for most of the next decade and lead to the publication in early 1917 of a paper entitled 'Cosmological Considerations on the General Theory of Relativity'. The Special Theory of Relativity of 1905 was a profound and important piece of work, of course; but, as C. P. Snow once observed, if Einstein hadn't thought of it when he did someone else would have, probably within five years; it was an idea waiting to happen. But the general theory was something else altogether. 'Without it,' wrote Snow in 1979, 'it is likely that we should still be waiting for the theory today.'

With his pipe, genially self-effacing manner and electrified hair, Einstein was too splendid a figure to remain permanently obscure and in 1919, the war over, the world suddenly discovered him. Almost at once his theories of relativity developed a reputation for being impossible for an ordinary person to grasp. Matters were not helped, as David Bodanis points out in his superb book $E = mc^2$, when the *New York Times* decided to do a story, and – for reasons that can never fail to excite wonder – sent the paper's golfing correspondent, one Henry Crouch, to conduct the interview.

Crouch was hopelessly out of his depth, and got nearly everything wrong. Among the more lasting errors in his report was the assertion that Einstein had found a publisher daring enough to publish a book that only twelve men 'in all the world could comprehend'. There was no such book, no such publisher, no such circle of learned men, but the notion stuck anyway. Soon the number of people who could grasp relativity had been reduced even further in the popular imagination – and the scientific establishment, it must be said, did little to disturb the myth.

When a journalist asked the British astronomer Sir Arthur Eddington if it was true that he was one of only three people in the world who could understand Einstein's relativity theories, Eddington considered deeply for a moment and replied: 'I am trying to think who the third person is.' In fact, the problem with relativity wasn't that it involved a lot of differential equations, Lorentz transformations and other complicated mathematics (though it did – even Einstein needed help with some of it), but that it was just so thoroughly non-intuitive.

In essence what relativity says is that space and time are not absolute, but relative both to the observer and to the thing being observed, and the faster one moves the more pronounced these effects become. We can never accelerate ourselves to the speed of light, and the harder we try (and the faster we go) the more distorted we will become, relative to an outside observer.

Almost at once, popularizers of science tried to come up with ways to make these concepts accessible to a general audience. One of the more successful attempts – commercially at least – was *The ABC of Relativity* by the mathematician and philosopher Bertrand Russell. In it, Russell employed an image that has been used many times since. He asked the reader to envision a train 100 yards long moving at 60 per cent of the speed of light. To someone standing on a platform watching it pass, the train would appear to be only 80 yards long and everything on it would be similarly compressed. If we could hear the passengers on the train speak, their voices would sound slurred and sluggish, like a record played at too slow a speed, and their movements would appear similarly ponderous. Even the clocks on the train would seem to be running at only four-fifths of their normal speed.

However – and here's the thing – people on the train would have no sense of these distortions. To them, everything on the train would seem quite normal. It would be us on the platform who looked weirdly compressed and slowed down. It is all to do, you see, with your position relative to the moving object.

This effect actually happens every time you move. Fly across the United States and you will step from the plane a quinzillionth of a second, or something, younger than those you left behind. Even in walking across the room you will very slightly alter your own experience of time and space. It has been calculated that a baseball thrown at 160 kilometres an hour will pick up 0.000000000002 grams of mass on its way to home plate. So the effects of relativity are real and have been measured. The problem is that such changes are much too small to make the tiniest detectable difference to us. But for other things in the universe – light, gravity, the universe itself – these are matters of consequence.

So if the ideas of relativity seem weird, it is only because we don't experience these sorts of interactions in normal life. However, to turn to Bodanis again, we all commonly encounter other kinds of relativity – for instance, with regard to sound. If you are in a park and someone is playing annoying music, you know that if you move to a more distant spot the music will seem quieter. That's not because the music *is* quieter, of course, but simply that your position relative to it has changed. To something too small or sluggish to duplicate this experience – a snail, say – the idea that a boom box could seem to two observers to produce two different volumes of music simultaneously might seem incredible.

The most challenging and non-intuitive of all the

concepts in the General Theory of Relativity is the idea that time is part of space. Our instinct is to regard time as eternal, absolute, immutable; to believe that nothing can disturb its steady tick. In fact, according to Einstein, time is variable and ever-changing. It even has shape. It is bound up – 'inextricably interconnected', in Stephen Hawking's expression – with the three dimensions of space in a curious dimension known as spacetime.

Spacetime is usually explained by asking you to imagine something flat but pliant – a mattress, say, or a sheet of stretched rubber – on which is resting a heavy round object, such as an iron ball. The weight of the iron ball causes the material on which it is sitting to stretch and sag slightly. This is roughly analogous to the effect that a massive object such as the Sun (the iron ball) has on spacetime (the material): it stretches and curves and warps it. Now, if you roll a smaller ball across the sheet, it tries to go in a straight line as required by Newton's laws of motion, but as it nears the massive object and the slope of the sagging fabric, it rolls downwards, ineluctably drawn to the more massive object. This is gravity – a product of the bending of spacetime.

Every object that has mass creates a little depression in the fabric of the cosmos. Thus the universe, as Dennis Overbye has put it, is 'the ultimate sagging mattress'. Gravity on this view is no longer so much a thing as an outcome – 'not a "force" but a byproduct of the warping of spacetime', in the words of the physicist Michio Kaku, who goes on: 'In some sense, gravity does not exist; what moves the planets and stars is the distortion of space and time.'

Of course, the sagging mattress analogy can take us only so far, because it doesn't incorporate the effect of time. But then, our brains can take us only so far, because it is so nearly impossible to envision a dimension comprising three parts space to one part time, all interwoven like the threads in a plaid fabric. At all events, I think we can agree that this was an awfully big thought for a young man staring out of the window of a patent office in the capital of Switzerland.

Among much else, Einstein's General Theory of Relativity suggested that the universe must be either expanding or contracting. But Einstein was not a cosmologist and he accepted the prevailing wisdom that the universe was fixed and eternal. More or less reflexively, he dropped into his equations something called the cosmological constant, which arbitrarily counterbalanced the effects of gravity, serving as a kind of mathematical pause button. Books on the history of science always forgive Einstein this lapse, but it was actually a fairly appalling piece of science and he knew it. He called it 'the biggest blunder of my life'.

Coincidentally, at about the time that Einstein was affixing a cosmological constant to his theory, at the Lowell Observatory in Arizona an astronomer with the cheerily intergalactic name of Vesto Slipher (who was in fact from Indiana) was taking spectrographic readings of distant stars and discovering that they appeared to be moving away from us. The universe wasn't static. The stars Slipher looked at showed unmistakable signs of a Doppler shift – the same mechanism behind that distinctive stretched-out *yee–yummm* sound cars make as they flash past on a racetrack.* The phenomenon also applies to light, and in the case of receding galaxies it is known as a red shift (because light moving away

---

*Named for Johann Christian Doppler, an Austrian physicist, who first noticed the effect in 1842. Briefly, what happens is that as a moving object approaches a stationary one its sound waves become bunched up as they cram up against whatever device is receiving them (your ears, say), just as you would expect of anything that is being pushed from behind towards an immobile object. This bunching is perceived by the listener as a kind of pinched and elevated sound (the *yee*). As the sound source passes, the sound waves spread out and lengthen, causing the pitch to drop abruptly (the *yummm*).

from us shifts towards the red end of the spectrum; approaching light shifts to blue).

Slipher was the first to notice this effect with light and to realize its potential importance for understanding the motions of the cosmos. Unfortunately, no-one much noticed him. The Lowell Observatory, as you will recall, was a bit of an oddity thanks to Percival Lowell's obsession with Martian canals, which in the 1910s made it, in every sense, an outpost of astronomical endeavour. Slipher was unaware of Einstein's theory of relativity and the world was equally unaware of Slipher. So his finding had no impact.

Glory instead would pass to a large mass of ego named Edwin Hubble. Hubble was born in 1889, ten years after Einstein, in a small Missouri town on the edge of the Ozarks, and grew up there and in Wheaton, Illinois, a suburb of Chicago. His father was a successful insurance executive, so life was always comfortable, and Edwin enjoyed a wealth of physical endowments, too. He was a strong and gifted athlete, charming, smart and immensely good-looking – 'handsome almost to a fault', in the description of William H. Cropper, 'an Adonis' in the words of another admirer. According to his own accounts, he also managed to fit into his life more or less constant acts of valour – rescuing drowning swimmers, leading frightened men to safety across the battlefields of France, embarrassing world-champion boxers with knockdown punches in exhibition bouts. It all seemed too good to be true. It was. For all his gifts, Hubble was also an inveterate liar.

This was more than a little odd, for Hubble's life was filled from an early age with a level of genuine distinction that was at times almost ludicrously golden. At a single high-school track meeting in 1906, he won the pole vault, shot-put, discus, hammer throw, standing high jump and running high jump, and was on the winning mile relay team – that is, seven first places in one meeting – and came third in the long jump. In the same year, he set a state record for the high jump in Illinois.

As a scholar he was equally proficient, and had no trouble gaining admission to study physics and astronomy at the University of Chicago (where, coincidentally, the head of the department was now Albert Michelson). There he was selected to be one of the first Rhodes Scholars at Oxford. Three years of English life evidently turned his head, for he returned to Wheaton in 1913 wearing an Inverness cape, smoking a pipe and talking with a peculiarly orotund accent – not quite British but not quite not – that would remain with him for life. Though he later claimed to have passed most of the second decade of the century practising law in Kentucky, in fact he worked as a high-school teacher and basketball coach in New Albany, Indiana, before belatedly attaining his doctorate and passing briefly through the Army. (He arrived in France one month before the armistice and almost certainly never heard a shot fired in anger.)

In 1919, now aged thirty, he moved to California and took up a position at the Mount Wilson Observatory near Los Angeles. Swiftly, and more than a little unexpectedly, he became the most outstanding astronomer of the twentieth century.

It is worth pausing for a moment to consider just how little was known of the cosmos at this time. Astronomers today believe there are perhaps 140 billion galaxies in the visible universe. That's a huge number, much bigger than merely saying it would lead you to suppose. If galaxies were frozen peas, it would be enough to fill a large auditorium – the old Boston Garden, say, or the Royal Albert Hall. (An astrophysicist named Bruce Gregory has actually computed this.) In 1919, when Hubble first put his head to the eyepiece, the number of these galaxies that were known to us was exactly one: the Milky Way. Everything else was thought to be either part of the Milky Way itself or one of many distant, peripheral puffs of gas. Hubble quickly demonstrated how wrong that belief was.

Over the next decade, Hubble tackled two of the most fundamental questions of the universe: how old is it, and how big? To answer both it is necessary to know two things – how far away certain galaxies are and how fast they are flying away from us (what is known as their recessional velocity). The red shift gives the speed at which galaxies are retiring, but doesn't tell us how far away they are to begin with. For that you need what are known as 'standard candles' – stars whose brightness can be reliably calculated and used as benchmarks to measure the brightness (and hence relative distance) of other stars.

Hubble's luck was to come along soon after an ingenious woman named Henrietta Swan Leavitt had figured out a way to find these stars. Leavitt worked at the Harvard College Observatory as a computer, as they were known. Computers spent their lives studying photographic plates of stars and making computations – hence the name. It was little more than drudgery by another name, but it was as close as women could get to real astronomy at Harvard – or, indeed, pretty much anywhere – in those days. The system, however unfair, did have certain unexpected benefits: it meant that half the finest minds available were directed to work that would otherwise have attracted little reflective attention and it ensured that women ended up with an appreciation of the fine structure of the cosmos that often eluded their male counterparts.

One Harvard computer, Annie Jump Cannon, used her repetitive acquaintance with the stars to devise a system of stellar classifications so practical that it is still in use today. Leavitt's contribution was even more profound. She noticed that a type of star known as a Cepheid variable (after the constellation Cepheus, where the first was identified) pulsated with a regular rhythm – a kind of stellar heartbeat.

Cepheids are quite rare, but at least one of them is well known to most of us. Polaris, the Pole Star, is a Cepheid.

We now know that Cepheids throb as they do because they are elderly stars that have moved past their 'main sequence phase', in the parlance of astronomers, and become red giants. The chemistry of red giants is a little weighty for our purposes here (it requires an appreciation for the properties of singly ionized helium atoms, among quite a lot else), but put simply it means that they burn their remaining fuel in a way that produces a very rhythmic, very reliable brightening and dimming. Leavitt's genius was to realize that by comparing the relative magnitudes of Cepheids at different points in the sky you could work out where they were in relation to each other. They could be used as standard candles – a term she coined and still in universal use. The method provided only relative distances, not absolute distances, but even so it was the first time that anyone had come up with a usable way to measure the large-scale universe.

(Just to put these insights into perspective, it is perhaps worth noting that at the time Leavitt and Cannon were inferring fundamental properties of the cosmos from dim smudges of distant stars on photographic plates, the Harvard astronomer William H. Pickering, who could of course peer into a first-class telescope as often as he wanted, was developing *his* seminal theory that dark patches on the Moon were caused by swarms of seasonally migrating insects.)

Combining Leavitt's cosmic yardstick with Vesto Slipher's handy red shifts, Hubble began to measure selected points in space with a fresh eye. In 1923 he showed that a puff of distant gossamer in the Andromeda constellation known as M31 wasn't a gas cloud at all, but a blaze of stars, a galaxy in its own right, a hundred thousand light years across and at least nine hundred thousand light years away. The universe was vaster – vastly vaster – than anyone had ever supposed. In 1924 Hubble produced a landmark paper, 'Cepheids in Spiral Nebulae' (nebulae, from the Latin for 'clouds', was his word for galaxies), showing that the universe consisted not just of the Milky Way but of lots of independent galaxies – 'island universes' – many of them bigger than the Milky Way and much more distant.

This finding alone would have ensured Hubble's reputation, but he now turned to the question of working out just how much vaster the universe was, and made an even more striking discovery. Hubble began to measure the spectra of distant galaxies – the business that Slipher had begun in Arizona. Using Mount Wilson's new 100-inch Hooker telescope and some clever inferences, by the early 1930s he had worked out that all the galaxies in the sky (except for our own local cluster) are moving away from us. Moreover their speed and distance were neatly proportional: the further away the galaxy, the faster it was moving.

This was truly startling. The universe was expanding, swiftly and evenly in all directions. It didn't take a huge amount of imagination to read backwards from this and realize that it must therefore have started from some central point. Far from being the stable, fixed, eternal void that everyone had always assumed, this was a universe that had a beginning. It might therefore also have an end.

The wonder, as Stephen Hawking has noted, is that no-one had hit on the idea of the expanding universe before. A static universe, as should have been obvious to Newton and every thinking astronomer since, would collapse in upon itself. There was also the problem that if stars had been burning indefinitely in a static universe they'd have made the whole intolerably hot – certainly much too hot for the likes of us. An expanding universe resolved much of this at a stroke.

Hubble was a much better observer than a thinker and

didn't immediately appreciate the full implications of what he had found. Partly this was because he was woefully ignorant of Einstein's General Theory of Relativity. This was quite remarkable because, for one thing, Einstein and his theory were world-famous by now. Moreover, in 1929 Albert Michelson – now in his twilight years but still one of the world's most alert and esteemed scientists – accepted a position at Mount Wilson to measure the velocity of light with his trusty interferometer, and must surely have at least mentioned to him the applicability of Einstein's theory to his own findings.

At all events, Hubble failed to make theoretical hay when the chance was there. Instead, it was left to a Belgian priest–scholar (with a PhD from MIT) named Georges Lemaître to bring together the two strands in his own 'fireworks theory', which suggested that the universe began as a geometrical point, a 'primeval atom', which burst into glory and had been moving apart ever since. It was an idea that very neatly anticipated the modern conception of the Big Bang, but was so far ahead of its time that Lemaître seldom gets more than the sentence or two that we have given him here. The world would need additional decades, and the inadvertent discovery of cosmic background radiation by Penzias and Wilson at their hissing antenna in New Jersey, before the Big Bang would begin to move from interesting idea to established theory.

Neither Hubble nor Einstein would be much of a part of that big story. Though no-one would have guessed it at the time, both men had done about as much as they were ever going to do.

In 1936 Hubble produced a popular book called *The Realm of the Nebulae*, which explained in flattering style his own considerable achievements. Here at last he showed that he had acquainted himself with Einstein's theory – up to a point, anyway: he gave it four pages out of about two hundred.

Hubble died of a heart attack in 1953. One last small oddity awaited him. For reasons cloaked in mystery, his wife declined to have a funeral and never revealed what she did with his body. Half a century later the whereabouts of the century's greatest astronomer remain unknown. For a memorial you must look to the sky and the Hubble Space Telescope, launched in 1990 and named in his honour.

9

THE MIGHTY ATOM

While Einstein and Hubble were productively unravelling the large-scale structure of the cosmos, others were struggling to understand something closer to hand but in its way just as remote: the tiny and ever-mysterious atom.

The great Caltech physicist Richard Feynman once observed that if you had to reduce scientific history to one important statement it would be: 'All things are made of atoms.' They are everywhere and they constitute everything. Look around you. It is all atoms. Not just the solid things like walls and tables and sofas, but the air in between. And they are there in numbers that you really cannot conceive.

The basic working arrangement of atoms is the molecule (from the Latin for 'little mass'). A molecule is simply two or more atoms working together in a more or less stable arrangement: add two atoms of hydrogen to one of oxygen and you have a molecule of water. Chemists tend to think in terms of molecules rather than elements in much the way that writers tend to think in terms of words and not letters, so it is molecules they count, and these are numerous to say the least. At sea level, at a temperature of 0 degrees Celsius, one cubic centimetre of air (that is, a space about the size of a sugar cube) will contain 45 billion billion molecules. And they are in every single cubic centimetre you see around you. Think how many cubic centimetres there are in the world outside your window – how many sugar cubes it would take to fill that view. Then think how many it would take to build a universe. Atoms, in short, are very abundant.

They are also fantastically durable. Because they are so long-lived, atoms really get around. Every atom you possess has almost certainly passed through several stars and been part of millions of organisms on its way to becoming you. We are each so atomically numerous and so vigorously recycled at death that a significant number of our atoms – up to a billion for each of us, it has been suggested – probably once belonged to Shakespeare. A billion more each came from Buddha and Genghis Khan and Beethoven, and any other historical figure you care to name. (The personages have to be historical, apparently, as it takes the atoms some decades to become thoroughly redistributed; however much you may wish it, you are not yet one with Elvis Presley.)

So we are all reincarnations – though short-lived ones. When we die, our atoms will disassemble and move off to find new uses elsewhere – as part of a leaf or other human being or drop of dew. Atoms themselves, however, go on practically for ever. Nobody actually knows how long an atom can survive, but according to Martin Rees it is probably about 10^{35} years – a number so big that even I am happy to express it in mathematical notation.

Above all, atoms are tiny – very tiny indeed. Half a million of them lined up shoulder to shoulder could hide behind a human hair. On such a scale an individual atom is essentially impossible to imagine, but we can of course try.

Start with a millimetre, which is a line this long: -. Now imagine that line divided into a thousand equal widths. Each of those widths is a micron. This is the scale of microorganisms. A typical paramecium, for instance – a tiny, single-celled, freshwater creature – is about 2 microns wide, 0.002 millimetres, which is really very small. If you wanted to see with your naked eye a paramecium swimming in a drop of water, you would have to enlarge the drop until it was some 12 metres across. However, if you wanted to see the atoms in the same drop, you would have to make the drop 24 *kilometres* across.

Atoms, in other words, exist on a scale of minuteness of another order altogether. To get down to the scale of atoms, you would need to take each one of those micron slices and shave it into ten thousand finer widths. *That*'s the scale of an atom: one ten-millionth of a millimetre. It is a degree of slenderness way beyond the capacity of our imaginations, but you can get some idea of the proportions if you bear in mind that one atom is to that millimetre line above as the thickness of a sheet of paper is to the height of the Empire State Building.

It is, of course, the abundance and extreme durability of atoms that make them so useful, and the tininess that makes them so hard to detect and understand. The realization that atoms are these three things – small, numerous, practically indestructible – and that all things are made from them first occurred not to Antoine-Laurent Lavoisier, as you might expect, or even to Henry Cavendish or Humphry Davy, but rather to a spare and lightly educated English Quaker named John Dalton, whom we first encountered in Chapter 7.

Dalton was born in 1766 on the edge of the Lake District, near Cockermouth, to a family of poor and devout Quaker weavers. (Four years later the poet William Wordsworth would also join the world at Cockermouth.) He was an exceptionally bright student – so very bright, indeed, that at the improbably youthful age of twelve he was put in charge of the local Quaker school. This perhaps says as much about the school as about Dalton's precocity, but perhaps not: we know from his diaries that at about this time he was reading

Newton's *Principia* – in the original Latin – and other works of a similarly challenging nature. At fifteen, still schoolmastering, he took a job in the nearby town of Kendal, and a decade after that he moved to Manchester, whence he scarcely stirred for the remaining fifty years of his life. In Manchester he became something of an intellectual whirlwind, producing books and papers on subjects ranging from meteorology to grammar. Colour blindness, a condition from which he suffered, was for a long time called Daltonism because of his studies. But it was a plump book called *A New System of Chemical Philosophy*, published in 1808, that established his reputation.

There, in a short chapter of just five pages (out of the book's more than nine hundred), people of learning first encountered atoms in something approaching their modern conception. Dalton's simple insight was that at the root of all matter are exceedingly tiny, irreducible particles. 'We might as well attempt to introduce a new planet into the solar system or annihilate one already in existence, as to create or destroy a particle of hydrogen,' he wrote.

Neither the idea of atoms nor the term itself was exactly new. Both had been developed by the ancient Greeks. Dalton's contribution was to consider the relative sizes and characters of these atoms and how they fit together. He knew, for instance, that hydrogen was the lightest element, so he gave it an atomic weight of 1. He believed also that water consisted of seven parts of oxygen to one of hydrogen, and so he gave oxygen an atomic weight of 7. By such means was he able to arrive at the relative weights of the known elements. He wasn't always terribly accurate – oxygen's atomic weight is actually 16, not 7 – but the principle was sound and formed the basis for all of modern chemistry and much of the rest of modern science.

The work made Dalton famous – albeit in a low-key, English Quaker sort of way. In 1826, the French chemist

P. J. Pelletier travelled to Manchester to meet the atomic hero. Pelletier expected to find him attached to some grand institution, so he was astounded to discover him teaching elementary arithmetic to boys in a small school on a back street. According to the scientific historian E. J. Holmyard, a confused Pelletier, upon beholding the great man, stammered:

> 'Est-ce que j'ai l'honneur de m'addresser à Monsieur Dalton?' for he could hardly believe his eyes that this was the chemist of European fame, teaching a boy his first four rules. 'Yes,' said the matter-of-fact Quaker. 'Wilt thou sit down whilst I put this lad right about his arithmetic?'

Although Dalton tried to avoid all honours, he was elected to the Royal Society against his wishes, showered with medals and given a handsome government pension. When he died in 1844, forty thousand people viewed the coffin and the funeral cortège stretched for two miles. His entry in the *Dictionary of National Biography* is one of the longest, rivalled in length among nineteenth-century men of science only by those of Darwin and Lyell.

For a century after Dalton made his proposal, it remained entirely hypothetical, and a few eminent scientists – notably the Viennese physicist Ernst Mach, for whom is named the speed of sound – doubted the existence of atoms at all. 'Atoms cannot be perceived by the senses . . . they are things of thought,' he wrote. Such was the scepticism with which the existence of atoms was viewed in the German-speaking world in particular that it was said to have played a part in the suicide of the great theoretical physicist and atomic enthusiast Ludwig Boltzmann in 1906.

It was Einstein who provided the first incontrovertible evidence of atoms' existence with his paper on Brownian motion in 1905, but this attracted little attention and in any case Einstein was soon to become consumed with his work on general relativity. So the first real hero of the atomic age, if not the first personage on the scene, was Ernest Rutherford.

Rutherford was born in 1871 in the 'back blocks' of New Zealand to parents who had emigrated from Scotland to raise a little flax and a lot of children (to paraphrase Steven Weinberg). Growing up in a remote part of a remote country, he was about as far from the mainstream of science as it was possible to be, but in 1895 he won a scholarship that took him to the Cavendish Laboratory at Cambridge University, which was about to become the hottest place in the world to do physics.

Physicists are notoriously scornful of scientists from other fields. When the great Austrian physicist Wolfgang Pauli's wife left him for a chemist, he was staggered with disbelief. 'Had she taken a bullfighter I would have understood,' he remarked in wonder to a friend. 'But a *chemist* . . .'

It was a feeling Rutherford would have understood. 'All science is either physics or stamp collecting,' he once said, in a line that has been used many times since. There is a certain engaging irony, therefore, that his award of the Nobel Prize in 1908 was in chemistry, not physics.

Rutherford was a lucky man – lucky to be a genius, but even luckier to live at a time when physics and chemistry were so exciting and so compatible (his own sentiments notwithstanding). Never again would they quite so comfortably overlap.

For all his success, Rutherford was not an especially brilliant man and was actually pretty terrible at mathematics. Often during lectures he would get so lost in his own equations that he would give up halfway through and tell the students to work it out for themselves. According to his longtime colleague James Chadwick, discoverer of the neutron, he wasn't even particularly clever at experimentation. He was simply tenacious and open-minded. For brilliance he substituted shrewdness and a kind of daring. His mind, in the words of one biographer, was 'always operating out towards the frontiers, as far as he could see, and that was a great deal further than most other men'. Confronted with an intractable problem, he was prepared to work at it harder and longer than most people and to be more receptive to unorthodox explanations. His greatest breakthrough came because he was prepared to spend immensely tedious hours sitting at a screen counting alpha particle scintillations, as they were known – the sort of work that would normally have been farmed out. He was one of the first – possibly the very first – to see that the power inherent in the atom could, if harnessed, make bombs powerful enough to 'make this old world vanish in smoke'.

Physically he was big and booming, with a voice that made the timid shrink. Once, when told that Rutherford was about to make a radio broadcast across the Atlantic, a colleague drily asked: 'Why use radio?' He also had a huge amount of good-natured confidence. When someone remarked to him that he seemed always to be at the crest of a wave, he responded, 'Well, after all, I made the wave, didn't I?' C. P. Snow recalled how, in a Cambridge tailor's, he overheard Rutherford remark: 'Every day I grow in girth. And in mentality.'

But both girth and fame were far ahead of him in 1895 when he fetched up at the Cavendish.* It was a singularly eventful period in science. In the year of Rutherford's arrival in Cambridge, Wilhelm Roentgen discovered X-rays at the

*The name comes from the same Cavendishes who produced Henry. This one was William Cavendish, seventh Duke of Devonshire, who was a gifted mathematician and steel baron in Victorian England. In 1870 he gave the university £6,300 to build an experimental laboratory.

University of Würzburg in Germany; the next year, Henri Becquerel discovered radioactivity. And the Cavendish itself was about to embark on a long period of greatness. In 1897, J. J. Thomson and colleagues would discover the electron there, in 1911 C. T. R. Wilson would produce the first particle detector there (as we shall see), and in 1932 James Chadwick would discover the neutron there. Further still in the future, in 1953, James Watson and Francis Crick would discover the structure of DNA at the Cavendish.

In the beginning Rutherford worked on radio waves, and with some distinction – he managed to transmit a crisp signal more than a mile, a very reasonable achievement for the time – but he gave it up when he was persuaded by a senior colleague that radio had little future. On the whole, however, Rutherford didn't thrive at the Cavendish, and after three years there, feeling he was going nowhere, he took a post at McGill University in Montreal, where he began his long and steady rise to greatness. By the time he received his Nobel Prize (for 'investigations into the disintegration of the elements, and the chemistry of radioactive substances', according to the official citation) he had moved on to Manchester University, and it was there, in fact, that he would do his most important work in determining the structure and nature of the atom.

By the early twentieth century it was known that atoms were made of parts – Thomson's discovery of the electron had established that – but it wasn't known how many parts there were or how they fitted together or what shape they took. Some physicists thought that atoms might be cube-shaped, because cubes can be packed together so neatly without any wasted space. The more general view, however, was that an atom was more like a currant bun or a plum pudding: a dense, solid object that carried a positive charge but that was studded with negatively charged electrons, like the currants in a currant bun.

In 1910, Rutherford (assisted by his student Hans Geiger, who would later invent the radiation detector that bears his name) fired ionized helium atoms, or alpha particles, at a sheet of gold foil.* To Rutherford's astonishment, some of the particles bounced back. It was as if, he said, he had fired a 15-inch shell at a sheet of paper and it rebounded into his lap. This was just not supposed to happen. After considerable reflection he realized there could be only one possible explanation: the particles that bounced back were striking something small and dense at the heart of the atom, while the other particles sailed through unimpeded. An atom, Rutherford realized, was mostly empty space, with a very dense nucleus at the centre. This was a most gratifying discovery, but it presented one immediate problem. By all the laws of conventional physics, atoms shouldn't therefore exist.

Let us pause for a moment and consider the structure of the atom as we know it now. Every atom is made from three kinds of elementary particles: protons, which have a positive electrical charge; electrons, which have a negative electrical charge; and neutrons, which have no charge. Protons and neutrons are packed into the nucleus, while electrons spin around outside. The number of protons is what gives an atom its chemical identity. An atom with one proton is an atom of hydrogen, one with two protons is helium, with three protons lithium, and so on up the scale. Each time you add a proton you get a new element. (Because the number of protons in an atom is always balanced by an equal number of electrons, you will sometimes see it written that it is the number of electrons that defines an element; it comes to the same thing. The way it was explained to me is that

*Geiger would also later become a loyal Nazi, unhesitatingly betraying Jewish colleagues, including many who had helped him.

protons give an atom its identity, electrons its personality.)

Neutrons don't influence an atom's identity, but they do add to its mass. The number of neutrons is generally about the same as the number of protons, but they can vary up and down slightly. Add or subtract a neutron or two and you get an isotope. The terms you hear in reference to dating techniques in archaeology refer to isotopes – carbon-14, for instance, which is an atom of carbon with six protons and eight neutrons (the fourteen being the sum of the two).

Neutrons and protons occupy the atom's nucleus. The nucleus of an atom is tiny – only one-millionth of a billionth of the full volume of the atom – but fantastically dense, since it contains virtually all the atom's mass. As Cropper has put it, if an atom were expanded to the size of a cathedral, the nucleus would be only about the size of a fly – but a fly many thousands of times heavier than the cathedral. It was this spaciousness – this resounding, unexpected roominess – that had Rutherford scratching his head in 1910.

It is still a fairly astounding notion to consider that atoms are mostly empty space, and that the solidity we experience all around us is an illusion. When two objects come together in the real world – billiard balls are most often used for illustration – they don't actually strike each other. 'Rather,' as Timothy Ferris explains, 'the negatively charged fields of the two balls repel each other . . . [W]ere it not for their electrical charges they could, like galaxies, pass right through each other unscathed.' When you sit in a chair, you are not actually sitting there, but levitating above it at a height of one angstrom (a hundred millionth of a centimetre), your electrons and its electrons implacably opposed to any closer intimacy.

The picture of an atom that nearly everybody has in mind is of an electron or two flying around a nucleus, like planets orbiting a sun. This image was created in 1904, based on little more than clever guesswork, by a Japanese physicist named Hantaro Nagaoka. It is completely wrong, but durable just the same. As Isaac Asimov liked to note, it inspired generations of science-fiction writers to create stories of worlds-within-worlds, in which atoms become tiny inhabited solar systems or our solar system turns out to be merely a mote in some much larger scheme. Even now CERN, the European Organization for Nuclear Research, uses Nagaoka's image as a logo on its website. In fact, as physicists were soon to realize, electrons are not like orbiting planets at all, but more like the blades of a spinning fan, managing to fill every bit of space in their orbits simultaneously (but with the crucial difference that the blades of a fan only *seem* to be everywhere at once; electrons *are*).

Needless to say, very little of this was understood in 1910 or for many years afterwards. Rutherford's finding presented some large and immediate problems, not least that no electron should be able to orbit a nucleus without crashing. Conventional electrodynamic theory demanded that a flying electron should run out of energy very quickly – in only an instant or so – and spiral into the nucleus, with disastrous consequences for both. There was also the problem of how protons, with their positive charges, could bundle together inside the nucleus without blowing themselves and the rest of the atom apart. Clearly, whatever was going on down there in the world of the very small was not governed by the laws that applied in the macro world where our expectations reside.

As physicists began to delve into this subatomic realm, they realized that it wasn't merely different from anything we knew, but different from anything ever imagined. 'Because atomic behaviour is so unlike ordinary experience,' Richard Feynman once observed, 'it is very difficult to get used to and it appears peculiar and mysterious to everyone, both to the novice and to the experienced physicist.' When

Feynman made that comment, physicists had had half a century to adjust to the strangeness of atomic behaviour. So think how it must have felt to Rutherford and his colleagues in the early 1910s when it was all brand new.

One of the people working with Rutherford was a mild and affable young Dane named Niels Bohr. In 1913, while puzzling over the structure of the atom, Bohr had an idea so exciting that he postponed his honeymoon to write what became a landmark paper.

Because physicists couldn't see anything so small as an atom, they had to try to work out its structure from how it behaved when they did things to it, as Rutherford had done by firing alpha particles at foil. Sometimes, not surprisingly, the results of these experiments were puzzling. One puzzle that had been around for a long time was to do with spectrum readings of the wavelengths of hydrogen. These produced patterns showing that hydrogen atoms emitted energy at certain wavelengths but not others. It was rather as if someone under surveillance kept turning up at particular locations but was never observed travelling between them. No-one could understand why this should be.

It was while puzzling over this problem that Bohr was struck by a solution and dashed off his famous paper. Called 'On the Constitutions of Atoms and Molecules', the paper explained how electrons could keep from falling into the nucleus by suggesting that they could occupy only certain well-defined orbits. According to the new theory, an electron moving between orbits would disappear from one and reappear instantaneously in another *without visiting the space between*. This idea – the famous 'quantum leap' – is of course utterly strange, but it was too good not to be true. It not only kept electrons from spiralling catastrophically into the nucleus, it also explained hydrogen's bewildering wavelengths. The electrons only appeared in certain orbits because they only existed in certain orbits. It was a dazzling insight and it won Bohr the 1922 Nobel Prize in physics, the year after Einstein received his.

Meanwhile the tireless Rutherford, now back at Cambridge having succeeded J. J. Thomson as head of the Cavendish Laboratory, came up with a model that explained why the nuclei didn't blow up. He saw that the positive charge of the protons must be offset by some type of neutralizing particles, which he called neutrons. The idea was simple and appealing, but not easy to prove. Rutherford's associate, James Chadwick, devoted eleven intensive years to hunting for neutrons before finally succeeding in 1932. He, too, was awarded a Nobel Prize in physics, in 1935. As Boorse and his colleagues point out in their history of the subject, the delay in discovery was probably a very good thing, as mastery of the neutron was essential to the development of the atomic bomb. (Because neutrons have no charge, they aren't repelled by the electrical fields at the heart of an atom and thus could be fired like tiny torpedoes into an atomic nucleus, setting off the destructive process known as fission.) Had the neutron been isolated in the 1920s, they note, it is 'very likely the atomic bomb would have been developed first in Europe, undoubtedly by the Germans'.

As it was, the Europeans had their hands full trying to understand the strange behaviour of the electron. The principal problem they faced was that the electron sometimes behaved like a particle and sometimes like a wave. This impossible duality drove physicists nearly mad. For the next decade all across Europe they furiously thought and scribbled and offered competing hypotheses. In France, Prince Louis-Victor de Broglie, the scion of a ducal family, found that certain anomalies in the behaviour of electrons disappeared when one regarded them as waves. The observation excited the attention of the Austrian Erwin Schrödinger, who made some deft refinements and devised a handy system called wave mechanics. At almost the same time, the German physicist Werner Heisenberg came up with a competing theory called matrix mechanics. This was so mathematically complex that hardly anyone really understood it, including Heisenberg himself ('I do not even know what a matrix *is*,' Heisenberg despaired to a friend at one point), but it did seem to solve certain problems that Schrödinger's waves failed to explain.

The upshot is that physics had two theories, based on conflicting premises, that produced the same results. It was an impossible situation.

Finally, in 1926, Heisenberg came up with a celebrated compromise, producing a new discipline that came to be known as quantum mechanics. At the heart of it was Heisenberg's Uncertainty Principle, which states that the electron is a particle but a particle that can be described in terms of waves. The uncertainty around which the theory is built is that we can know the path an electron takes as it moves through a space or we can know where it is at a given instant, but we cannot know both.* Any attempt to measure one will unavoidably disturb the other. This isn't a matter of simply needing more precise instruments; it is an immutable property of the universe.

What this means in practice is that you can never predict where an electron will be at any given moment. You can only list its probability of being there. In a sense, as Dennis Overbye has put it, an electron doesn't exist until it is observed. Or, put slightly differently, until it is observed an

*There is a little uncertainty about the use of the word uncertainty in regard to Heisenberg's principle. Michael Frayn, in an afterword to his play *Copenhagen*, notes that several words in German – *Unsicherheit, Unschärfe, Ungenauigkeit* and *Unbestimmtheit* – have been used by various translators, but that none quite equates to the English *uncertainty*. Frayn suggests that *indeterminacy* would be a better word for the principle and *indeterminability* would be better still. Heisenberg himself generally used *Unbestimmtheit*.

electron must be regarded as being 'at once everywhere and nowhere'.

If this seems confusing, you may take some comfort in knowing that it was confusing to physicists, too. Overbye notes: 'Bohr once commented that a person who wasn't outraged on first hearing about quantum theory didn't understand what had been said.' Heisenberg, when asked how one could envision an atom, replied: 'Don't try.'

So the atom turned out to be quite unlike the image that most people had created. The electron doesn't fly around the nucleus like a planet around its sun, but instead takes on the more amorphous aspect of a cloud. The 'shell' of an atom isn't some hard, shiny casing, as illustrations sometimes encourage us to suppose, but simply the outermost of these fuzzy electron clouds. The cloud itself is essentially just a zone of statistical probability marking the area beyond which the electron only very seldom strays. Thus an atom, if you could see it, would look more like a very fuzzy tennis ball than a hard-edged metallic sphere (but not much like either or, indeed, like anything you've ever seen; we are, after all, dealing here with a world very different from the one we see around us).

It seemed as if there was no end of strangeness. For the first time, as James Trefil has put it, scientists had encountered 'an area of the universe that our brains just aren't wired to understand'. Or, as Feynman expressed it, 'things on a small scale behave *nothing* like things on a large scale.' As physicists delved deeper, they realized they had found a world not only where electrons could jump from one orbit to another without travelling across any intervening space, but where matter could pop into existence from nothing at all – 'provided', in the words of Alan Lightman of MIT, 'it disappears again with sufficient haste.'

Perhaps the most arresting of quantum improbabilities is the idea, arising from Wolfgang Pauli's Exclusion Principle of 1925, that certain pairs of subatomic particles, even when separated by the most considerable distances, can each instantly 'know' what the other is doing. Particles have a quality known as spin and, according to quantum theory, the moment you determine the spin of one particle, its sister particle, no matter how distant away, will immediately begin spinning in the opposite direction and at the same rate.

It is as if, in the words of the science writer Lawrence Joseph, you had two identical pool balls, one in Ohio and the other in Fiji, and that the instant you sent one spinning the other would immediately spin in a contrary direction at precisely the same speed. Remarkably, the phenomenon was proved in 1997 when physicists at the University of Geneva sent photons seven miles in opposite directions and demonstrated that interfering with one provoked an instantaneous response in the other.

Things reached such a pitch that at one conference Bohr remarked of a new theory that the question was not whether it was crazy, but whether it was crazy enough. To illustrate the non-intuitive nature of the quantum world, Schrödinger offered a famous thought experiment in which a hypothetical cat was placed in a box with one atom of a radioactive substance attached to a vial of hydrocyanic acid. If the particle degraded within an hour, it would trigger a mechanism that would break the vial and poison the cat. If not, the cat would live. But we could not know which was the case, so there was no choice, scientifically, but to regard the cat as 100 per cent alive and 100 per cent dead at the same time. This means, as Stephen Hawking has observed with a touch of understandable excitement, that one cannot 'predict future events exactly if one cannot even measure the present state of the universe precisely!'

Because of its oddities, many physicists disliked quantum theory, or at least certain aspects of it, and none more so than Einstein. This was more than a little ironic since it was he, in his *annus mirabilis* of 1905, who had so persuasively explained how photons of light could sometimes behave like particles and sometimes like waves – the notion at the very heart of the new physics. 'Quantum theory is very worthy of regard,' he observed politely, but he really didn't like it. 'God doesn't play dice,' he said.*

Einstein couldn't bear the notion that God could create a universe in which some things were for ever unknowable. Moreover, the idea of action at a distance – that one particle could instantaneously influence another trillions of miles away – was a stark violation of the special theory of relativity. Nothing could outrace the speed of light and yet here were physicists insisting that, somehow, at the subatomic level, information could. (No-one, incidentally, has ever explained how the particles achieve this feat. Scientists have dealt with this problem, according to the physicist Yakir Aharanov, 'by not thinking about it'.)

Above all, there was the problem that quantum physics introduced a level of untidiness that hadn't previously existed. Suddenly you needed two sets of laws to explain the behaviour of the universe – quantum theory for the world of the very small and relativity for the larger universe beyond. The gravity of relativity theory was brilliant at explaining why planets orbited suns or why galaxies tended to cluster, but turned out to have no influence at all at the particle level. To explain what kept atoms together, other forces were needed and in the 1930s two were discovered: the strong nuclear force and the weak nuclear force. The strong force binds atoms together; it's what allows protons to bed down together in the nucleus. The weak force engages

*Or at least, that is how it is nearly always rendered. The actual quote was: 'It seems hard to sneak a look at God's cards. But that He plays dice and uses "telepathic" methods . . . is something that I cannot believe for a single moment.'

in more miscellaneous tasks, mostly to do with controlling the rates of certain sorts of radioactive decay.

The weak nuclear force, despite its name, is ten billion billion billion times stronger than gravity, and the strong nuclear force is more powerful still – vastly so, in fact – but their influence extends to only the tiniest distances. The grip of the strong force reaches out only to about one-hundred-thousandth of the diameter of an atom. That's why the nuclei of atoms are so compacted and dense, and why elements with big, crowded nuclei tend to be so unstable: the strong force just can't hold on to all the protons.

The upshot of all this is that physics ended up with two bodies of laws – one for the world of the very small, one for the universe at large – leading quite separate lives. Einstein disliked that, too. He devoted the rest of his life to searching for a way to tie up these loose ends by finding a Grand Unified Theory, and always failed. From time to time he thought he had it, but it always unravelled on him in the end. As time passed he became increasingly marginalized and even a little pitied. Almost without exception, wrote Snow, 'his colleagues thought, and still think, that he wasted the second half of his life.'

Elsewhere, however, real progress was being made. By the mid-1940s scientists had reached a point where they understood the atom at an extremely profound level – as they all too effectively demonstrated in August 1945 by exploding a pair of atomic bombs over Japan.

By this point physicists could be excused for thinking that they had just about conquered the atom. In fact, everything in particle physics was about to get a whole lot more complicated. But before we take up that slightly exhausting story, we must bring another strand of our history up to date by considering an important and salutary tale of avarice, deceit, bad science, several needless deaths and the final determination of the age of the Earth.

10

GETTING THE LEAD OUT

In the late 1940s, a graduate student at the University of Chicago named Clair Patterson (who was, first name notwithstanding, an Iowa farm boy by origin) was using a new method of lead isotope measurement to try to get a definitive age for the Earth at last. Unfortunately, all his rock samples became contaminated – usually wildly so. Most contained something like two hundred times the levels of lead that would normally be expected to occur. Many years would pass before Patterson realized that the reason for this lay with a regrettable Ohio inventor named Thomas Midgley, Junior.

Midgley was an engineer by training and the world would no doubt have been a safer place if he had stayed so. Instead, he developed an interest in the industrial applications of chemistry. In 1921, while working for the General Motors Research Corporation in Dayton, Ohio, he investigated a compound called tetraethyl lead (also known, confusingly, as lead tetraethyl), and discovered that it significantly reduced the juddering condition known as engine knock.

Even though lead was widely known to be dangerous, by the early years of the twentieth century it could be found in all manner of consumer products. Food came in cans sealed with lead solder. Water was often stored in lead-lined tanks. Lead arsenate was sprayed onto fruit as a pesticide. Lead even came as part of the composition of toothpaste tubes. Hardly a product existed that didn't bring a little lead into consumers' lives. However, nothing gave it a greater and more lasting intimacy than its addition to motor fuel.

Lead is a neurotoxin. Get too much of it and you can irreparably damage the brain and central nervous system. Among the many symptoms associated with over-exposure are blindness, insomnia, kidney failure, hearing loss, cancer, palsies and convulsions. In its most acute form it produces abrupt and terrifying hallucinations, disturbing to victims and onlookers alike, which generally then give way to coma and death. You really don't want to get too much lead into your system.

On the other hand, lead was easy to extract and work, and almost embarrassingly profitable to produce industrially – and tetraethyl lead did indubitably stop engines from knocking. So in 1923 three of America's largest corporations, General Motors, Du Pont and Standard Oil of New Jersey, formed a joint enterprise called the Ethyl Gasoline Corporation (later shortened to simply Ethyl Corporation) with a view to making as much tetraethyl lead as the world was willing to buy, and that proved to be a very great deal. They called their additive 'ethyl' because it sounded friendlier and less toxic than 'lead', and introduced it for public consumption (in more ways than most people realized) on 1 February 1923.

Almost at once production workers began to exhibit the staggered gait and confused faculties that mark the recently poisoned. Also almost at once, the Ethyl Corporation embarked on a policy of calm but unyielding denial that would serve it well for decades. As Sharon Bertsch McGrayne notes in her absorbing history of industrial chemistry, *Prometheans in the Lab*, when employees at one plant developed irreversible delusions, a spokesman blandly informed reporters: 'These men probably went insane because they worked too hard.' Altogether, at least fifteen workers died in the early days of production of leaded gasoline, and untold numbers of others became ill, often violently so; the exact numbers are unknown because the company nearly always managed to hush up news of embarrassing leakages, spills and poisonings. At times, however, suppressing the news became impossible – most notably in 1924 when, in a matter of days, five production workers died and thirty-five more were turned into permanent staggering wrecks at a single ill-ventilated facility.

As rumours circulated about the dangers of the new product, ethyl's ebullient inventor, Thomas Midgley, decided to hold a demonstration for reporters to allay their concerns. As he chatted away about the company's commitment to safety, he poured tetraethyl lead over his hands, then held a beaker of it to his nose for sixty seconds, claiming all the while that he could repeat the procedure daily without harm. In fact, Midgley knew only too well the perils of lead poisoning: he had himself been made seriously ill from overexposure a few months earlier and now, except when reassuring journalists, never went near the stuff if he could help it.

Buoyed by the success of leaded petrol, Midgley now turned to another technological problem of the age. Refrigerators in the 1920s were often appallingly risky because they used insidious and dangerous gases that sometimes seeped out. One leak from a refrigerator at a hospital in Cleveland, Ohio, in 1929 killed more than a hundred people. Midgley set out to create a gas that was stable, non-flammable, non-corrosive and safe to breathe. With an instinct for the regrettable that was almost uncanny, he invented chlorofluorocarbons, or CFCs.

Seldom has an industrial product been more swiftly or unfortunately embraced. CFCs went into production in the early 1930s and found a thousand applications in everything from car air-conditioners to deodorant sprays before it was noticed, half a century later, that they were devouring the ozone in the stratosphere. As you will be aware, this was not a good thing.

Ozone is a form of oxygen in which each molecule bears three atoms of oxygen instead of the normal two. It is a bit of a chemical oddity in that at ground level it is a pollutant, while way up in the stratosphere it is beneficial since it soaks up dangerous ultraviolet radiation. Beneficial ozone is not terribly abundant, however. If it were distributed evenly throughout the stratosphere, it would form a layer just 2 millimetres or so thick. That is why it is so easily disturbed.

Chlorofluorocarbons are also not very abundant – they constitute only about one part per billion of the atmosphere as a whole – but they are extravagantly destructive. A single kilogram of CFCs can capture and annihilate 70,000 kilograms of atmospheric ozone. CFCs also hang around for a long time – about a century on average – wreaking havoc all the while. And they are great heat sponges. A single CFC molecule is about ten thousand times more efficient at exacerbating greenhouse effects than a molecule of carbon dioxide – and carbon dioxide is of course no slouch itself as a greenhouse gas. In short, chlorofluorocarbons may ultimately prove to be just about the worst invention of the twentieth century.

Midgley never knew this because he died long before anyone realized how destructive CFCs were. His death was itself memorably unusual. After becoming crippled with polio, Midgley invented a contraption involving a series of motorized pulleys that automatically raised or turned him in bed. In 1944, he became entangled in the cords as the machine went into action and was strangled.

* * *

If you were interested in finding out the ages of things, the University of Chicago in the 1940s was the place to be. Willard Libby was in the process of inventing radiocarbon dating, allowing scientists to get an accurate reading of the age of bones and other organic remains, something they had never been able to do before. Up to this time, the oldest reliable dates went back no further than the First Dynasty in Egypt – about 3000 BC. No-one could confidently say, for instance, when the last ice sheets had retreated or at what time in the past the Cro-Magnon people had decorated the caves of Lascaux in France.

Libby's idea was so useful that he would be awarded a Nobel Prize for it in 1960. It was based on the realization that all living things have within them an isotope of carbon called carbon-14, which begins to decay at a measurable rate the instant they die. Carbon-14 has a half-life – that is, the time it takes for half of any sample to disappear – of about 5,600 years, so by working out how much of a given sample of carbon had decayed, Libby could get a good fix on the age of an object – though only up to a point. After eight half-lives, only 0.39 per cent of the original radioactive carbon remains, which is too little to make a reliable measurement, so radiocarbon dating works only for objects up to forty thousand or so years old.

Curiously, just as the technique was becoming widespread, certain flaws within it became apparent. To begin with, it was discovered that one of the basic components of Libby's formula, known as the decay constant, was out by about 3 per cent. By this time, however, thousands of measurements had been taken throughout the world. Rather than restate every one, scientists decided to keep the inaccurate constant. 'Thus,' Tim Flannery notes, 'every raw radiocarbon date you read today is given as too young by around 3 per cent.' The problems didn't quite stop there. It was also quickly discovered that carbon-14 samples can be easily contaminated with carbon from other sources – a tiny scrap of vegetable matter, for instance, that has been collected with the sample and not noticed. For younger samples – those under twenty thousand years or so – slight contamination does not always matter so much, but for older samples it can be a serious problem because so few remaining atoms are being counted. In the first instance, to borrow from Flannery, it is like miscounting by a dollar when counting to a thousand; in the second it is more like miscounting by a dollar when you have only two dollars to count.

Libby's method was also based on the assumption that the amount of carbon-14 in the atmosphere, and the rate at which it has been absorbed by living things, has been consistent throughout history. In fact it hasn't been. We now know that the volume of atmospheric carbon-14 varies depending on how well or not the Earth's magnetism is deflecting cosmic rays, and that that can vary significantly over time. This means that some carbon-14 dates are more dubious than others. Among the more dubious are dates just around the time that people first came to the Americas, which is one of the reasons the matter is so perennially in dispute.

Finally, and perhaps a little unexpectedly, readings can be thrown out by seemingly unrelated external factors – such as the diets of those whose bones are being tested. One recent case involved the long-running debate over whether syphilis originated in the New World or the Old. Archaeologists in Hull found that monks in a monastery graveyard had suffered from syphilis, but the initial conclusion that the monks had done so before Columbus's voyage was cast into doubt by the realization that they had eaten a lot of fish, which could make their bones appear to be older than in fact they were. The monks may well have had syphilis, but how it got to them, and when, remain tantalizingly unresolved.

Because of the accumulated shortcomings of carbon-14, scientists devised other methods of dating ancient materials, among them thermoluminescence, which measures electrons trapped in clays, and electron spin resonance, which involves bombarding a sample with electromagnetic waves and measuring the vibrations of the electrons. But even the best of these could not date anything older than about two hundred thousand years, and they couldn't date inorganic materials like rocks at all, which is of course what you need to do if you wish to determine the age of your planet.

The problems of dating rocks were such that at one point almost everyone in the world had given up on them. Had it not been for a determined English professor named Arthur Holmes, the quest might well have fallen into abeyance altogether.

Holmes was heroic as much for the obstacles he overcame as for the results he achieved. By the 1920s, when he was in the prime of his career, geology had slipped out of fashion – physics was the new excitement of the age – and had become severely underfunded, particularly in Britain, its spiritual birthplace. At Durham University, Holmes was for many years the entire geology department. Often he had to borrow or patch together equipment in order to pursue his radiometric dating of rocks. At one point, his calculations were effectively held up for a year while he waited for the university to provide him with a simple adding machine. Occasionally, he had to drop out of academic life altogether to earn enough to support his family – for a time he ran a curio shop in Newcastle upon Tyne – and sometimes he could not even afford the £5 annual membership fee for the Geological Society.

The technique Holmes used in his work was theoretically straightforward and arose directly from the process first observed by Ernest Rutherford in 1904 by which some atoms decay from one element into another at a rate predictable enough that you can use them as clocks. If you know how long it takes for potassium-40 to become argon-40, and you measure the amounts of each in a sample, you can work out how old a material is. Holmes's contribution was to measure the decay rate of uranium into lead to calculate the age of rocks, and thus – he hoped – of the Earth.

But there were many technical difficulties to overcome. Holmes also needed – or at least would very much have appreciated – sophisticated gadgetry of a sort that could make very fine measurements from tiny samples, and, as we have seen, it was all he could do to get a simple adding machine. So it was quite an achievement when in 1946 he was able to announce with some confidence that the Earth was at least three billion years old and possibly rather more. Unfortunately, he now met yet another formidable impediment to acceptance: the conservativeness of his fellow scientists. Although happy to praise his methodology, many maintained that he had found not the age of the Earth but merely the age of the materials from which the Earth had been formed.

It was just at this time that Harrison Brown of the University of Chicago developed a new method for counting lead isotopes in igneous rocks (which is to say those that were created through heating, as opposed to the laying down of sediments). Realizing that the work would be exceedingly tedious, he assigned it to young Clair Patterson as his dissertation project. Famously, he promised Patterson that determining the age of the Earth with his new method would be 'duck soup'. In fact, it would take years.

Patterson began work on the project in 1948. Compared with Thomas Midgley's colourful contributions to the march of progress, Patterson's discovery of the age of the Earth feels more than a touch anti-climactic. For seven years, first at the University of Chicago and then at the California Institute of Technology (where he moved in 1952), he worked in a sterile lab, making very precise measurements of the lead/uranium ratios in carefully selected samples of old rock.

The problem with measuring the age of the Earth was that you needed rocks that were extremely ancient, containing lead- and uranium-bearing crystals that were about as old as the planet itself – anything much younger would obviously give you misleadingly youthful dates – but really ancient rocks are only rarely found on Earth. In the late 1940s no-one altogether understood why this should be. Indeed, and rather extraordinarily, we would be well into the space age before anyone could plausibly account for where all the Earth's old rocks went. (The answer was plate tectonics, which we shall of course get to.) Patterson, meanwhile, was left to try to make sense of things with very limited materials. Eventually, and ingeniously, it occurred to him that he could circumvent the rock shortage by using rocks from beyond Earth. He turned to meteorites.

The assumption he made – rather a large one, but correct as it turned out – was that many meteorites are essentially left-over building materials from the early days of the solar system, and thus have managed to preserve a more or less pristine interior chemistry. Measure the age of these wandering rocks and you would have the age also (near enough) of the Earth.

As always, however, nothing was quite as straightforward as such a breezy description makes it sound. Meteorites are not abundant and meteoritic samples not especially easy to get hold of. Moreover, Brown's measurement technique proved finicky in the extreme and needed much refinement. Above all, there was the problem that Patterson's samples were continuously and unaccountably contaminated with large doses of atmospheric lead whenever they were exposed to air. It was this that eventually led him to create a sterile laboratory – the world's first, according to at least one account.

It took Patterson seven years of patient work just to find and measure suitable samples for final testing. In the spring of 1953 he took his specimens to the Argonne National Laboratory in Illinois, where he was granted time on a late-model mass spectrograph, a machine capable of detecting and measuring the minute quantities of uranium and lead locked up in ancient crystals. When at last he had his results, Patterson was so excited that he drove straight to his boyhood home in Iowa and had his mother check him into a hospital because he thought he was having a heart attack.

Soon afterwards, at a meeting in Wisconsin, Patterson announced a definitive age for the Earth of 4,550 million years (plus or minus 70 million years) – 'a figure that stands unchanged 50 years later', as McGrayne admiringly notes. After two hundred years of attempts, the Earth finally had an age.

Almost at once, Patterson turned his attention to the question of all that lead in the atmosphere. He was astounded to find that what little was known about the effects of lead on humans was almost invariably wrong or misleading – and not surprisingly, since for forty years every study of lead's effects had been funded exclusively by manufacturers of lead additives.

In one such study, a doctor who had no specialized training in chemical pathology undertook a five-year programme in which volunteers were asked to breathe in or swallow lead in elevated quantities. Then their urine and faeces were tested. Unfortunately, as the doctor appears not to have known, lead is not excreted as a waste product. Rather, it accumulates in the bones and blood – that's what makes it so dangerous – and neither bone nor blood was tested. In consequence, lead was given a clean bill of health.

Patterson quickly established that we had a lot of lead in the atmosphere – still do, in fact, since lead never goes away – and that about 90 per cent of it appeared to come from car exhaust pipes; but he couldn't prove it. What he needed was a way to compare lead levels in the atmosphere now with the levels that existed before 1923, when tetraethyl lead began to be commercially produced. It occurred to him that ice cores could provide the answer.

It was known that snowfall in places like Greenland accumulates into discrete annual layers (because seasonal temperature differences produce slight changes in coloration from winter to summer). By counting back through these layers and measuring the amount of lead in each, he could work out global atmospheric lead concentrations at any time for hundreds, or even thousands, of years. The notion became the foundation of ice core studies, on which much modern climatological work is based.

What Patterson found was that before 1923 there was almost no lead in the atmosphere, and that since that time lead levels had climbed steadily and dangerously. He now made it his life's quest to get lead taken out of petrol. To that end, he became a constant and often vocal critic of the lead industry and its interests.

It would prove to be a hellish campaign. Ethyl was a powerful global corporation with many friends in high places. (Among its directors have been Supreme Court Justice Lewis Powell and Gilbert Grosvenor of the National Geographic Society.) Patterson suddenly found research funding withdrawn or difficult to acquire. The American Petroleum Institute cancelled a research contract with him, as did the United States Public Health Service, a supposedly neutral government body.

As Patterson increasingly became a liability to his institution, the Caltech trustees were repeatedly pressed by lead industry officials to shut him up or let him go. According to Jamie Lincoln Kitman, writing in *The Nation* in 2000, Ethyl executives allegedly offered to endow a chair at Caltech 'if Patterson was sent packing'. Absurdly, he was excluded from a 1971 National Research Council panel appointed to investigate the dangers of atmospheric lead poisoning, even though he was by then unquestionably America's leading expert on atmospheric lead.

To his great credit, Patterson never wavered. Eventually his efforts led to the introduction of the Clean Air Act of 1970 and finally to the removal from sale of all leaded petrol in the United States in 1986. Almost immediately lead levels in the blood of Americans fell by 80 per cent. But because lead is for ever, Americans alive today each have about 625 times more lead in their blood than people did a century ago. The amount of lead in the atmosphere also continues to grow, quite legally, by about a hundred thousand tonnes a year, mostly from mining, smelting and industrial activities. The United States also banned lead in indoor paint, '44 years after most of Europe', as McGrayne notes. Remarkably, considering its startling toxicity, lead solder was not removed from American food containers until 1993.

As for the Ethyl Corporation, it's still going strong, though GM, Standard Oil and Du Pont no longer have stakes in the company. (They sold out to a company called Albemarle Paper in 1962.) According to McGrayne, as late as February 2001 Ethyl continued to contend 'that research has failed to show that leaded gasoline poses a threat to human health or the environment'. On its website, a history of the company makes no mention of lead – or indeed of Thomas Midgley – but simply refers to the original product as containing 'a certain combination of chemicals'.

Ethyl no longer makes leaded petrol, although, according to its 2001 company accounts, tetraethyl lead (or TEL as it calls it) still accounted for $25.1 million sales in 2000 (out of overall sales of $795 million), up from $24.1 million in 1999, but down from $117 million in 1998. The company stated in its report its determination to 'maximize the cash generated by TEL as its usage continues to phase down around the world'. Ethyl markets TEL worldwide through an agreement with Associated Octel Ltd of England.

As for the other scourge left to us by Thomas Midgley, chlorofluorocarbons, they were banned in 1974 in the United States, but they are tenacious little devils and any that were loosed into the atmosphere before then (in deodorants or hairsprays, for instance) will almost certainly be around and devouring ozone long after you and I have shuffled off. Worse, we are still introducing huge amounts of CFCs into the atmosphere every year. According to Wayne Biddle, over 27 million kilograms of the stuff, worth $1.5 billion, still finds its way onto the market every year. So who is making it? We are – that is to say, many large corporations are still making it at their plants overseas. It will not be banned in third world countries until 2010.

Clair Patterson died in 1995. He didn't win a Nobel Prize for his work. Geologists never do. Nor, more puzzlingly, did he gain any fame or even much attention from half a century of consistent and increasingly selfless achievement. A good case could be made that he was the most influential geologist of the twentieth century. Yet who has ever heard of Clair Patterson? Most geology textbooks don't mention him. Two recent popular books on the history of the dating of the Earth actually manage to misspell his name. In early 2001, a reviewer of one of these books in the journal *Nature* made the additional, rather astounding error of thinking Patterson was a woman.

At all events, thanks to the work of Clair Patterson, by 1953 the Earth at last had an age everyone could agree on. The only problem now was that it was older than the universe that contained it.

11

MUSTER MARK'S QUARKS

In 1911, a British scientist named C. T. R. Wilson was studying cloud formations by tramping regularly to the summit of Ben Nevis, a famously damp Scottish mountain, when it occurred to him that there must be an easier way. Back in the Cavendish Lab in Cambridge he built an artificial cloud chamber – a simple device in which he could cool and moisten the air, creating a reasonable model of a cloud in laboratory conditions.

The device worked very well, but had an additional, unexpected benefit. When he accelerated an alpha particle through the chamber to seed his make-believe clouds, it left a visible trail – like the contrails of a passing airliner. He had just invented the particle detector. It provided convincing evidence that subatomic particles did indeed exist.

Eventually two other Cavendish scientists invented a more powerful proton-beam device, while in California Ernest Lawrence at Berkeley produced his famous and impressive cyclotron, or atom-smasher as such devices were long excitingly known. All of these contraptions worked – and indeed still work – on more or less the same principle, the idea being to accelerate a proton or other charged particle to an extremely high speed along a track (sometimes circular, sometimes linear), then bang it into another particle and see what flies off. That's why they were called atom-smashers. It wasn't science at its subtlest, but it was generally effective.

As physicists built bigger and more ambitious machines, they began to find or postulate particles or particle families seemingly without number: muons, pions, hyperons, mesons, K-mesons, Higgs bosons, intermediate vector bosons, baryons, tachyons. Even physicists began to grow a little uncomfortable. 'Young man,' Enrico Fermi replied when a student asked him the name of a particular particle, 'if I could remember the names of these particles, I would have been a botanist.'

Today accelerators have names that sound like something Flash Gordon would use in battle: the Super Proton Synchrotron, the Large Electron-Positron Collider, the Large Hadron Collider, the Relativistic Heavy Ion Collider. Using huge amounts of energy (some operate only at night so that people in neighbouring towns don't have to witness their lights fading when the apparatus is fired up), they can whip particles into such a state of liveliness that a single electron can do 47,000 laps around a 7-kilometre tunnel in under a second. Fears have been raised that in their enthusiasm scientists might inadvertently create a black hole or even something called 'strange quarks', which could, theoretically, interact with other subatomic particles and propagate uncontrollably. If you are reading this, that hasn't happened.

Finding particles takes a certain amount of concentration. They are not just tiny and swift but often also tantalizingly evanescent. Particles can come into being and be gone again in as little as 0.000000000000000000000001 of a second (10^{-24} seconds). Even the most sluggish of unstable particles hang around for no more than 0.0000001 of a second (10^{-7} seconds).

Some particles are almost ludicrously slippery. Every second the Earth is visited by ten thousand trillion trillion tiny, all-but-massless neutrinos (mostly shot out by the nuclear broilings of the Sun) and virtually all of them pass right through the planet and everything that is on it, including you and me, as if it weren't there. To trap just a few of them, scientists need tanks holding up to 57,000 cubic metres of heavy water (that is, water with a relative abundance of deuterium in it) in underground chambers (old mines, usually) where they can't be interfered with by other types of radiation.

Very occasionally, a passing neutrino will bang into one of the atomic nuclei in the water and produce a little puff of energy. Scientists count the puffs and by such means take us very slightly closer to understanding the fundamental properties of the universe. In 1998, Japanese observers reported that neutrinos do have mass, but not a great deal – about one ten-millionth that of an electron.

What it really takes to find particles these days is money and lots of it. There is a curious inverse relationship in modern physics between the tininess of the thing being sought and the scale of the facilities required to do the searching. CERN, the European Organization for Nuclear Research, is like a little city. Straddling the border of France and Switzerland, it employs three thousand people and occupies a site that is measured in square kilometres. CERN boasts a string of magnets that weigh more than the Eiffel Tower and an underground tunnel some 26 kilometres around.

Breaking up atoms, as James Trefil has noted, is easy; you do it each time you switch on a fluorescent light. Breaking up atomic nuclei, however, requires quite a lot of money and a generous supply of electricity. Getting down to the level of quarks – the particles that make up particles – requires still more: trillions of volts of electricity and the budget of a small Central American state. CERN's new Large Hadron Collider, scheduled to begin operations in 2005, will achieve 14 trillion volts of energy and cost something over $1.5 billion to construct.*

But these numbers are as nothing compared with what could have been achieved by, and spent upon, the vast and now unfortunately never-to-be Superconducting Supercollider, which began construction near Waxahachie, Texas, in the 1980s, before experiencing a supercollision of its own with the United States Congress. The intention of the collider was to let scientists probe 'the ultimate nature of matter', as it is always put, by recreating as nearly as possible the conditions in the universe during its first ten thousand billionths of a second. The plan was to fling particles through a tunnel 84 kilometres long, achieving a truly staggering 99 trillion volts of energy. It was a grand scheme, but would have cost $8 billion to build (a figure that eventually rose to $10 billion) and hundreds of millions of dollars a year to run.

In perhaps the finest example in history of pouring money into a hole in the ground, Congress spent $2 billion on the project, then cancelled it in 1993 after 22 kilometres of tunnel had been dug. So Texas now boasts the most expensive hole in the universe. The site is, I am told by my friend Jeff Guinn of the *Fort Worth Star-Telegram*, 'essentially a vast, cleared field dotted along the circumference by a series of disappointed small towns'.

Since the supercollider debacle, particle physicists have set their sights a little lower, but even comparatively modest projects can be quite breathtakingly costly when compared with, well, almost anything. A proposed neutrino observatory

*There are practical side-effects to all this costly effort. The World Wide Web is a CERN offshoot. It was invented by a CERN scientist, Tim Berners-Lee, in 1989.

at the old Homestake Mine in Lead, South Dakota, would cost $500 million to build – this in a mine that is already dug – before even looking at the annual running costs. There would also be $281 million of 'general conversion costs'. A particle accelerator at Fermilab in Illinois, meanwhile, cost $260 million merely to refit.

Particle physics, in short, is a hugely expensive enterprise – but it is a productive one. Today the particle count is well over 150, with a further 100 or so suspected, but unfortunately, in the words of Richard Feynman, 'it is very difficult to understand the relationships of all these particles, and what nature wants them for, or what the connections are from one to another.' Inevitably, each time we manage to unlock a box, we find that there is another locked box inside. Some people think there are particles called tachyons, which can travel faster than the speed of light. Others long to find gravitons – the seat of gravity. At what point we reach the irreducible bottom is not easy to say. Carl Sagan in *Cosmos* raised the possibility that if you travelled downwards into an electron, you might find that it contained a universe of its own, recalling all those science-fiction stories of the 1950s. 'Within it, organized into the local equivalent of galaxies and smaller structures, are an immense number of other, much tinier elementary particles, which are themselves universes at the next level and so on forever – an infinite downward regression, universes within universes, endlessly. And upward as well.'

For most of us it is a world that surpasses understanding. To read even an elementary guide to particle physics nowadays you must find your way through lexical thickets such as this: 'The charged pion and antipion decay respectively into a muon plus antineutrino and an antimuon plus neutrino with an average lifetime of 2.603×10^{-8} seconds, the neutral pion decays into two photons with an average lifetime of about 0.8×10^{-16} seconds, and the muon and antimuon decay respectively into . . .' And so it runs on – and this from a book for the general reader by one of the (normally) most lucid of interpreters, Steven Weinberg.

In the 1960s, in an attempt to bring just a little simplicity to matters, the Caltech physicist Murray Gell-Mann invented a new class of particles, essentially, in the words of Steven Weinberg, 'to restore some economy to the multitude of hadrons' – a collective term used by physicists for protons, neutrons and other particles governed by the strong nuclear force. Gell-Mann's theory was that all hadrons were made up of still smaller, even more fundamental particles. His colleague Richard Feynman wanted to call these new basic particles *partons*, as in Dolly, but was over-ruled. Instead they became known as *quarks*.

Gell-Mann took the name from a line in *Finnegans Wake*: 'Three quarks for Muster Mark!' (Discriminating physicists rhyme the word with *storks*, not *larks*, even though the latter is almost certainly the pronunciation Joyce had in mind.) The fundamental simplicity of quarks was not long-lived. As they became better understood it was necessary to introduce subdivisions. Although quarks are much too small to have colour or taste or any other physical characteristics we would recognize, they became clumped into six categories – up, down, strange, charm, top and bottom – which physicists oddly refer to as their 'flavours', and these are further divided into the colours red, green and blue. (One suspects that it was not altogether coincidental that these terms were first applied in California during the age of psychedelia.)

Eventually, out of all this emerged what is called the Standard Model, which is essentially a sort of parts kit for the subatomic world. The Standard Model consists of six quarks, six leptons, five known bosons and a postulated sixth, the Higgs boson (named for a Scottish scientist, Peter Higgs), plus three of the four physical forces: the strong and weak nuclear forces and electromagnetism.

The arrangement essentially is that among the basic building blocks of matter are quarks; these are held together by particles called gluons; and together quarks and gluons form protons and neutrons, the stuff of the atom's nucleus. Leptons are the source of electrons and neutrinos. Quarks and leptons together are called fermions. Bosons (named for the Indian physicist S. N. Bose) are particles that produce and carry forces, and include photons and gluons. The Higgs boson may or may not actually exist; it was invented simply as a way of endowing particles with mass.

It is all, as you can see, just a little unwieldy, but it is the simplest model that can explain all that happens in the world of particles. Most particle physicists feel, as Leon Lederman remarked in a 1985 television documentary, that the Standard Model lacks elegance and simplicity. 'It is too complicated. It has too many arbitrary parameters,' Lederman said. 'We don't really see the creator twiddling twenty knobs to set twenty parameters to create the universe as we know it.' Physics is really nothing more than a search for ultimate simplicity, but so far all we have is a kind of elegant messiness – or as Lederman put it: 'There is a deep feeling that the picture is not beautiful.'

The Standard Model is not only ungainly but incomplete. For one thing, it has nothing at all to say about gravity. Search through the Standard Model as you will and you won't find anything to explain why when you place a hat on a table it doesn't float up to the ceiling. Nor, as we've just noted, can it explain mass. In order to give particles any mass at all we have to introduce the notional Higgs boson; whether it actually exists is a matter for twenty-first century physics. As Feynman cheerfully observed: 'So we are stuck with a theory, and we do not know whether it is right or wrong, but we do know that it is a *little* wrong, or at least incomplete.'

In an attempt to draw everything together, physicists have

come up with something called superstring theory. This postulates that all those little things like quarks and leptons that we had previously thought of as particles are actually 'strings' – vibrating strands of energy that oscillate in eleven dimensions, consisting of the three we know already plus time and seven other dimensions that are, well, unknowable to us. The strings are very tiny – tiny enough to pass for point particles.

By introducing extra dimensions, superstring theory enables physicists to pull together quantum laws and gravitational ones into one comparatively tidy package; but it also means that anything scientists say about the theory begins to sound worryingly like the sort of thoughts that would make you edge away if conveyed to you by a stranger on a park bench. Here, for example, is the physicist Michio Kaku explaining the structure of the universe from a superstring perspective:

> The heterotic string consists of a closed string that has two types of vibrations, clockwise and counterclockwise, which are treated differently. The clockwise vibrations live in a ten-dimensional space. The counterclockwise live in a 26-dimensional space, of which 16 dimensions have been compactified. (We recall that in Kaluza's original five-dimensional, the fifth dimension was compactified by being wrapped up into a circle.)

And so it goes, for some 350 pages.

String theory has further spawned something called M theory, which incorporates surfaces known as membranes – or simply branes to the hipper souls of the world of physics. This, I'm afraid, is the stop on the knowledge highway where most of us must get off. Here is a sentence from the *New York Times*, explaining this as simply as possible to a general audience:

> The ekpyrotic process begins far in the indefinite past with a pair of flat empty branes sitting parallel to each other in a warped five-dimensional space . . . The two branes, which form the walls of the fifth dimension, could have popped out of nothingness as a quantum fluctuation in the even more distant past and then drifted apart.

No arguing with that. No understanding it either. *Ekpyrotic*, incidentally, comes from the Greek word for conflagration.

Matters in physics have now reached such a pitch that, as Paul Davies noted in *Nature*, it is 'almost impossible for the non-scientist to discriminate between the legitimately weird and the outright crackpot'. The question came interestingly to a head in the autumn of 2002 when two French physicists, twin brothers Igor and Grichka Bogdanov, produced a theory of ambitious density involving such concepts as 'imaginary time' and the 'Kubo–Schwinger–Martin condition', and purporting to describe the nothingness that was the universe before the Big Bang – a period that was always assumed to be unknowable (since it predated the birth of physics and its properties).

Almost at once the Bogdanov theory excited debate among physicists as to whether it was twaddle, a work of genius or a hoax. 'Scientifically, it's clearly more or less complete nonsense,' Columbia University physicist Peter Woit told the *New York Times*, 'but these days that doesn't much distinguish it from a lot of the rest of the literature.'

Karl Popper, whom Steven Weinberg has called 'the dean of modern philosophers of science', once suggested that there may not in fact be an ultimate theory for physics – that, rather, every explanation may require a further explanation, producing 'an infinite chain of more and more fundamental principles'. A rival possibility is that such knowledge may simply be beyond us. 'So far, fortunately,' writes Weinberg in *Dreams of a Final Theory*, 'we do not seem to be coming to the end of our intellectual resources.'

Almost certainly this is an area that will see further developments of thought, and almost certainly again these thoughts will be beyond most of us.

While physicists in the middle decades of the twentieth century were looking perplexedly into the world of the very small, astronomers were finding no less arresting an incompleteness of understanding in the universe at large.

When we last met Edwin Hubble, he had determined that nearly all the galaxies in our field of view are flying away from us, and that the speed and distance of this retreat are neatly proportional: the further away the galaxy, the faster it is moving. Hubble realized that this could be expressed with a simple equation, Ho = *v*/*d* (where Ho is the constant, *v* is the recessional velocity of a flying galaxy and *d* its distance away from us). Ho has been known ever since as the Hubble constant and the whole as Hubble's Law. Using his formula, Hubble calculated that the universe was about two billion years old, which was a little awkward because even by the late 1920s it was increasingly evident that many things within the universe – including, probably, the Earth itself – were older than that. Refining this figure has been an ongoing preoccupation of cosmology.

Almost the only thing constant about the Hubble constant has been the amount of disagreement over what value to give it. In 1956, astronomers discovered that Cepheid variables were more variable than they had thought; they came in two varieties, not one. This allowed them to rework their calculations and come up with a new age for the universe of between seven billion and twenty billion years – not terribly precise, but at least old enough, at last, to embrace the formation of the Earth.

In the years that followed there erupted a dispute that would run and run, between Allan Sandage, heir to Hubble at Mount Wilson, and Gérard de Vaucouleurs, a French-born astronomer based at the University of Texas. Sandage, after years of careful calculations, arrived at a value for the Hubble constant of 50, giving the universe an age of twenty billion years. De Vaucouleurs was equally certain that the Hubble constant was 100.* This would mean that the universe was only half the size and age that Sandage believed – ten billion years. Matters took a further lurch into uncertainty when in 1994 a team from the Carnegie Observatories in California, using measures from the Hubble Space Telescope, suggested that the universe could be as little as eight billion years old – an age even they conceded was younger than some of the stars within the universe. In February 2003, a team from NASA and the Goddard Space Flight Center in Maryland, using a new, far-reaching type of satellite called the Wilkinson Microwave Anistropy Probe, announced with some confidence that the age of the universe is 13.7 billion years, give or take a hundred million years or so. There matters rest, at least for the moment.

The difficulty in making final determinations is that there are often acres of room for interpretation. Imagine standing in a field at night and trying to decide how far away two distant electric lights are. Using fairly straightforward tools of astronomy you can easily enough determine that the bulbs

---

*You are of course entitled to wonder what is meant exactly by 'a constant of 50' or 'a constant of 100'. The answer lies in astronomical units of measure. Except conversationally, astronomers don't use light years. They use a distance called the *parsec* (a contraction of *parallax* and *second*), based on a universal measure called the stellar parallax and equivalent to 3.26 light years. Really big measures, like the size of a universe, are measured in megaparsecs: 1 megaparsec = 1 million parsecs. The constant is expressed in terms of kilometres per second per megaparsec. Thus when astronomers refer to a Hubble constant of 50, what they really mean is '50 kilometres per second per megaparsec'. For most of us that is of course an utterly meaningless measure; but then, with astronomical measures most distances are so huge as to be utterly meaningless.

are of equal brightness and that one is, say, 50 per cent more distant than the other. But what you can't be certain of is whether the nearer light is, let us say, a 58-watt bulb that is 37 metres away or a 61-watt light that is 36.5 metres away. On top of that you must make allowances for distortions caused by variations in the Earth's atmosphere, by intergalactic dust, by contaminating light from foreground stars and many other factors. The upshot is that your computations are necessarily based on a series of nested assumptions, any of which could be a source of contention. There is also the problem that access to telescopes is always at a premium and historically measuring red shifts has been notably costly in telescope time. It could take all night to get a single exposure. In consequence, astronomers have sometimes been compelled (or willing) to base conclusions on notably scanty evidence. In cosmology, as the journalist Geoffrey Carr has suggested, we have 'a mountain of theory built on a molehill of evidence'. Or as Martin Rees has put it: 'Our present satisfaction [with our state of understanding] may reflect the paucity of the data rather than the excellence of the theory'.

This uncertainty applies, incidentally, to relatively nearby things as much as to the distant edges of the universe. As Donald Goldsmith notes, when astronomers say that the galaxy M87 is 60 million light years away, what they really mean ('but do not often stress to the general public') is that it is somewhere between 40 million and 90 million light years away – not quite the same thing. For the universe at large, matters are naturally magnified. For all the éclat surrounding the latest pronouncements, we remain a long way from unanimity.

One interesting theory recently suggested is that the universe is not nearly as big as we thought; that when we peer into the distance some of the galaxies we see may simply be reflections, ghost images created by rebounded light.

The fact is, there is a great deal, even at quite a fundamental level, that we don't know – not least what the universe is made of. When scientists calculate the amount of matter needed to hold things together, they always come up desperately short. It appears that at least 90 per cent of the universe, and perhaps as much as 99 per cent, is composed of Fritz Zwicky's 'dark matter' – stuff that is by its nature invisible to us. It is slightly galling to think that we live in a universe that for the most part we can't even see, but there you are. At least the names for the two main possible culprits are entertaining: they are said to be either WIMPs (for Weakly Interacting Massive Particles, which is to say specks of invisible matter left over from the Big Bang) or MACHOs (for MAssive Compact Halo Objects – really just another name for black holes, brown dwarfs and other very dim stars).

Particle physicists have tended to favour the particle explanation of WIMPs, astrophysicists the stellar explanation of MACHOs. For a time MACHOs had the upper hand, but not nearly enough of them were detected, so sentiment swung back towards WIMPs – with the problem that no WIMP has ever been found. Because they are weakly interacting, they are (assuming they even exist) very hard to identify. Cosmic rays would cause too much interference. So scientists must go deep underground. One kilometre underground cosmic bombardments would be one-millionth what they would be on the surface. But even when all these are added in, 'two-thirds of the universe is still missing from the balance sheet,' as one commentator has put it. For the moment we might very well call them DUNNOS (for Dark Unknown Nonreflective Nondetectable Objects Somewhere).

Recent evidence suggests not only that the galaxies of the universe are racing away from us, but that they are doing so at a rate that is accelerating. This is counter to all expectations. It appears that the universe may be filled not only with dark matter, but with dark energy. Scientists sometimes also call it vacuum energy or quintessence. Whatever it is, it seems to be driving an expansion that no-one can altogether account for. The theory is that empty space isn't so empty at all – that there are particles of matter and anti-matter popping into existence and popping out again – and that these are pushing the universe outwards at an accelerating rate. Improbably enough, the one thing that resolves all this is Einstein's cosmological constant – the little piece of maths he dropped into the General Theory of Relativity to stop the universe's presumed expansion and that he called 'the biggest blunder of my life'. It now appears that he may have got things right after all.

The upshot of all this is that we live in a universe whose age we can't quite compute, surrounded by stars whose distances from us and each other we don't altogether know, filled with matter we can't identify, operating in conformance with physical laws whose properties we don't truly understand.

And on that rather unsettling note, let's return to Planet Earth and consider something that we *do* understand – though by now you perhaps won't be surprised to hear that we don't understand it completely and what we do understand we haven't understood for long.

12

THE EARTH MOVES

In one of his last professional acts before his death in 1955, Albert Einstein wrote a short but glowing foreword to a book by a geologist named Charles Hapgood entitled *Earth's Shifting Crust: A Key to Some Basic Problems of Earth Science*. Hapgood's book was a steady demolition of the idea that continents were in motion. In a tone that all but invited the reader to join him in a tolerant chuckle, Hapgood observed that a few gullible souls had noticed 'an apparent correspondence in shape between certain continents'. It would appear, he went on, 'that South America might be fitted together with Africa, and so on . . . It is even claimed that rock formations on opposite sides of the Atlantic match.'

Mr Hapgood briskly dismissed any such notions, noting that the geologists K. E. Caster and J. C. Mendes had done extensive fieldwork on both sides of the Atlantic and had established beyond question that no such similarities existed. Goodness knows what outcrops Messrs Caster and Mendes had looked at, because in fact many of the rock formations on both sides of the Atlantic *are* the same – not just very similar but the same.

This was not an idea that flew with Mr Hapgood, or many other geologists of his day. The theory Hapgood alluded to was one first propounded in 1908 by an amateur American geologist named Frank Bursley Taylor. Taylor came from a wealthy family and had both the means and the freedom from academic constraints to pursue unconventional lines of enquiry. He was one of those struck by the similarity in shape between the facing coastlines of Africa and South America, and from this observation he developed the idea that the continents had once slid around. He suggested – presciently, as it turned out – that the crunching together of continents could have thrust up the world's mountain chains. He failed, however, to produce much in the way of evidence, and the theory was considered too crackpot to merit serious attention.

In Germany, however, Taylor's idea was picked up, and effectively appropriated, by a theorist named Alfred Wegener, a meteorologist at the University of Marburg. Wegener investigated the many plant and fossil anomalies that did not fit comfortably into the standard model of Earth history and realized that very little of it made sense if conventionally interpreted. Animal fossils repeatedly turned up on opposite sides of oceans that were clearly too wide to swim. How, he wondered, did marsupials travel from South America to Australia? How did identical snails turn up in Scandinavia and New England? And how, come to that, did one account for coal seams and other semi-tropical remnants in frigid spots like Spitsbergen, over 600 kilometres north of Norway, if they had not somehow migrated there from warmer climes?

Wegener developed the theory that the world's continents had once existed as a single land mass he called Pangaea, where flora and fauna had been able to mingle, before splitting apart and floating off to their present positions. He set the idea out in a book called *Die Entstehung der Kontinente und Ozeane*, or *The Origin of Continents and Oceans*, which was published in German in 1912 and – despite the outbreak of the First World War in the meantime – in English three years later.

Because of the war, Wegener's theory didn't attract much notice at first, but by 1920, when he produced a revised and expanded edition, it quickly became a subject of discussion. Everyone agreed that continents moved – but up and down, not sideways. The process of vertical movement, known as isostasy, was a foundation of geological belief for generations, though no-one had any really good theories as to how or why it happened. One idea, which remained in textbooks well into my own schooldays, was the 'baked apple' theory propounded by the Austrian Eduard Suess just before the turn of the century. This suggested that as the molten Earth had cooled, it had become wrinkled in the manner of a baked apple, creating ocean basins and mountain ranges. Never mind that James Hutton had shown long before that any such static arrangement would eventually result in a featureless spheroid as erosion levelled the bumps and filled in the divots. There was also the problem, demonstrated by Rutherford and Soddy early in the century, that earthly elements hold huge reserves of heat – much too much to allow for the sort of cooling and shrinking Suess suggested. And anyway, if Suess's theory were correct, then mountains should be evenly distributed across the face of the Earth, which patently they were not, and of more or less the same ages; yet by the early 1900s it was already evident that some ranges, like the Urals and Appalachians, were hundreds of millions of years older than others, like the Alps and Rockies. Clearly the time was ripe for a new theory. Unfortunately, Alfred Wegener was not the man geologists wished to provide it.

For a start, his radical notions questioned the foundations of their discipline, seldom an effective way to generate warmth in an audience. Such a challenge would have been painful enough coming from a geologist, but Wegener had no background in geology. He was a meteorologist, for goodness' sake. A weatherman – a German weatherman. These were not remediable deficiencies.

And so geologists took every pain they could to dismiss his evidence and belittle his suggestions. To get around the problems of fossil distributions, they posited ancient 'land bridges' wherever they were needed. When an ancient horse named *Hipparion* was found to have lived in France and Florida at the same time, a land bridge was drawn across the Atlantic. When it was realized that ancient tapirs had existed simultaneously in South America and Southeast Asia a land bridge was drawn there, too. Soon maps of prehistoric seas were almost solid with hypothesized land bridges – from North America to Europe, from Brazil to Africa, from Southeast Asia to Australia, from Australia to Antarctica. These connective tendrils had not only conveniently appeared whenever it was necessary to move a living organism from one land mass to another, but then had obligingly vanished without leaving a trace of their former existence. None of this, of course, was supported by so much as a grain of evidence – nothing so wrong could be – yet it was geological orthodoxy for the next half-century.

Even land bridges couldn't explain some things. One species of trilobite that was well known in Europe was also found to have lived on Newfoundland – but only on one side. No-one could persuasively explain how it had managed to cross 3,000 kilometres of hostile ocean but then failed to find its way around the corner of an island 300 kilometres wide. Even more awkwardly anomalous was another species of trilobite found in Europe and the Pacific northwest of America but nowhere in between, which would have required not so much a land bridge as a flyover. Yet as late as 1964, when the *Encyclopaedia Britannica* discussed the rival theories it was Wegener's that was held to be full of 'numerous grave theoretical difficulties'. To be sure,

Wegener made mistakes. He asserted that Greenland is drifting west by about 1.6 kilometres a year, a clear nonsense. (It's more like a centimetre.) Above all, he could offer no convincing explanation for how the land masses moved about. To believe in his theory you had to accept that massive continents somehow pushed through solid crust, like a farm plough through soil, without leaving any furrow in their wake. Nothing then known could plausibly explain what motored these massive movements.

It was Arthur Holmes, the English geologist who did so much to determine the age of the Earth, who came up with a suggestion. Holmes was the first scientist to understand that radioactive warming could produce convection currents within the Earth. In theory, these could be powerful enough to slide continents around on the surface. In his popular and influential textbook *Principles of Physical Geology*, first published in 1944, Holmes laid out a continental drift theory that was, in its fundamentals, the theory that prevails today. It was still a radical proposition for the time and widely criticized, particularly in the United States, where resistance to drift lasted longer than elsewhere. One reviewer there fretted, without any sense of irony, that Holmes presented his arguments so clearly and compellingly that students might actually come to believe them. Elsewhere, however, the new theory drew steady if cautious support. In 1950, a vote at the annual meeting of the British Association for the Advancement of Science showed that about half of those present now embraced the idea of continental drift. (Hapgood soon after cited this figure as proof of how tragically misled British geologists had become.) Curiously, Holmes himself sometimes wavered in his conviction. In 1953 he confessed: 'I have never succeeded in freeing myself from a nagging prejudice against continental drift; in my geological bones, so to speak, I feel the hypothesis is a fantastic one.'

Continental drift was not entirely without support in the United States. Reginald Daly of Harvard spoke for it, but he, you may recall, was the man who suggested that the Moon had been formed by a cosmic impact and his ideas tended to be considered interesting, even worthy, but a touch too exuberant for serious consideration. And so most American academics stuck to the belief that the continents had occupied their present positions for ever and that their surface features could be attributed to something other than lateral motions.

Interestingly, oil company geologists had known for years that if you wanted to find oil you had to allow for precisely the sort of surface movements that were implied by plate tectonics. But oil geologists didn't write academic papers; they just found oil.

There was one other major problem with Earth theories that no one had resolved, or even come close to resolving. That was the question of where all the sediments went. Every year the Earth's rivers carried massive volumes of eroded material – 500 million tonnes of calcium, for instance – to the seas. If you multiplied the rate of deposition by the number of years it had been going on, you arrived at a disturbing figure: there should be about 20 kilometres of sediments on the ocean bottoms – or, put another way, the ocean bottoms should by now be well above the ocean tops. Scientists dealt with this paradox in the handiest possible way. They ignored it. But eventually there came a point when they could ignore it no longer.

In the Second World War, a Princeton University mineralogist named Harry Hess was put in charge of an attack transport ship, the USS *Cape Johnson*. Aboard this vessel was a fancy new depth sounder called a fathometer, which was designed to facilitate inshore manoeuvres during beach landings, but Hess realized that it could equally well be used for scientific purposes and never switched it off, even when far out at sea, even in the heat of battle. What he found was entirely unexpected. If the ocean floors were ancient, as everyone assumed, they should be thickly blanketed with sediments, like the mud on the bottom of a river or lake. But Hess's readings showed that the ocean floor offered anything but the gooey smoothness of ancient silts. It was scored everywhere with canyons, trenches and crevasses and dotted with volcanic seamounts that he called guyots after an earlier Princeton geologist named Arnold Guyot. All this was a puzzle, but Hess had a war to take part in, and put such thoughts to the back of his mind.

After the war, Hess returned to Princeton and the preoccupations of teaching, but the mysteries of the sea floor continued to occupy a space in his thoughts. Meanwhile, throughout the 1950s oceanographers were undertaking more and more sophisticated surveys of the ocean floors. In so doing, they found an even bigger surprise: the mightiest and most extensive mountain range on Earth was – mostly – under water. It traced a continuous path along the world's seabeds, rather like the pattern on a tennis ball. If you began at Iceland and travelled south, you could follow it down the centre of the Atlantic Ocean, around the bottom of Africa, and across the Indian and Southern oceans and into the Pacific just below Australia; there it angled across the Pacific as if making for Baja California before shooting up the west coast of the United States to Alaska. Occasionally its higher peaks poked above the water as an island or archipelago – the Azores and Canaries in the Atlantic, Hawaii in the Pacific, for instance – but mostly it was buried under thousands of fathoms of salty sea, unknown and unsuspected. When all its branches were added together, the network extended to 75,000 kilometres.

A very little of this had been known for some time. People laying ocean-floor cables in the nineteenth century had realized that there was some kind of mountainous intrusion in the mid-Atlantic from the way the cables ran, but the continuous nature and overall scale of the chain was a stunning surprise. Moreover, it contained physical anomalies that couldn't be explained. Down the middle of the mid-Atlantic ridge was a canyon – a rift – up to 20 kilometres wide for its entire 19,000-kilometre length. This seemed to suggest that the Earth was splitting apart at the seams, like a nut bursting out of its shell. It was an absurd and unnerving notion, but the evidence couldn't be denied.

Then in 1960 core samples showed that the ocean floor was quite young at the mid-Atlantic ridge but grew progressively older as you moved away from it to east or west. Harry Hess considered the matter and realized that this could mean only one thing: new ocean crust was being formed on either side of the central rift, then being pushed away from it as more new crust came along behind. The Atlantic floor was effectively two large conveyor belts, one carrying crust towards North America, the other carrying crust towards Europe. The process became known as sea-floor spreading.

When the crust reached the end of its journey at the boundary with continents, it plunged back into the Earth in a process known as subduction. That explained where all the sediment went. It was being returned to the bowels of the Earth. It also explained why ocean floors everywhere were so comparatively youthful. None had ever been found to be older than about 175 million years, which was a puzzle because continental rocks were often billions of years old. Now Hess could see why. Ocean rocks lasted only as long as it took them to travel to shore. It was a beautiful theory that explained a great deal. Hess elaborated his arguments in an important paper, which was almost universally ignored. Sometimes the world just isn't ready for a good idea.

Meanwhile, two researchers, working independently,

were making some startling findings by drawing on a curious fact of Earth history that had been discovered several decades earlier. In 1906, a French physicist named Bernard Brunhes had found that the planet's magnetic field reverses itself from time to time, and that the record of these reversals is permanently fixed in certain rocks at the time of their birth. Specifically, tiny grains of iron ore within the rocks point to wherever the magnetic poles happen to be at the time of their formation, then stay pointing in that direction as the rocks cool and harden. In effect, they 'remember' where the magnetic poles were at the time of their creation. For years this was little more than a curiosity, but in the 1950s Patrick Blackett of the University of London and S. K. Runcorn of the University of Newcastle studied the ancient magnetic patterns frozen in British rocks and were startled, to say the very least, to find them indicating that at some time in the distant past Britain had spun on its axis and travelled some distance to the north, as if it had somehow come loose from its moorings. Moreover, they also discovered that if you placed a map of Europe's magnetic patterns alongside an American one from the same period, they fit together as neatly as two halves of a torn letter. It was uncanny. Their findings were ignored, too.

It finally fell to two men from Cambridge University, a geophysicist named Drummond Matthews and a graduate student of his named Fred Vine, to draw all the strands together. In 1963, using magnetic studies of the Atlantic Ocean floor, they demonstrated conclusively that the sea floors were spreading in precisely the manner Hess had suggested and that the continents were in motion, too. An unlucky Canadian geologist named Lawrence Morley came up with the same conclusion at the same time, but couldn't find anyone to publish his paper. In what has become a famous snub, the editor of the *Journal of Geophysical Research* told him: 'Such speculations make interesting talk at cocktail parties, but it is not the sort of thing that ought to be published under serious scientific aegis.' One geologist later described it as 'probably the most significant paper in the earth sciences ever to be denied publication'.

At all events, mobile crust was an idea whose time had finally come. A symposium of many of the most important figures in the field was convened in London under the auspices of the Royal Society in 1964, and suddenly, it seemed, everyone was a convert. The Earth, the meeting agreed, was a mosaic of interconnected segments whose various stately jostlings accounted for much of the planet's surface behaviour.

The name 'continental drift' was fairly swiftly discarded when it was realized that the whole crust was in motion and not just the continents, but it took a while to settle on a name for the individual segments. At first people called them 'crustal blocks' or sometimes 'paving stones'. Not until late 1968, with the publication of an article by three American seismologists in the *Journal of Geophysical Research*, did the segments receive the name by which they have since been known: plates. The same article called the new science plate tectonics.

Old ideas die hard and not everyone rushed to embrace the exciting new theory. Well into the 1970s, one of the most popular and influential geological textbooks, *The Earth* by the venerable Harold Jeffreys, strenuously insisted that plate tectonics was a physical impossibility, just as it had in the first edition way back in 1924. It was equally dismissive of convection and sea-floor spreading. And in *Basin and Range*, published in 1980, John McPhee noted that even then one American geologist in eight still didn't believe in plate tectonics.

Today we know that the Earth's surface is made up of eight to twelve big plates (depending on how you define big) and twenty or so smaller ones, and that they all move in different directions and at different speeds. Some plates are large and comparatively inactive, others small but energetic. They bear only an incidental relationship to the land masses that sit upon them. The North American plate, for instance, is much larger than the continent with which it is associated. It roughly traces the outline of the continent's western coast (which is why that area is so seismically active, because of the bump and crush of the plate boundary), but ignores the eastern seaboard altogether and instead extends halfway across the Atlantic to the mid-ocean ridge. Iceland is split down the middle, which makes it tectonically half American and half European. New Zealand, meanwhile, is part of the immense Indian Ocean plate even though it is nowhere near the Indian Ocean. And so it goes for most plates.

The connections between modern land masses and those of the past were found to be infinitely more complex than anyone had imagined. Kazakhstan, it turns out, was once attached to Norway and New England. One corner of Staten Island, but only a corner, is European. So is part of Newfoundland. Pick up a pebble from a Massachusetts beach and its nearest kin will now be in Africa. The Scottish Highlands and much of Scandinavia are substantially American. Some of the Shackleton Range of Antarctica, it is thought, may once have belonged to the Appalachians of the eastern US. Rocks, in short, get around.

The constant turmoil keeps the plates from fusing into a single immobile plate. Assuming things continue much as at present, the Atlantic Ocean will expand until eventually it is much bigger than the Pacific. Much of California will float off and become a kind of Madagascar of the Pacific. Africa will push northward into Europe, squeezing the Mediterranean out of existence and thrusting up a chain of mountains of Himalayan majesty running from Paris to Calcutta. Australia will colonize the islands to its north and connect by some isthmian umbilicus to Asia. These are future outcomes, but not future events. The events are happening now. As we sit here, continents are adrift, like leaves on a pond. Thanks to Global Positioning Systems we can see that Europe and North America are parting at about the speed a fingernail grows – roughly two metres in a human lifetime. If you were prepared to wait long enough, you could ride from Los Angeles all the way up to San Francisco. It is only the brevity of lifetimes that keeps us from appreciating the changes. Look at a globe and what you are seeing really is a snapshot of the continents as they have been for just one-tenth of 1 per cent of the Earth's history.

Earth is alone among the rocky planets in having tectonics and why this should be is a bit of a mystery. It is not simply a matter of size or density – Venus is nearly a twin of Earth in these respects and yet has no tectonic activity – but it may be that we have just the right materials in just the right measures to keep the Earth bubbling away. It is thought – though it is really nothing more than a thought – that tectonics is an important part of the planet's organic well-being. As the physicist and writer James Trefil has put it, 'It would be hard to believe that the continuous movement of tectonic plates has no effect on the development of life on earth.' He suggests that the challenges induced by tectonics – changes in climate, for instance – were an important spur to the development of intelligence. Others believe the driftings of the continents may have produced at least some of the Earth's various extinction events. In November 2002 Tony Dickson of Cambridge University produced a report, published in the journal *Science*, strongly suggesting that there may well be a relationship between the history of rocks and the history of life. What Dickson established was that the chemical composition of the world's oceans has altered abruptly and dramatically at times throughout the past half-billion years and that these changes often correlate with important events in biological history – the huge outburst of tiny organisms that created the chalk cliffs of England's south coast, the sudden fashion for shells among marine organisms during the Cambrian period, and so on. No-one can say what causes the oceans' chemistry to change so dramatically from time to time, but the opening and shutting of ocean ridges would be an obvious possible culprit.

At all events, plate tectonics explained not only the surface dynamics of the Earth – how an ancient *Hipparion* got from France to Florida, for example – but also many of its internal actions. Earthquakes, the formation of island chains, the carbon cycle, the locations of mountains, the coming of ice ages, the origins of life itself – there was hardly a matter that wasn't directly influenced by this remarkable new theory. Geologists, as McPhee has noted, found themselves in the giddying position where 'the whole earth suddenly made sense.'

But only up to a point. The distribution of continents in former times is much less neatly resolved than most people outside geophysics think. Although textbooks give confident-looking representations of ancient land masses with names like Laurasia, Gondwana, Rodinia and Pangaea, these are sometimes based on conclusions that don't altogether hold up. As George Gaylord Simpson observes in *Fossils and the History of Life*, species of plants and animals from the ancient world have a habit of appearing inconveniently where they shouldn't and failing to be where they ought.

The outline of Gondwana, a once-mighty continent connecting Australia, Africa, Antarctica and South America, was based in large part on the distributions of a genus of ancient tongue fern called *Glossopteris*, which was found in all the right places. However, much later *Glossopteris* was also discovered in parts of the world that had no known connection to Gondwana. This troubling discrepancy was – and continues to be – mostly ignored. Similarly, a Triassic reptile called lystrosaurus has been found from Antarctica all the way to Asia, supporting the idea of a former connection between those continents, but it has never turned up in South America or Australia, which are believed to have been part of the same continent at the same time.

There are also many surface features that tectonics can't explain. Take Denver. It is, as everyone knows, a mile high, but that rise is comparatively recent. When dinosaurs roamed the Earth, Denver was part of an ocean bottom, many thousands of metres lower. Yet the rocks on which Denver sits are not fractured or deformed in the way they would be if Denver had been pushed up by colliding plates, and anyway Denver was too far from the plate edges to be susceptible to their actions. It would be as if you pushed against the edge of a rug hoping to raise a ruck at the opposite end. Mysteriously and over millions of years, it appears that Denver has been rising, like baking bread. So, too, has much of southern Africa; a portion of it 1,600 kilometres across has risen about one and a half kilometres in a hundred million years without any known associated tectonic activity. Australia, meanwhile, has been tilting and sinking. Over the past hundred million years, as it has drifted north towards Asia, its leading edge has sunk by nearly 200 metres. It appears that Indonesia is very slowly drowning, and dragging Australia down with it. Nothing in the theories of tectonics can explain any of this.

Alfred Wegener never lived to see his ideas vindicated. On an expedition to Greenland in 1930, he set out alone, on his fiftieth birthday, to check out a supply drop. He never returned. He was found a few days later, frozen to death on the ice. He was buried on the spot and lies there yet, but about a metre closer to North America than on the day he died.

Einstein also failed to live long enough to see that he had backed the wrong horse. In fact, he died at Princeton, New Jersey, in 1955, before Charles Hapgood's rubbishing of continental drift theories was even published.

The other principal player in the emergence of tectonics theory, Harry Hess, was also at Princeton at the time, and would spend the rest of his career there. One of his students was a bright young fellow named Walter Alvarez, who would eventually change the world of science in a quite different way.

As for geology itself, its cataclysms had only just begun, and it was young Alvarez who helped to start the process.

IV

DANGEROUS PLANET

The history of any one part of the Earth, like the life of a soldier, consists of long periods of boredom and short periods of terror.

British geologist Derek V. Ager

13

BANG!

People knew for a long time that there was something odd about the earth beneath Manson, Iowa. In 1912, a man drilling a well for the town water supply reported bringing up a lot of strangely deformed rock – 'crystalline clast breccia with a melt matrix' and 'overturned ejecta flap', as it was later described in an official report. The water was odd, too. It was almost as soft as rainwater. Naturally occurring soft water had never been found in Iowa before.

Though Manson's strange rocks and silken waters were matters of curiosity, forty-one years would pass before a team from the University of Iowa got around to making a trip to the community, then as now a town of about two thousand people in the northwest part of the state. In 1953, after sinking a series of experimental bores, university geologists agreed that the site was indeed anomalous and attributed the deformed rocks to some ancient, unspecified volcanic action. This was in keeping with the wisdom of the day, but it was also about as wrong as a geological conclusion can get.

The trauma to Manson's geology had come not from within the Earth, but from at least one hundred million miles beyond. Some time in the very ancient past, when

Manson stood on the edge of a shallow sea, a rock about a mile and a half across, weighing 10 billion tons and travelling at perhaps two hundred times the speed of sound, ripped through the atmosphere and punched into the Earth with a violence and suddenness that we can scarcely imagine. Where Manson now stands became in an instant a hole three miles deep and more than 20 miles across. The limestone that elsewhere gives Iowa its hard, mineralized water was obliterated and replaced by the shocked basement rocks that so puzzled the water driller in 1912.

The Manson impact was the biggest thing that has ever occurred on the mainland United States. Of any type. Ever. The crater it left behind was so colossal that if you stood on one edge you would only just be able to see the other side on a good day. It would make the Grand Canyon look quaint and trifling. Unfortunately for lovers of spectacle, 2.5 million years of passing ice sheets filled the Manson crater right to the top with rich glacial till, then graded it smooth, so that today the landscape at Manson, and for miles around, is as flat as a table top. Which is of course why no one has ever heard of the Manson crater.

At the library in Manson they are delighted to show you a collection of newspaper articles and a box of core samples from a 1991–2 drilling programme – indeed, they positively bustle to produce them – but you have to ask to see them. Nothing permanent is on display and nowhere in the town is there any historical marker.

To most people in Manson the biggest thing ever to happen was a tornado that rolled up Main Street in 1979, tearing apart the business district. One of the advantages of all that surrounding flatness is that you can see danger from a long way off. Virtually the whole town turned out at one end of Main Street and watched for half an hour as the tornado came towards them, hoping it would veer off, then prudently scampered when it did not. Four of them, alas, didn't move quite fast enough and were killed. Every June now Manson has a week-long event called Crater Days, which was dreamed up as a way of helping people forget that unhappy anniversary. It doesn't really have anything to do with the crater. Nobody's figured out a way to capitalize on an impact site that isn't visible.

'Very occasionally we get people coming in and asking where they should go to see the crater and we have to tell them that there is nothing to see,' says Anna Schlapkohl, the town's friendly librarian. 'Then they go away kind of disappointed.' However, most people, including most Iowans, have never heard of the Manson crater. Even for geologists it barely rates a footnote. But for one brief period in the 1980s, Manson was the most geologically exciting place on Earth.

The story begins in the early 1950s when a bright young geologist named Eugene Shoemaker paid a visit to Meteor Crater in Arizona. Today Meteor Crater is the most famous impact site on Earth and a popular tourist attraction. In those days, however, it didn't receive many visitors and was still often referred to as Barringer Crater, after a wealthy mining engineer named Daniel M. Barringer who had staked a claim on it in 1903. Barringer believed that the crater had been formed by a 10 million tonne meteor, heavily freighted with iron and nickel, and it was his confident expectation that he would make a fortune digging it out. Unaware that the meteor and everything in it would have been vaporized on impact, he wasted a fortune, and the next twenty-six years, cutting tunnels that yielded nothing.

By the standards of today, crater research in the early 1900s was a trifle unsophisticated, to say the least. The leading early investigator, G. K. Gilbert of Columbia University, modelled the effects of impacts by flinging marbles into pans of oatmeal. (For reasons I cannot supply, Gilbert conducted these experiments not in a laboratory at Columbia but in a hotel room.) Somehow, from this Gilbert concluded that the Moon's craters were indeed formed by impacts – in itself quite a radical notion for the time – but that the Earth's were not. Most scientists refused to go even that far. To them, the Moon's craters were evidence of ancient volcanoes and nothing more. The few craters that remained evident on the Earth (most had been eroded away) were generally attributed to other causes or treated as fluky rarities.

By the time Shoemaker came along, a common view was that Meteor Crater had been formed by an underground steam explosion. Shoemaker knew nothing about underground steam explosions – he couldn't: they don't exist – but he did know all about blast zones. One of his first jobs out of college had been to study explosion rings at the Yucca Flats nuclear test site in Nevada. He concluded, as Barringer had before him, that there was nothing at Meteor Crater to suggest volcanic activity, but that there were huge distributions of other stuff – anomalous fine silicas and magnetites principally – that suggested an impact from space. Intrigued, he began to study the subject in his spare time.

Working first with his colleague Eleanor Helin and later with his wife, Carolyn, and associate David Levy, Shoemaker began a systematic survey of the inner solar system. They spent one week each month at the Palomar Observatory in California looking for objects, asteroids primarily, whose trajectories carried them across the Earth's orbit.

'At the time we started, only slightly more than a dozen of these things had ever been discovered in the entire course of astronomical observation,' Shoemaker recalled some years later in a television interview. 'Astronomers in the twentieth century essentially abandoned the solar system,' he added. 'Their attention was turned to the stars, the galaxies.'

What Shoemaker and his colleagues found was that there was more risk out there – a great deal more – than anyone had ever imagined.

* * *

Asteroids, as most people know, are rocky objects orbiting in loose formation in a belt between Mars and Jupiter. In illustrations they are always shown as existing in a jumble, but in fact the solar system is quite a roomy place and the average asteroid actually will be about one and a half million kilometres from its nearest neighbour. Nobody knows even approximately how many asteroids there are tumbling through space, but the number is thought to be probably not less than a billion. They are presumed to be a planet that never quite made it, owing to the unsettling gravitational pull of Jupiter, which kept – and keeps – them from coalescing.

When asteroids were first detected in the 1800s – the very first was discovered on the first day of the century by a Sicilian named Giuseppi Piazzi – they were thought to be planets, and the first two were named Ceres and Pallas. It took some inspired deductions by the astronomer William Herschel to work out that they were nowhere near planet-sized but much smaller. He called them asteroids – Latin for 'starlike' – which was slightly unfortunate as they are not like stars at all. Sometimes now they are more accurately called planetoids.

Finding asteroids became a popular activity in the 1800s and by the end of the century about a thousand were known. The problem was that no-one was systematically recording them. By the early 1900s, it had often become impossible to know whether an asteroid that popped into view was new or simply one that had been noted earlier and then lost track of. By this time, too, astrophysics had moved on so much that few astronomers wanted to devote their lives to anything as mundane as rocky planetoids. Only a few, notably Gerard Kuiper, the Dutch-born astronomer for whom is named the Kuiper belt of comets, took any interest in the solar system at all. Thanks to his work at the

McDonald Observatory in Texas, followed later by work done by others at the Minor Planet Center in Cincinnati and the Spacewatch project in Arizona, a long list of lost asteroids was gradually whittled down until by the close of the twentieth century only one known asteroid was unaccounted for – an object called 719 Albert. Last seen in October 1911, it was finally tracked down in 2000 after being missing for eighty-nine years.

So, from the point of view of asteroid research the twentieth century was essentially just a long exercise in book-keeping. It is really only in the last few years that astronomers have begun to count and keep an eye on the rest of the asteroid community. As of July 2001, 26,000 asteroids had been named and identified – half in just the previous two years. With up to a billion to identify, the count obviously has barely begun.

In a sense it hardly matters. Identifying an asteroid doesn't make it safe. Even if every asteroid in the solar system had a name and known orbit, no one could say what perturbations might send any of them hurtling towards us. We can't forecast rock disturbances on our own surface. Put those rocks adrift in space and what they might do is beyond guessing. Any asteroid out there that has our name on it is very likely to have no other.

Think of the Earth's orbit as a kind of motorway on which we are the only vehicle, but which is crossed regularly by pedestrians who don't know enough to look before stepping off the verge. At least 90 per cent of these pedestrians are quite unknown to us. We don't know where they live, what sort of hours they keep, how often they come our way. All we know is that at some point, at uncertain intervals, they trundle across the road down which we are cruising at over 100,000 kilometres an hour. As Steven Ostro of the Jet Propulsion Laboratory has put it, 'Suppose that there was a button you could push and you could light up all the Earth-crossing asteroids larger than about ten metres, there would be over a hundred million of these objects in the sky.' In short, you would see not a couple of thousand distant twinkling stars, but millions upon millions upon millions of nearer, randomly moving objects – 'all of which are capable of colliding with the Earth and all of which are moving on slightly different courses through the sky at different rates. It would be deeply unnerving.' Well, be unnerved, because it is there. We just can't see it.

Altogether it is thought – though it is really only a guess, based on extrapolating from cratering rates on the moon – that some two thousand asteroids big enough to imperil civilized existence regularly cross our orbit. But even a small asteroid – the size of a house, say – could destroy a city. The number of these relative tiddlers in Earth-crossing orbits is almost certainly in the hundreds of thousands and possibly in the millions, and they are nearly impossible to track.

The first one wasn't spotted until 1991, and that was after it had already gone by. Named 1991 BA, it was noticed as it sailed past us at a distance of 170,000 kilometres – in cosmic terms the equivalent of a bullet passing through one's sleeve without touching the arm. Two years later, another, somewhat larger asteroid missed us by just 145,000 kilometres – the closest pass yet recorded. It, too, was not seen until it had passed and would have arrived without warning. According to Timothy Ferris, writing in the *New Yorker*, such near misses probably happen two or three times a week and go unnoticed.

An object a hundred metres across couldn't be picked up by any Earth-based telescope until it was within just a few days of us, and that is only if a telescope happened to be trained on it, which is unlikely because even now the number of people searching for such objects is modest. The arresting analogy that is always made is that the number of people in the world who are actively searching for asteroids is fewer than the staff of a typical McDonald's restaurant. (It is actually somewhat higher now. But not much.)

While Gene Shoemaker was trying to get people galvanized about the potential dangers of the inner solar system, another development – wholly unrelated on the face of it – was quietly unfolding in Italy with the work of a young geologist from the Lamont Doherty Laboratory at Columbia University. In the early 1970s, Walter Alvarez was doing fieldwork in a comely defile known as the Bottaccione Gorge, near the Umbrian hill town of Gubbio, when he grew curious about a thin band of reddish clay that divided two ancient layers of limestone – one from the Cretaceous period, the other from the Tertiary. This is a point known to geology as the KT boundary* and it marks the time, 65 million years ago, when the dinosaurs and roughly half the world's other species of animals abruptly vanish from the fossil record. Alvarez wondered what it was about a thin lamina of clay, barely 6 millimetres thick, that could account for such a dramatic moment in the Earth's history.

At the time, the conventional wisdom about the dinosaur extinction was the same as it had been in Charles Lyell's day a century earlier – namely, that the dinosaurs had died out over millions of years. But the thinness of the clay layer clearly suggested that in Umbria, if nowhere else, something rather more abrupt had happened. Unfortunately, in the 1970s no tests existed for determining how long such a deposit might have taken to accumulate.

In the normal course of things, Alvarez almost certainly would have had to leave the problem at that; but luckily he

*It is KT rather than CT because *C* had already been appropriated for *Cambrian*. Depending on which source you credit, the *K* comes from either the Greek *kreta* or the German *Kreide*. Both conveniently mean chalk, which is also what *Cretaceous* means.

had an impeccable connection to someone outside his discipline who could help – his father, Luis. Luis Alvarez was an eminent nuclear physicist; he had won the Nobel Prize for physics the previous decade. He had always been mildly scornful of his son's attachment to rocks, but this problem intrigued him. It occurred to him that the answer might lie in dust from space.

Every year the Earth accumulates some 30,000 tonnes of 'cosmic spherules' – space dust, in plainer language – which would be quite a lot if you swept it into one pile, but is infinitesimal when spread across the globe. Scattered through this thin dusting are exotic elements not normally much found on Earth. Among these is the element iridium, which is a thousand times more abundant in space than in the Earth's crust (because, it is thought, most of the iridium on Earth sank to the core when the planet was young).

Luis Alvarez knew that a colleague of his at the Lawrence Berkeley Laboratory in California, Frank Asaro, had developed a technique for measuring very precisely the chemical composition of clays using a process called neutron activation analysis. This involved bombarding samples with neutrons in a small nuclear reactor and carefully counting the gamma rays that were emitted; it was extremely finicky work. Previously Asaro had used the technique to analyse pieces of pottery, but Alvarez reasoned that if they measured the amount of one of the exotic elements in his son's soil samples and compared that with its annual rate of deposition, they would know how long it had taken the samples to form. On an October afternoon in 1977, Luis and Walter Alvarez dropped in on Asaro and asked him if he would run the necessary tests for them.

It was really quite a presumptuous request. They were asking Asaro to devote months to making the most painstaking measurements of geological samples merely to confirm what seemed entirely self-evident to begin with –

that the thin layer of clay had been formed as quickly as its thinness suggested. Certainly no-one expected his survey to yield any dramatic breakthroughs.

'Well, they were very charming, very persuasive,' Asaro recalled in an interview in 2002. 'And it seemed an interesting challenge, so I agreed to try. Unfortunately, I had a lot of other work on, so it was eight months before I could get to it.' He consulted his notes from the period. 'On June 21, 1978, at 1.45 p.m., we put a sample in the detector. It ran for 224 minutes and we could see we were getting interesting results, so we stopped it and had a look.'

The results were so unexpected, in fact, that the three scientists at first thought they had to be wrong. The amount of iridium in the Alvarez sample was more than three hundred times normal levels – far beyond anything they might have predicted. Over the following months Asaro and his colleague Helen Michel worked up to thirty hours at a stretch ('Once you started you couldn't stop,' Asaro explained) analysing samples, always with the same results. Tests on other samples – from Denmark, Spain, France, New Zealand, Antarctica – showed that the iridium deposit was worldwide and greatly elevated everywhere, sometimes by as much as five hundred times normal levels. Clearly something big and abrupt, and probably cataclysmic, had produced this arresting spike.

After much thought, the Alvarezes concluded that the most plausible explanation – plausible to them, at any rate – was that the Earth had been struck by an asteroid or comet.

The idea that the Earth might be subjected to devastating impacts from time to time was not quite as new as is now sometimes suggested. As far back as 1942, a Northwestern University astrophysicist named Ralph B. Baldwin had suggested such a possibility in an article in *Popular Astronomy* magazine. (He published the article there because no academic publisher was prepared to run it.) And at least two well-known scientists, the astronomer Ernst Öpik and the chemist and Nobel laureate Harold Urey, had also voiced support for the notion at various times. Even among palaeontologists it was not unknown. In 1956 a professor at Oregon State University, M.W. de Laubenfels, writing in the *Journal of Paleontology*, had actually anticipated the Alvarez theory by suggesting that the dinosaurs may have been dealt a death blow by an impact from space, and in 1970 the president of the American Paleontological Society, Dewey J. McLaren, proposed at the group's annual conference the possibility that an extraterrestrial impact may have been the cause of an earlier event known as the Frasnian extinction.

As if to underline just how un-novel the idea had become by this time, in 1979 a Hollywood studio actually produced a movie called *Meteor* ('It's five miles wide . . . It's coming at 30,000 m.p.h. – and there's no place to hide!') starring Henry Fonda, Natalie Wood, Karl Malden and a very large rock.

So when, in the first week of 1980, at a meeting of the American Association for the Advancement of Science, the Alvarezes announced their belief that the dinosaur extinction had not taken place over millions of years as part of some slow inexorable process, but suddenly in a single explosive event, it shouldn't have come as a shock.

But it did. It was received everywhere, but particularly in the palaeontological world, as an outrageous heresy.

'Well, you have to remember,' Asaro recalls, 'that we were amateurs in this field. Walter was a geologist specializing in palaeomagnetism, Luis was a physicist and I was a nuclear chemist. And now here we were telling palaeontologists that we had solved a problem that had eluded them for over a century. It's not terribly surprising that they didn't embrace it immediately.' As Luis Alvarez joked: 'We were caught practising geology without a licence.'

But there was also something much deeper and more fundamentally abhorrent in the impact theory. The belief that terrestrial processes were gradual had been elemental in natural history since the time of Lyell. By the 1980s, catastrophism had been out of fashion for so long that it had become literally unthinkable. For most geologists the idea of a devastating impact was, as Eugene Shoemaker noted, 'against their scientific religion'.

Nor did it help that Luis Alvarez was openly contemptuous of palaeontologists and their contributions to scientific knowledge. 'They're really not very good scientists. They're more like stamp collectors,' he wrote in the *New York Times*, in an article that stings yet.

Opponents of the Alvarez theory produced any number of alternative explanations for the iridium deposits – for instance, that they were generated by prolonged volcanic eruptions in India called the Deccan Traps ('trap' comes from a Swedish word for a type of lava; 'Deccan' is the name of the area today) – and above all insisted that there was no proof that the dinosaurs disappeared abruptly from the fossil record at the iridium boundary. One of the most vigorous opponents was Charles Officer of Dartmouth College. He insisted that the iridium had been deposited by volcanic action even while conceding in a newspaper interview that he had no actual evidence of it. As late as 1988, more than half of all American palaeontologists contacted in a survey continued to believe that the extinction of the dinosaurs was in no way related to an asteroid or cometary impact.

The one thing that would most obviously support the Alvarezes' theory was the one thing they didn't have – an impact site. Enter Eugene Shoemaker. Shoemaker had an Iowa connection – his daughter-in-law taught at the University of Iowa – and he was familiar with the Manson crater from his own studies. Thanks to him, all eyes now turned to Iowa.

★ ★ ★

Geology is a profession that varies from place to place. In Iowa, a state that is flat and stratigraphically uneventful, it tends to be comparatively serene. There are no alpine peaks or grinding glaciers, no great deposits of oil or precious metals, not a hint of a pyroclastic flow. If you are a geologist employed by the state of Iowa, a big part of the work you do is to evaluate Manure Management Plans, which all the state's 'animal confinement operators' – pig farmers, to the rest of us – are required to file periodically. There are 15 million pigs in Iowa, so a lot of manure to manage. I'm not mocking this at all – it's vital and enlightened work; it keeps Iowa's water clean – but with the best will in the world it's not exactly dodging lava bombs on Mount Pinatubo or scrabbling over crevasses on the Greenland ice sheet in search of ancient life-bearing quartzes. So we may well imagine the flutter of excitement that swept through the Iowa Department of Natural Resources when in the mid-1980s the world's geological attention focused on Manson and its crater.

Trowbridge Hall in Iowa City is a turn-of-the-century pile of red brick that houses the University of Iowa's Earth Sciences Department and – way up in a kind of garret – the geologists of the Iowa Department of Natural Resources. No-one now can remember quite when, still less why, the state geologists were placed in an academic facility, but you get the impression that the space was conceded grudgingly, for the offices are cramped and low-ceilinged and not very accessible. When being shown the way, you half expect to be taken out onto a roof ledge and helped in through a window.

Ray Anderson and Brian Witzke spend their working lives up here amid disordered heaps of papers, journals, furled charts and hefty specimen stones. (Geologists are never at a loss for paperweights.) It's the kind of space where if you want to find anything – an extra chair, a coffee cup, a ringing telephone – you have to move stacks of documents around.

'Suddenly we were at the centre of things,' Anderson told me, gleaming at the memory of it, when I met him and Witzke in their offices on a dismal, rainy morning in June. 'It was a wonderful time.'

I asked them about Gene Shoemaker, a man who seems to have been universally revered. 'He was just a great guy,' Witzke replied without hesitation. 'If it hadn't been for him, the whole thing would never have gotten off the ground. Even with his support, it took two years to get it up and running. Drilling's an expensive business – about thirty-five dollars a foot back then, more now, and we needed to go down three thousand feet.'

'Sometimes more than that,' Anderson added.

'Sometimes more than that,' Witzke agreed. 'And at several locations. So you're talking a lot of money. Certainly more than our budget would allow.'

So a collaboration was formed between the Iowa Geological Survey and the US Geological Survey.

'At least, we *thought* it was a collaboration,' said Anderson, producing a small pained smile.

'It was a real learning curve for us,' Witzke went on. 'There was actually quite a lot of bad science going on throughout the period – people rushing in with results that didn't always stand up to scrutiny.' One of those moments came at the annual meeting of the American Geophysical Union in 1985, when Glenn Izett and C. L. Pillmore of the US Geological Survey announced that the Manson crater was of the right age to have been involved with the dinosaurs' extinction. The declaration attracted a good deal of press attention but was unfortunately premature. A more careful examination of the data revealed that Manson was not only too small, but also nine million years too early.

The first Anderson or Witzke learned of this setback to their careers was when they arrived at a conference in South Dakota, and found people coming up to them with sympathetic looks and saying: 'We hear you lost your crater.' It was news to them that Izett and the other USGS scientists had just announced refined figures revealing that Manson couldn't after all have been the extinction crater.

'It was pretty stunning,' recalls Anderson. 'I mean, we had this thing that was really important and then suddenly we didn't have it any more. But even worse was the realization that the people we thought we'd been collaborating with hadn't bothered to share with us their new findings.'

'Why not?'

He shrugged. 'Who knows? Anyway, it was a pretty good insight into how unattractive science can get when you're playing at a certain level.'

The search moved elsewhere. By chance, in 1990 one of the searchers, Alan Hildebrand of the University of Arizona, met a reporter from the *Houston Chronicle* who happened to know about a large, unexplained ring formation, 193 kilometres wide and 48 kilometres deep, under Mexico's Yucatán Peninsula at Chicxulub, near the city of Progreso, about 950 kilometres due south of New Orleans. The formation had been found by Pemex, the Mexican oil company, in 1952 – the year, coincidentally, that Gene Shoemaker first visited Meteor Crater in Arizona – but the company's geologists had concluded that it was volcanic, in line with the thinking of the day. Hildebrand travelled to the site and decided fairly swiftly that they had their crater. By early 1991 it had been established to nearly everyone's satisfaction that Chicxulub was the impact site.

Still, many people didn't quite grasp what an impact could do. As Stephen Jay Gould recalled in one of his essays: 'I remember harboring some strong initial doubts about the efficacy of such an event . . . [W]hy should an object only six miles across wreak such havoc upon a planet with a diameter of eight thousand miles?'

Conveniently, a natural test of the theory arose soon after when the Shoemakers and Levy discovered Comet Shoemaker–Levy 9, which they soon realized was headed for Jupiter. For the first time, humans would be able to witness a cosmic collision – and witness it very well, thanks to the new Hubble Space Telescope. Most astronomers, according to Curtis Peebles, expected little, particularly as the comet was not a coherent sphere but a string of twenty-one fragments. 'My sense', wrote one, 'is that Jupiter will swallow these comets up without so much as a burp.' One week before the impact, *Nature* ran an article, 'The Big Fizzle Is Coming', predicting that the impact would constitute nothing more than a meteor shower.

The impacts began on 16 July 1994, went on for a week and were bigger by far than anyone – with the possible exception of Gene Shoemaker – expected. One fragment, known as Nucleus G, struck with the force of about six million megatonnes – seventy-five times all the nuclear weaponry in existence. Nucleus G was only about the size of a small mountain, but it created wounds in the Jovian surface the size of Earth. It was the final blow for critics of the Alvarez theory.

Luis Alvarez never knew of the discovery of the Chicxulub crater or of the Shoemaker–Levy comet, as he died in 1988. Shoemaker also died early. On the third anniversary of the Jupiter collision, he and his wife were in the Australian outback, where they went every year to search for impact sites. On a dirt track in the Tanami Desert – normally one of the emptiest places on Earth – they came over a slight rise just as another vehicle was approaching. Shoemaker was killed instantly, his wife injured. Some of his ashes were sent to the Moon aboard the Lunar Prospector spacecraft. The rest were scattered around Meteor Crater.

★ ★ ★

Anderson and Witzke no longer had the crater that killed the dinosaurs, 'but we still had the largest and most perfectly preserved impact crater in the mainland United States', Anderson said. (A little verbal dexterity is required to keep Manson's superlative status. Other craters are larger – notably, Chesapeake Bay, which was recognized as being an impact site in 1994 – but they are either offshore or deformed.) 'Chicxulub is buried under two to three kilometres of limestone and mostly offshore, which makes it difficult to study,' Anderson went on, 'while Manson is really quite accessible. It's because it is buried that it is actually comparatively pristine.'

I asked them how much warning we would receive if a similar hunk of rock were coming towards us today.

'Oh, probably none,' said Anderson breezily. 'It wouldn't be visible to the naked eye until it warmed up and that wouldn't happen until it hit the atmosphere, which would be about one second before it hit the Earth. You're talking about something moving many tens of times faster than the fastest bullet. Unless it had been seen by someone with a telescope, and that's by no means a certainty, it would take us completely by surprise.'

How hard an impactor hits depends on a lot of variables – angle of entry, velocity and trajectory, whether the collision is head-on or from the side, and the mass and density of the impacting object, among much else – none of which we can know so many millions of years after the fact. But what scientists can do – and Anderson and Witzke have done – is measure the impact site and calculate the amount of energy released. From that they can work out plausible scenarios of what it must have been like – or, more chillingly, would be like if it happened now.

An asteroid or comet travelling at cosmic velocities would enter the Earth's atmosphere at such a speed that the air beneath it couldn't get out of the way and would be compressed, as in a bicycle pump. As anyone who has used such a pump knows, compressed air grows swiftly hot, and the temperature below it would rise to some 60,000 Kelvin, or ten times the surface temperature of the Sun. In this instant of its arrival in our atmosphere, everything in the meteor's path – people, houses, factories, cars – would crinkle and vanish like cellophane in a flame.

One second after entering the atmosphere, the meteorite would slam into the Earth's surface where the people of Manson had a moment before been going about their business. The meteorite itself would vaporize instantly, but the blast would blow out 1,000 cubic kilometres of rock, earth and superheated gases. Every living thing within 250 kilometres that hadn't been killed by the heat of entry would now be killed by the blast. Radiating outwards at almost the speed of light would be the initial shock wave, sweeping everything before it.

For those outside the zone of immediate devastation, the first inkling of catastrophe would be a flash of blinding light – the brightest ever seen by human eyes – followed an instant to a minute or two later by an apocalyptic sight of unimaginable grandeur: a roiling wall of darkness reaching high into the heavens, filling one entire field of view and travelling at thousands of kilometres an hour. Its approach would be eerily silent since it would be moving far beyond the speed of sound. Anyone in a tall building in Omaha or Des Moines, say, who chanced to look in the right direction would see a bewildering veil of turmoil followed by instantaneous oblivion.

Within minutes, over an area stretching from Denver to Detroit and encompassing what had once been Chicago, St Louis, Kansas City, the Twin Cities – the whole of the Midwest, in short – nearly every standing thing would be flattened or on fire, and nearly every living thing would be dead. People up to 1,500 kilometres away would be knocked off their feet and sliced or clobbered by a blizzard of flying projectiles. Beyond 1,500 kilometres the devastation from the blast would gradually diminish.

But that's just the initial shock wave. No-one can do more than guess what the associated damage would be, other than that it would be brisk and global. The impact would almost certainly set off a chain of devastating earthquakes. Volcanoes across the globe would begin to rumble and spew. Tsunamis would rise up and head devastatingly for distant shores. Within an hour, a cloud of blackness would cover the Earth and burning rock and other debris would be pelting down everywhere, setting much of the planet ablaze. It has been estimated that at least a billion and a half people would be dead by the end of the first day. The massive disturbances to the ionosphere would knock out communications systems everywhere, so survivors would have no idea what was happening elsewhere or where to turn. It would hardly matter. As one commentator has put it, fleeing would mean 'selecting a slow death over a quick one. The death toll would be very little affected by any plausible relocation effort, since Earth's ability to support life would be universally diminished.'

The amount of soot and floating ash from the impact and following fires would blot out the sun certainly for months, possibly for years, disrupting growing cycles. In 2001 researchers at the California Institute of Technology analysed helium isotopes from sediments left from the later KT impact and concluded that it affected the Earth's climate for about ten thousand years. This was actually used as evidence to support the notion that the extinction of dinosaurs was swift and emphatic – and so it was, in geological terms. We can only guess how well, or whether, humanity would cope with such an event.

And in all likelihood, remember, this would come without warning, out of a clear sky.

But let's suppose we did see the object coming. What would we do? Everyone assumes we would send up a nuclear warhead and blast it to smithereens. There are some problems with that idea, however. First, as John S. Lewis notes, our missiles are not designed for space work. They haven't the oomph to escape Earth's gravity, and even if they did there are no mechanisms to guide them across tens of millions of kilometres of space. Still less could we send up a shipload of space cowboys to do the job for us, as in the movie *Armageddon*; we no longer possess a rocket powerful enough to send humans even as far as the Moon. The last rocket that could, Saturn 5, was retired years ago and has never been replaced. Nor could we quickly build a new one because, amazingly, the plans for Saturn launchers were destroyed as part of a NASA spring-cleaning exercise.

Even if we did manage somehow to get a warhead to the asteroid and blast it to pieces, the chances are that we would simply turn it into a string of rocks that would slam into us one after the other in the manner of Comet Shoemaker–Levy on Jupiter – but with the difference that now the rocks would be intensely radioactive. Tom Gehrels, an asteroid hunter at the University of Arizona, thinks that even a year's warning would probably be insufficient to take appropriate action. The greater likelihood, however, is that we wouldn't see any object – even a comet – until it was about six months away, which would be much too late. Shoemaker–Levy 9 had been orbiting Jupiter in a fairly conspicuous manner since 1929, but it was over half a century before anyone noticed.

Because these things are so difficult to compute and must incorporate such a significant margin of error, even if we knew an object was heading our way we wouldn't know until nearly the end – the last couple of weeks anyway – whether collision was certain. For most of the time of the object's approach we would exist in a kind of cone of uncertainty. It would certainly be the most interesting few months in the history of the world. And imagine the party if it passed safely.

'So how often does something like the Manson impact happen?' I asked Anderson and Witzke before leaving.

'Oh, about once every million years on average,' said Witzke.

'And remember,' added Anderson, 'this was a relatively minor event. Do you know how many extinctions were associated with the Manson impact?'

'No idea,' I replied.

'None,' he said, with a strange air of satisfaction. 'Not one.'

Of course, Witzke and Anderson added hastily and more or less in unison, there would have been terrible devastation across much of the Earth, as just described, and complete annihilation for hundreds of miles around ground zero. But life is hardy, and when the smoke cleared there were enough lucky survivors from every species that none permanently perished.

The good news, it appears, is that it takes an awful lot to extinguish a species. The bad news is that the good news can never be counted on. Worse still, it isn't actually necessary to look to space for petrifying danger. As we are about to see, Earth can provide plenty of danger of its own.

14

THE FIRE BELOW

In the summer of 1971, a young geologist named Mike Voorhies was scouting around on some grassy farmland in eastern Nebraska, not far from the little town of Orchard where he had grown up. Passing through a steep-sided gully, he spotted a curious glint in the brush above and clambered up to have a look. What he had seen was the perfectly preserved skull of a young rhinoceros, which had been washed out by recent heavy rains.

A few yards beyond, it turned out, was one of the most extraordinary fossil beds ever discovered in North America: a dried-up waterhole that had served as a mass grave for scores of animals – rhinoceroses, zebra-like horses, sabre-toothed deer, camels, turtles. All had died from some mysterious cataclysm just under twelve million years ago in the time known to geology as the Miocene. In those days Nebraska stood on a vast, hot plain very like the Serengeti of Africa today. The animals had been found buried under volcanic ash up to 3 metres deep. The puzzle of it was that there were not, and never had been, any volcanoes in Nebraska.

Today, the site of Voorhies' discovery is called Ashfall Fossil Beds State Park. It has a stylish new visitors' centre and museum, with thoughtful displays on the geology of Nebraska and the history of the fossil beds. The centre incorporates a lab with a glass wall through which visitors can watch palaeontologists cleaning bones. Working alone in the lab on the morning I passed through was a cheerfully grizzled-looking fellow in a blue workshirt whom I recognized as Mike Voorhies from a BBC *Horizon* documentary in which he had featured. They don't get a huge number of visitors to Ashfall Fossil Beds State Park – it's slightly in the middle of nowhere – and Voorhies seemed pleased to show me around. He took me to the spot atop a 6-metre-high ravine where he had made his find.

'It was a dumb place to look for bones,' he said happily. 'But I wasn't looking for bones. I was thinking of making a geological map of eastern Nebraska at the time, and really just kind of poking around. If I hadn't gone up this ravine or the rains hadn't just washed out that skull, I'd have walked on by and this would never have been found.' He indicated a roofed enclosure nearby, which had become the main excavation site. There, some two hundred animals had been found lying together in a jumble.

I asked him in what way it was a dumb place to hunt for bones. 'Well, if you're looking for bones, you really need exposed rock. That's why most palaeontology is done in hot, dry places. It's not that there are more bones there. It's just that you have some chance of spotting them. In a setting like this' – he made a sweeping gesture across the vast and unvarying prairie – 'you wouldn't know where to begin. There could be really magnificent stuff out there, but there's no surface clues to show you where to start looking.'

At first they thought the animals were buried alive and Voorhies stated as much in a *National Geographic* article in 1981. 'The article called the site a "Pompeii of prehistoric animals",' he told me, 'which was unfortunate because just afterwards we realized that the animals hadn't died suddenly at all. They were all suffering from something called hypertrophic pulmonary osteodystrophy, which is what you would get if you were breathing a lot of abrasive ash – and they must have been breathing a lot of it because the ash was feet thick for hundreds of miles.' He picked up a chunk of greyish, claylike dirt and crumbled it into my hand. It was powdery but slightly gritty. 'Nasty stuff to have to breathe,' he went on, 'because it's very fine but also quite sharp. So anyway they came here to this watering hole, presumably seeking relief, and died in some misery. The ash would have ruined everything. It would have buried all the grass and coated every leaf and turned the water into an undrinkable grey sludge. It couldn't have been very agreeable at all.'

The *Horizon* documentary had suggested that the existence of so much ash in Nebraska was a surprise. In fact, Nebraska's huge ash deposits had been known about for a long time. For almost a century they had been mined to make household cleaning powders like Comet and Ajax. But, curiously, no-one had ever thought to wonder where all the ash came from.

'I'm a little embarrassed to tell you,' Voorhies said, smiling briefly, 'that the first I thought about it was when an editor at the *National Geographic* asked me the source of all the ash and I had to confess that I didn't know. Nobody knew.'

Voorhies sent samples to colleagues all over the western United States asking if there was anything about it that they recognized. Several months later a geologist named Bill Bonnichsen from the Idaho Geological Survey got in touch and told him that the ash matched a volcanic deposit from a place called Bruneau-Jarbidge in southwest Idaho. The event that killed the plains animals of Nebraska was a volcanic explosion on a scale previously unimagined – but big enough to leave an ash layer 3 metres deep some 1,600 kilometres away in eastern Nebraska. It turned out that under the western United States there was a huge cauldron of magma, a colossal volcanic hot spot, which erupted cataclysmically every six hundred thousand years or so. The last such eruption was just over six hundred thousand years ago. The hot spot is still there. These days we call it Yellowstone National Park.

We know amazingly little about what happens beneath our feet. It is fairly remarkable to think that Ford has been building cars and Nobel committees awarding prizes for longer than we have known that the Earth has a core. And of course the idea that the continents move about on the surface like lily pads has been common wisdom for much less than a generation. 'Strange as it may seem,' wrote Richard Feynman, 'we understand the distribution of matter in the interior of the sun far better than we understand the interior of the earth.'

The distance from the surface of Earth to the middle is 6,370 kilometres, which isn't so very far. It has been calculated that if you sunk a well to the centre and dropped a brick down it, it would take only forty-five minutes for it to hit the bottom (though at that point it would be weightless since all the Earth's gravity would be above and around it rather than beneath it). Our own attempts to penetrate towards the middle have been modest indeed. One or two South African gold mines reach to a depth of over 3 kilometres, but most mines on Earth go no more than about 400 metres beneath the surface. If the planet were an apple, we wouldn't yet have broken through the skin. Indeed, we haven't even come close.

Until slightly under a century ago, what the best-informed scientific minds knew about Earth's interior was not much more than what a coal miner knew – namely, that you could dig down through soil for a distance and then you'd hit rock, and that was about it. Then, in 1906, an Irish geologist named R. D. Oldham, while examining some seismograph readings from an earthquake in Guatemala, noticed that certain shock waves had penetrated to a point deep within the Earth and then bounced off at an angle, as if they had encountered some kind of barrier. From this he deduced that the Earth has a core. Three years later, a Croatian seismologist named Andrija Mohorovičić was studying graphs from an earthquake in Zagreb when he noticed a similar odd deflection, but at a shallower level. He had discovered the boundary between the crust and the layer immediately below, the mantle; this zone has been known ever since as the Mohorovičić discontinuity, or Moho for short.

We were beginning to get a vague idea of the Earth's layered interior – though it really was only vague. Not until 1936 did a Danish scientist named Inge Lehmann, studying seismographs of earthquakes in New Zealand, discover that there were two cores – an inner one, which we now believe to be solid, and an outer one (the one that Oldham had detected), which is thought to be liquid and the seat of magnetism.

At just about the time that Lehmann was refining our basic understanding of the Earth's interior by studying the seismic waves of earthquakes, two geologists at Caltech in California were devising a way to make comparisons between one earthquake and the next. They were Charles Richter and Beno Gutenberg, though for reasons that have nothing to do with fairness the scale became known almost at once as Richter's alone. (They were nothing to do with Richter, either. A modest fellow, he never referred to the scale by his own name, but always called it 'the Magnitude Scale'.)

The Richter scale has always been widely misunderstood by non-scientists, though it is perhaps a little less so now than in its early days when visitors to Richter's office often asked to see his celebrated scale, thinking it was some kind of machine. The scale is, of course, more an idea than a thing, an arbitrary measure of the Earth's tremblings based on surface measurements. It rises exponentially, so that a 7.3 quake is fifty times more powerful than a 6.3 earthquake and 2,500 times more powerful than a 5.3 earthquake.

Theoretically, at least, there is no upper limit for an earthquake – nor, come to that, a lower limit. The scale is a simple measure of force, but says nothing about damage. A magnitude 7 quake happening deep in the mantle – say, 650 kilometres down – might cause no surface damage at all, while a significantly smaller one happening just 6 or 7 kilometres under the surface could wreak widespread devastation. Much, too, depends on the nature of the subsoil, the quake's duration, the frequency and severity of aftershocks, and the physical setting of the affected area. All this means that the most fearsome quakes are not necessarily the most forceful, though force obviously counts for a lot.

The largest earthquake since the scale's invention was (depending on which source you credit) either one centred on Prince William Sound in Alaska in March 1964, which measured 9.2 on the Richter scale, or one in the Pacific Ocean off the coast of Chile in 1960, which was initially logged at 8.6 magnitude but later revised upwards by some authorities (including the US Geological Survey) to a truly grand-scale 9.5. As you will gather from this, measuring earthquakes is not always an exact science, particularly when it involves interpreting readings from remote locations. At all events, both quakes were whopping. The 1960 quake not only caused widespread damage across coastal South America, but also set off a giant tsunami that rolled nearly ten thousand kilometres across the Pacific and slapped away much of downtown Hilo, Hawaii, destroying five hundred buildings and killing sixty people. Similar wave surges claimed yet more victims as far away as Japan and the Philippines.

For pure, focused devastation, however, probably the most intense earthquake in recorded history was one that struck – and essentially shook to pieces – Lisbon, Portugal, on All Saints Day (1 November), 1755. Just before ten in the morning, the city was hit by a sudden sideways lurch now estimated at magnitude 9.0 and shaken ferociously for seven full minutes. The convulsive force was so great that the water rushed out of the city's harbour and returned in a wave over 15 metres high, adding to the destruction. When at last the motion ceased, survivors enjoyed just three minutes of calm before a second shock came, only slightly less severe than the first. A third and final shock followed two hours later. At the end of it all, sixty thousand people were dead and virtually every building for miles reduced to rubble. The San Francisco earthquake of 1906, for comparison, measured an estimated 7.8 on the Richter scale and lasted less than thirty seconds.

Earthquakes are fairly common. Every day on average somewhere in the world there are two of magnitude 2.0 or greater – that's enough to give anyone nearby a pretty good jolt. Although they tend to cluster in certain places – notably around the rim of the Pacific – they can occur almost anywhere. In the United States, only Florida, eastern Texas and the upper Midwest seem – so far – to be almost entirely immune. New England has had two quakes of magnitude 6.0 or greater in the last two hundred years. In April 2002, the region experienced a 5.1 magnitude shaking in a quake near Lake Champlain on the New York–Vermont border, causing extensive local damage and (I can attest) knocking pictures from walls and children from beds as far away as New Hampshire.

The most common types of earthquakes are those where two plates meet, as in California along the San Andreas Fault. As the plates push against each other, pressures build up until one or the other gives way. In general, the longer the interval between quakes, the greater the pent-up pressure and thus the greater the scope for a really big jolt. This is a particular worry for Tokyo, which Bill McGuire, a hazards specialist at University College London, describes as 'the city waiting to die' (not a motto you will find on many tourism leaflets). Tokyo stands on the meeting point of three tectonic plates in a country already well known for its seismic instability. In 1995, as you will remember, the city of Kobe, nearly 500 kilometres to the west, was struck by a magnitude 7.2 quake, which killed 6,394 people. The damage was estimated at $99 billion. But that was as nothing – well, as comparatively little – compared with what may await Tokyo.

Tokyo has already suffered one of the most devastating earthquakes in modern times. On 1 September 1923, just before midday, the city was hit by what is known as the Great Kanto quake – an event over ten times as powerful as Kobe's earthquake. Two hundred thousand people were killed. Since that time, Tokyo has been eerily quiet, so the strain beneath the surface has been building for eighty years. Eventually it is bound to snap. In 1923, Tokyo had a population of about three million. Today it is approaching thirty million. Nobody cares to guess how many people might die, but the potential economic cost has been put as high as $7 trillion.

Even more unnerving, because they are less well understood and capable of occurring anywhere at any time, are the rarer shakings of the type known as intraplate quakes. These happen away from plate boundaries, which makes them wholly unpredictable. And because they come from a much greater depth, they tend to propagate over much wider areas. The most notorious such quakes ever to hit the United States were a series of three in New Madrid, Missouri, in the winter of 1811–12. The adventure started just after midnight on 16 December when people were awakened first by the noise of panicking farm animals (the restiveness of animals before quakes is not an old wives' tale, but is in fact well established, though not at all understood) and then by an almighty rupturing noise from deep within the Earth. Emerging from their houses, locals found the land rolling in waves up to a metre high and opening up in fissures several metres deep. A strong smell of sulphur filled the air. The shaking lasted for four minutes, with the usual devastating effects to property. Among the witnesses was the artist John James Audubon, who happened to be in the area. The quake radiated outwards with such force that it knocked down chimneys in Cincinnati over 600 kilometres away and, according to at least one account, 'wrecked boats in East Coast harbors and . . . even collapsed scaffolding erected around the Capitol Building in Washington, D.C.' On 23 January and 4 February further quakes of similar magnitude followed. New Madrid has been silent ever since – but not surprisingly, since such episodes have never been known to happen in the same place twice. As far as we know, they are as random as lightning. The next one could be under Chicago or Paris or Kinshasa. No-one can even begin to guess. And what causes these massive intraplate rupturings? Something deep within the Earth. More than that, we don't know.

By the 1960s scientists had grown sufficiently frustrated by how little they understood of the Earth's interior that they decided to try to do something about it. Specifically, they got the idea to drill through the ocean floor (the continental crust was too thick) to the Moho discontinuity and to extract a piece of the Earth's mantle for examination at leisure. The thinking was that if they could understand the nature of the rocks inside the Earth, they might begin to understand how they interacted, and thus possibly be able to predict earthquakes and other unwelcome events.

The project became known, all but inevitably, as the Mohole, and it was pretty well disastrous. The hope was to lower a drill through over 4,000 metres of Pacific Ocean water off the coast of Mexico and drill some 5,000 metres through relatively thin crustal rock. Drilling from a ship in open waters is, in the words of one oceanographer, 'like trying to drill a hole in the sidewalks of New York from atop the Empire State Building using a strand of spaghetti'. Every attempt ended in failure. The deepest they penetrated was only about 180 metres. The Mohole became known as the No Hole. In 1966, exasperated with ever-rising costs and no results, Congress killed the project.

Four years later, Soviet scientists decided to try their luck on dry land. They chose a spot on Russia's Kola Peninsula, near the Finnish border, and set to work with the hope of drilling to a depth of 15 kilometres. The work proved harder than expected, but the Soviets were commendably persistent. When at last they gave up, nineteen years later, they had drilled to a depth of 12,262 metres. Bearing in mind that the crust of the Earth represents only about 0.3 per cent of the planet's volume and that the Kola hole had not cut even one-third of the way through the crust, we can hardly claim to have conquered the interior.

Even though the hole was modest, nearly everything about what it revealed surprised the researchers. Seismic wave studies had led the scientists to predict, and pretty confidently, that they would encounter sedimentary rock to a depth of 4,700 metres, followed by granite for the next 2,300 metres and basalt from there on down. In the event, the sedimentary layer was 50 per cent deeper than expected and the basaltic layer was never found at all. Moreover, the world down there was far warmer than anyone had expected, with a temperature at 10,000 metres of 180 degrees Celsius, nearly twice the forecast level. Most surprising of all was that the rock at depth was saturated with water – something that had not been thought possible.

Because we can't see into the Earth, we have to use other techniques, which mostly involve reading waves as they travel through the interior, to find out what is there. We know a little bit about the mantle from what are known as kimberlite pipes, where diamonds are formed. What happens is that deep in the Earth there is an explosion that fires, in effect, a cannonball of magma to the surface at supersonic speeds. It is a totally random event. A kimberlite pipe could explode in your back garden as you read this. Because they come up from such depths – up to 200 kilometres down – kimberlite pipes bring up all kinds of things not normally found on or near the surface: a rock called peridotite, crystals of olivine and – just occasionally, in about one pipe in a hundred – diamonds. Lots of carbon comes up with kimberlite ejecta, but most is vaporized or turns to graphite. Only occasionally does a hunk of it shoot up at just the right speed and cool down with the necessary swiftness to become a diamond. It was such a pipe that made South Africa the most productive diamond-mining country in the world, but there may be others even bigger that we don't know about. Geologists know that somewhere in the vicinity of northeastern Indiana there is evidence of a pipe or group of pipes that may be truly colossal. Diamonds up to 20 carats or more have been found at scattered sites throughout the region. But no-one has ever found the source. As John McPhee notes, it may be buried under glacially deposited soil, like the Manson crater in Iowa, or under the Great Lakes.

So how much do we know about what's inside the Earth? Very little. Scientists are generally agreed that the world beneath us is composed of four layers – a rocky outer crust, a mantle of hot, viscous rock, a liquid outer core and a solid inner core.* We know that the surface is dominated by silicates, which are relatively light and not heavy enough to account for the planet's overall density. Therefore there must be heavier stuff inside. We know that to generate our magnetic field somewhere in the interior there must be a concentrated belt of metallic elements in a liquid state. That much is universally accepted. Almost everything beyond that – how the layers interact, what causes them to behave in the way they do, what they will do at any time in the future – is a matter of at least some uncertainty, and generally quite a lot of uncertainty.

Even the one part of it we can see, the crust, is a matter of some fairly strident debate. Nearly all geology texts tell you that continental crust is 5 to 10 kilometres thick under the oceans, about 40 kilometres thick under the continents and 65–95 kilometres thick under big mountain chains, but there are many puzzling variabilities within these generalizations. The crust beneath the Sierra Nevada Mountains, for instance, is only about 30–40 kilometres thick, and no one knows why. By all the laws of geophysics the Sierra Nevadas should be sinking, as if into quicksand. (Some people think they may be.)

How and when the Earth got its crust are questions that divide geologists into two broad camps – those who think it happened abruptly, early in the Earth's history, and those who think it happened gradually and rather later. Strength of feeling runs deep on such matters. Richard Armstrong of

*For those who crave a more detailed picture of the Earth's interior, here are the dimensions of the various layers, using average figures: From 0 to 40 kilometres is the crust. From 40 to 400 kilometres is the upper mantle. From 400 to 650 kilometres is a transition zone between the upper and lower mantle. From 650 to 2,700 kilometres is the lower mantle. From 2,700 to 2,890 kilometres is the 'D' layer. From 2,890 to 5,150 kilometres is the outer core, and from 5,150 to 6,370 kilometres is the inner core.

Yale proposed an early-burst theory in the 1960s, then spent the rest of his career fighting those who did not agree with him. He died of cancer in 1991, but shortly before his death he 'lashed out at his critics in a polemic in an Australian earth science journal that charged them with perpetuating myths', according to a report in *Earth* magazine in 1998. 'He died a bitter man,' reported a colleague.

The crust and part of the outer mantle together are called the lithosphere (from the Greek *lithos*, meaning stone), which in turn floats on top of a layer of softer rock called the asthenosphere (from Greek words meaning 'without strength'), but such terms are never entirely satisfactory. To say that the lithosphere floats on top of the asthenosphere suggests a degree of easy buoyancy that isn't quite right. Similarly, it is misleading to think of the rocks as flowing in anything like the way we think of materials flowing on the surface. The rocks are viscous, but only in the same way that glass is. It may not look it, but all the glass on Earth is flowing downwards under the relentless drag of gravity. Remove a pane of really old glass from the window of a European cathedral and it will be noticeably thicker at the bottom than at the top.* That is the sort of 'flow' we are talking about. The hour hand on a clock moves about ten thousand times faster than the 'flowing' rocks of the mantle.

The movements occur not just laterally, as the Earth's plates move across the surface, but up and down too, as rocks rise and fall under the churning process known as convection. Convection as a process was first deduced by the eccentric Count von Rumford at the end of the eighteenth century. Sixty years later an English vicar named

*Or so it has often been written. However, in the summer of 2003, after this book came out, Science News reported a study by Prof. E.D. Zanotto of Brazil suggesting that the flow of glass, however venerable the pane, is actually much too slow to be detectable by the naked eye.

Osmond Fisher presciently suggested that the Earth's interior might well be fluid enough for the contents to move about, but that idea took a very long time to gain support.

In about 1970, when geophysicists realized just how much turmoil was going on down there, it came as a considerable shock. As Shawna Vogel put it in the book *Naked Earth: The New Geophysics*: 'It was as if scientists had spent decades figuring out the layers of the Earth's atmosphere – troposphere, stratosphere and so forth – and then had suddenly found out about wind.'

How deep the convection process goes has been a matter of controversy ever since. Some say it begins 650 kilometres down, others more than 3,000 kilometres below us. The problem, as James Trefil has observed, is that 'there are two sets of data, from two different disciplines, that cannot be reconciled.' Geochemists say that certain elements on the planet's surface cannot have come from the upper mantle, but must have come from deeper within the Earth. Therefore, the materials in the upper and lower mantle must at least occasionally mix. Seismologists insist that there is no evidence to support such a thesis.

So all that can be said is that at some slightly indeterminate point as we head towards the centre of the Earth we leave the asthenosphere and plunge into pure mantle. Considering that it accounts for 82 per cent of the Earth's volume and 65 per cent of its mass, the mantle doesn't attract a great deal of attention, largely because the things that interest earth scientists and general readers alike happen either deeper down (as with magnetism) or nearer the surface (as with earthquakes). We know that to a depth of about 150 kilometres the mantle consists predominantly of a type of rock known as peridotite, but what fills the next 2,650 kilometres is uncertain. According to a *Nature* report, it seems not to be peridotite. More than this we do not know.

Beneath the mantle are the two cores, a solid inner core

and a liquid outer one. Needless to say, our understanding of the nature of these cores is indirect, but scientists can make some reasonable assumptions. They know that the pressures at the centre of the Earth are sufficiently high – something over three million times those found at the surface – to turn any rock there solid. They also know from the Earth's history (among other clues) that the inner core is very good at retaining its heat. Although it is little more than a guess, it is thought that in over four billion years the temperature at the core has fallen by no more than 110 degrees Celsius. No one knows exactly how hot the Earth's core is, but estimates range from something over 4,000 degrees to over 7,000 degrees Celsius – about as hot as the surface of the Sun.

The outer core is in many ways even less well understood, though everyone is in agreement that it is fluid and that it is the seat of magnetism. The theory was put forward by E. C. Bullard of Cambridge University in 1949 that this fluid part of the Earth's core revolves in a way that makes it, in effect, an electrical motor, creating the Earth's magnetic field. The assumption is that the convecting fluids in the Earth act somehow like the currents in wires. Exactly what happens isn't known, but it is felt pretty certain that it is connected with the core spinning and with its being liquid. Bodies that don't have a liquid core – the Moon and Mars, for instance – don't have magnetism.

We know that the Earth's magnetic field changes in power from time to time: during the age of the dinosaurs, it was up to three times as strong as it is now. We also know that it reverses itself every five hundred thousand years or so on average, though that average hides a huge degree of unpredictability. The last reversal was about seven hundred and fifty thousand years ago. Sometimes it stays put for millions of years – 37 million years appears to be the longest stretch – and at other times it has reversed after as little as twenty thousand years. Altogether in the last hundred million years it has reversed itself about two hundred times, and we don't have any real idea why. This has been called 'the greatest unanswered question in the geological sciences'.

We may be going through a reversal now. The Earth's magnetic field has diminished by perhaps as much as 6 per cent in the last century alone. Any diminution in magnetism is likely to be bad news, because magnetism, apart from holding notes to refrigerators and keeping our compasses pointing the right way, plays a vital role in keeping us alive. Space is full of dangerous cosmic rays which, in the absence of magnetic protection, would tear through our bodies, leaving much of our DNA in useless shreds. When the magnetic field is working, these rays are safely herded away from the Earth's surface and into two zones in near space called the Van Allen belts. They also interact with particles in the upper atmosphere to create the bewitching veils of light known as the auroras.

A big part of the reason for our ignorance is that traditionally there has been little effort to co-ordinate what's happening on top of the Earth with what's going on inside it. According to Shawna Vogel: 'Geologists and geophysicists rarely go to the same meetings or collaborate on the same problems.'

Perhaps nothing better demonstrates our inadequate grasp of the dynamics of the Earth's interior than how badly we are caught out when it plays up, and it would be hard to come up with a more salutary reminder of the limitations of our understanding than the eruption of Mount St Helens in Washington state in 1980.

At that time, the lower forty-eight states of the Union had not seen a volcanic eruption for over sixty-five years. Therefore, most of the government volcanologists called in to monitor and forecast St Helens' behaviour had seen only

Hawaiian volcanoes in action, and they, it turned out, were not the same thing at all.

St Helens started its ominous rumblings on 20 March. Within a week it was erupting magma, albeit in modest amounts, up to a hundred times a day, and being constantly shaken with earthquakes. People were evacuated to what was assumed to be a safe distance of 13 kilometres. As the mountain's rumblings grew, St Helens became a tourist attraction for the world. Newspapers gave daily reports on the best places to get a view. Television crews repeatedly flew in helicopters to the summit and people were even seen climbing over the mountain. On one day, more than seventy copters and light aircraft circled the peak. But as the days passed and the rumblings failed to develop into anything dramatic, people grew restless and the view became general that the volcano wasn't going to blow after all.

On 19 April the northern flank of the mountain began to bulge conspicuously. Remarkably, no-one in a position of responsibility saw that this strongly signalled a lateral blast. The seismologists resolutely based their conclusions on the behaviour of Hawaiian volcanoes, which don't blow out sideways. Almost the only person who believed that something really bad might happen was Jack Hyde, a geology professor at a community college in Tacoma. He pointed out that St Helens didn't have an open vent, as Hawaiian volcanoes have, so any pressure building up inside was bound to be released dramatically and probably catastrophically. However, Hyde was not part of the official team and his observations attracted little notice.

We all know what happened next. At 8.32 a.m. on a Sunday morning, 18 May, the north side of the volcano collapsed, sending an enormous avalanche of dirt and rock rushing down the mountain slope at nearly 250 kilometres an hour. It was the biggest landslide in human history and carried enough material to bury the whole of Manhattan to a depth of 120 metres. A minute later, its flank severely weakened, St Helens exploded with the force of five hundred Hiroshima-sized atomic bombs, shooting out a murderous hot cloud at up to 1,050 kilometres an hour – much too fast, clearly, for anyone nearby to outrace it. Many people who were thought to be in safe areas, often far out of sight of the volcano, were overtaken. Fifty-seven people were killed. Twenty-three of the bodies were never found. The toll would have been much higher had it not been a Sunday. On any weekday, many lumber workers would have been working within the death zone. As it was, people were killed 30 kilometres away.

The luckiest person on that day was a graduate student named Harry Glicken. He had been manning an observation post 9 kilometres from the mountain, but he had a college placement interview on 18 May in California, and so had left the site the day before the eruption. His place was taken by David Johnston. Johnston was the first to report the volcano exploding; moments later he was dead. His body was never found. Glicken's luck, alas, was temporary. Eleven years later he was one of forty-three scientists and journalists fatally caught up in a lethal outpouring of superheated ash, gases and molten rock – what is known as a pyroclastic flow – at Mount Unzen in Japan when yet another volcano was catastrophically misread.

Volcanologists may or may not be the worst scientists in the world at making predictions, but they are without question the worst in the world at realizing how bad their predictions are. Less than two years after the Unzen catastrophe another group of volcano-watchers, led by Stanley Williams of the University of Arizona, descended into the rim of an active volcano called Galeras in Colombia. Despite the deaths of recent years, only two of the sixteen members of Williams's party wore safety helmets or other protective gear. The volcano erupted, killing six of the scientists, along with three tourists who had followed them, and seriously injuring several others, including Williams himself.

In an extraordinarily unselfcritical book called *Surviving Galeras*, Williams said he could 'only shake my head in wonder' when he learned afterwards that his colleagues in the world of volcanology had suggested that he had overlooked or disregarded important seismic signals and behaved recklessly. 'How easy it is to snipe after the fact, to apply the knowledge we have now to the events of 1993,' he wrote. He was guilty of nothing worse, he believed, than unlucky timing when Galeras 'behaved capriciously, as natural forces are wont to do. I was fooled, and for that I will take responsibility. But I do not feel guilty about the deaths of my colleagues. There is no guilt. There was only an eruption.'

But to return to Washington. Mount St Helens lost 400 metres of peak, and 600 square kilometres of forest were devastated. Enough trees to build 150,000 homes (or 300,000 according to some reports) were blown away. The damage was placed at $2.7 billion. A giant column of smoke and ash rose to a height of 18,000 metres in less than ten minutes. An airliner some 48 kilometres away reported being pelted with rocks.

Ninety minutes after the blast, ash began to rain down on Yakima, Washington, a community of fifty thousand people about 130 kilometres away. As you would expect, the ash turned day to night and got into everything, clogging motors, generators and electrical switching equipment, choking pedestrians, blocking filtration systems and generally bringing things to a halt. The airport shut down and highways in and out of the city were closed.

All this was happening, you will note, just downwind of a volcano that had been rumbling menacingly for two months. Yet Yakima had no volcano emergency procedures.

The city's emergency broadcast system, which was supposed to swing into action during a crisis, did not go on the air because 'the Sunday-morning staff did not know how to operate the equipment'. For three days, Yakima was paralysed and cut off from the world, its airport closed, its approach roads impassable. Altogether the city received just over 1.5 centimetres of ash after the eruption of Mount St Helens. Now bear that in mind, please, as we consider what a Yellowstone blast would do.

15

DANGEROUS BEAUTY

In the 1960s, while studying the volcanic history of Yellowstone National Park, Bob Christiansen of the United States Geological Survey became puzzled about something that, oddly, had not troubled anyone before: he couldn't find the park's volcano. It had been known for a long time that Yellowstone was volcanic in nature – that's what accounted for all its geysers and other steamy features – and the one thing about volcanoes is that they are generally pretty conspicuous. But Christiansen couldn't find the Yellowstone volcano anywhere. In particular, what he couldn't find was a structure known as a caldera.

Most of us, when we think of volcanoes, think of the classic cone shape of a Fuji or a Kilimanjaro, which is created when erupting magma accumulates in a symmetrical mound. These can form remarkably quickly. In 1943 at Paricutín in Mexico a farmer was startled to see smoke rising from a patch on his land. In one week he was the bemused owner of a cone 152 metres high. Within two years it had topped out at almost 430 metres and was more than 800 metres across. Altogether there are some ten thousand of these intrusively visible volcanoes on Earth, all but a few hundred of them extinct. But there is a second, less celebrated type of volcano that doesn't involve mountain-building. These are volcanoes so explosive that they burst open in a single mighty rupture, leaving behind a vast subsided pit, the caldera (from a Latin word for cauldron). Yellowstone obviously was of this second type, but Christiansen couldn't find the caldera anywhere.

By coincidence, just at this time NASA decided to test some new high-altitude cameras by taking photographs of Yellowstone, copies of which a thoughtful official passed on to the park authorities on the assumption that they might make a nice display for one of the visitor centres. As soon as Christiansen saw the photos he realized why he had failed to spot the caldera: virtually the whole park – 9,000 square kilometres – was caldera. The explosion had left a crater nearly 65 kilometres across – much too huge to be perceived from anywhere at ground level. At some time in the past Yellowstone must have blown up with a violence far beyond the scale of anything known to humans.

Yellowstone, it turns out, is a supervolcano. It sits on top of an enormous hot spot, a reservoir of molten rock that begins at least 200 kilometres down in the Earth and rises to near the surface, forming what is known as a superplume. The heat from the hot spot is what powers all of Yellowstone's vents, geysers, hot springs and popping mud pots. Beneath the surface is a magma chamber that is about 72 kilometres across – roughly the same dimensions as the park – and about 13 kilometres thick at its thickest point. Imagine a pile of TNT about the size of an English county and reaching 13 kilometres into the sky, to about the height of the highest cirrus clouds, and you have some idea of what visitors to Yellowstone are shuffling around on top of. The pressure that such a pool of magma exerts on the crust above has lifted Yellowstone and its surrounding territory about half a kilometre higher than they would otherwise be. If it blew, the cataclysm is pretty well beyond imagining.

According to Professor Bill McGuire of University College London, 'you wouldn't be able to get within a thousand kilometres of it' while it was erupting. The consequences that followed would be even worse.

Superplumes of the type on which Yellowstone sits are rather like martini glasses – thin on the way up, but spreading out as they near the surface to create vast bowls of unstable magma. Some of these bowls can be up to 1,900 kilometres across. According to current theories, they don't always erupt explosively, but sometimes burst forth in a vast, continuous outpouring – a flood – of molten rock, as happened with the Deccan Traps in India 65 million years ago. These covered an area of over 500,000 square kilometres and probably contributed to the demise of the dinosaurs – they certainly didn't help – with their noxious outpourings of gases. Superplumes may also be responsible for the rifts that cause continents to break up.

Such plumes are not all that rare. There are about thirty active ones on the Earth at the moment and they are responsible for many of the world's best-known islands and island chains – Iceland, Hawaii, the Azores, Canaries and Galápagos archipelagos, little Pitcairn in the middle of the South Pacific, and many others – but apart from Yellowstone they are all oceanic. No-one has the faintest idea how or why Yellowstone's ended up beneath a continental plate. Only two things are certain: that the crust at Yellowstone is thin and that the world beneath it is hot. But whether the crust is thin because of the hot spot or whether the hot spot is there because the crust is thin is a matter of heated (as it were) debate. The continental nature of the crust makes a huge difference to its eruptions. Whereas the other supervolcanoes tend to bubble away steadily and in a comparatively benign fashion, Yellowstone blows explosively. It doesn't happen often, but when it does you want to stand well back.

Since its first known eruption 16.5 million years ago, it has blown up about a hundred times, but the most recent three eruptions are the ones that get written about. The last eruption was a thousand times as big as that of Mount St Helens; the one before that was 280 times as big, and the one before *that* was so big nobody knows exactly how big it was. It was at least 2,500 times as big as St Helens, but perhaps 8,000 times as monstrous.

We have absolutely nothing to compare it to. The biggest blast in recent times was that of Krakatau in Indonesia in August 1883, which made a bang that reverberated around the world for nine days, and made water slosh as far away as the English Channel. But if you imagine the volume of ejected material from Krakatau as being about the size of a golf ball, then that from the biggest of the Yellowstone blasts would be the size of a sphere you could just about hide behind. On this scale, the Mount St Helens eruption would be no more than a pea.

The Yellowstone eruption of two million years ago put out enough ash to bury New York State to a depth of 20 metres or California to a depth of 6 metres. This was the ash that made Mike Voorhies' fossil beds in eastern Nebraska. That blast occurred in what is now Idaho, but over millions of years, at a rate of about 2.5 centimetres a year, the Earth's crust has travelled over it, so that today it is directly under northwest Wyoming. (The hot spot itself stays in one place, like an acetylene torch aimed at a ceiling.) In its wake it leaves the sort of rich volcanic plains that are ideal for growing potatoes, as Idaho's farmers long ago discovered. In another two million years, geologists like to joke, Yellowstone will be producing French fries for McDonald's and the people of Billings, Montana, will be stepping around geysers.

The ash fall from the last Yellowstone eruption covered all or parts of nineteen western states (plus parts of Canada and Mexico) – nearly the whole of the United States west of the Mississippi. This, bear in mind, is the breadbasket of America, an area that produces roughly half the world's cereals. And ash, it is worth remembering, is not like a big snowfall that will melt in the spring. If you wanted to grow crops again, you would have to find some place to put all the ash. It took thousands of workers eight months to clear 1.8 billion tonnes of debris from the 6.5 hectares of the World Trade Center site in New York. Imagine what it would take to clear Kansas.

And that's not even to consider the climatic consequences. The last supervolcano eruption on Earth was at Toba, in northern Sumatra, 74,000 years ago. No-one knows quite how big it was, but it was a whopper. Greenland ice cores show that the Toba blast was followed by at least six years of 'volcanic winter' and goodness knows how many poor growing seasons after that. The event, it is thought, may have carried humans right to the brink of extinction, reducing the global population to no more than a few thousand individuals. That would mean that all modern humans arose from a very small population base, which would explain our lack of genetic diversity. At all events, there is some evidence to suggest that for the next twenty thousand years the total number of people on Earth was never more than a few thousand at any time. That is, needless to say, a long time to spend recovering from a single volcanic blast.

All this was hypothetically interesting until 1973, when an odd occurrence made it suddenly momentous: water in Yellowstone Lake, in the heart of the park, began to run over the banks at the lake's southern end, flooding a meadow, while at the opposite end of the lake the water mysteriously flowed away. Geologists did a hasty survey and discovered that a large area of the park had developed an ominous bulge. This was lifting up one end of the lake and causing the water to run out at the other, as would happen if you lifted one side of a child's paddling pool. By 1984, the whole central region of the park – over 100 square kilometres – was more than a metre higher than it had been in 1924, when the park was last formally surveyed. Then, in 1985, the central part of the park subsided by 20 centimetres (about 8 inches). It now seems to be swelling again.

The geologists realized that only one thing could cause this – a restless magma chamber. Yellowstone wasn't the site of an ancient supervolcano; it was the site of an active one. It was also at about this time that they were able to work out that the cycle of Yellowstone's eruptions averaged one massive blow every 600,000 years. The last one was 630,000 years ago. Yellowstone, it appears, is due.

'It may not feel like it, but you're standing on the largest active volcano in the world,' Paul Doss, Yellowstone National Park geologist, told me soon after climbing off an enormous Harley-Davidson motorcycle and shaking hands when we met at the park headquarters at Mammoth Hot Springs early on a lovely morning in June. A native of Indiana, Doss is an amiable, soft-spoken, extremely thoughtful man who looks nothing like a National Park Service employee. He has a greying beard and hair tied back in a long ponytail. A small sapphire stud graces one ear. A slight paunch strains against his crisp Park Service uniform. He looks more like a blues musician than a government employee. In fact, he is a blues musician (harmonica). But he sure knows and loves geology. 'And I've got the best place in the world to do it,' he says as we set off in a bouncy, battered four-wheel-drive vehicle in the general direction of Old Faithful. He has agreed to let me accompany him for a day as he goes about doing whatever it is a park geologist does. The first assignment today is to give an introductory talk to a new crop of tour guides.

Yellowstone, I hardly need point out, is sensationally beautiful, with plump, stately mountains, bison-specked meadows, tumbling streams, a sky-blue lake, wildlife beyond counting. 'It really doesn't get any better than this if you're a geologist,' Doss says. 'You've got rocks up at Beartooth Gap that are nearly three billion years old – three-quarters of the way back to the Earth's beginning – and then you've got mineral springs here' – he points at the sulphurous hot springs from which Mammoth takes its title – 'where you can see rocks as they are being born. And in between there's everything you could possibly imagine. I've never been any place where geology is more evident – or prettier.'

'So you like it?' I say.

'Oh, no, I love it,' he answers with profound sincerity. 'I mean I really love it here. The winters are tough and the pay's not too hot, but when it's good, it's just—'

He interrupted himself to point out a distant gap in a range of mountains to the west, which had just come into view over a rise. The mountains, he told me, were known as the Gallatins. 'That gap is sixty or maybe seventy miles across. For a long time nobody could understand why that gap was there, and then Bob Christiansen realized that it had to be because the mountains were just blown away. When you've got sixty miles of mountains just obliterated, you know you're dealing with something pretty potent. It took Christiansen six years to figure it all out.'

I asked him what caused Yellowstone to blow when it did.

'Don't know. Nobody knows. Volcanoes are strange things. We really don't understand them at all. Vesuvius, in Italy, was active for three hundred years until an eruption in 1944 and then it just stopped. It's been silent ever since. Some volcanologists think that it is recharging in a big way, which is a little worrying because two million people live on or around it. But nobody knows.'

'And how much warning would you get if Yellowstone was going to go?'

He shrugged. 'Nobody was around the last time it blew, so nobody knows what the warning signs are. Probably you would have swarms of earthquakes and some surface uplift and possibly some changes in the patterns of behaviour of the geysers and steam vents, but nobody really knows.'

'So it could just blow without warning?'

He nodded thoughtfully. The trouble, he explained, is that nearly all the things that would constitute warning signs already exist in some measure at Yellowstone. 'Earthquakes are generally a precursor of volcanic eruptions, but the park already has lots of earthquakes – twelve hundred and sixty of them last year. Most of them are too small to be felt, but they are earthquakes nonetheless.'

A change in the pattern of geyser eruptions might also be taken as a clue, he said, but these too vary unpredictably. Once the most famous geyser in the park was Excelsior Geyser. It used to erupt regularly and spectacularly to heights of 100 metres, but in 1888 it just stopped. Then in 1985 it erupted again, though only to a height of 25 metres. Steamboat Geyser is the biggest geyser in the world when it blows, shooting water 120 metres into the air, but the intervals between its eruptions have ranged from as little as four days to almost fifty years. 'If it blew today and again next week, that wouldn't tell us anything at all about what it might do the following week or the week after or twenty years from now,' Doss says. 'The whole park is so volatile that it's essentially impossible to draw conclusions from almost anything that happens.'

Evacuating Yellowstone would never be easy. The park gets some three million visitors a year, mostly in the three peak months of summer. The park's roads are comparatively few and they are kept intentionally narrow, partly to slow traffic, partly to preserve an air of picturesqueness, and partly because of topographical constraints. At the height of summer, it can easily take half a day to cross the park and hours to get anywhere within it. 'Whenever people see animals, they just stop, wherever they are,' Doss says. 'We get bear jams. We get bison jams. We get wolf jams.'

In the autumn of 2000, representatives from the US Geological Survey and National Park Service, along with some academics, met and formed something called the Yellowstone Volcanic Observatory. Four of these bodies were in existence already – in Hawaii, California, Alaska and Washington – but, oddly, there was none in the largest volcanic zone in the world. The YVO is not actually a thing so much as an idea – an agreement to co-ordinate efforts at studying and analysing the park's diverse geology. One of its first tasks, Doss told me, was to draw up an 'earthquake and volcano hazards plan' – a plan of action in the event of a crisis.

'There isn't one already?' I said.

'No. Afraid not. But there will be soon.'

'Isn't that just a little tardy?'

He smiled. 'Well, let's just say that it's not any too soon.'

Once it is in place, the idea is that three people – Christiansen in Menlo Park, California, Professor Robert B. Smith at the University of Utah and Doss in the park – would assess the degree of danger of any potential cataclysm and advise the park superintendent. The superintendent would take the decision whether to evacuate the park. As for surrounding areas, there are no plans. You would be on your own once you left the park gates – not much help if Yellowstone were going to blow in a really big way.

Of course, it may be tens of thousands of years before that day comes. Doss thinks such a day may not come at all. 'Just because there was a pattern in the past doesn't mean that it still holds true,' he says. 'There is some evidence to suggest that the pattern may be a series of catastrophic explosions, then a long period of quiet. We may be in that now. The evidence now is that most of the magma chamber is cooling and crystallizing. It is releasing its volatiles; you need to *trap* volatiles for an explosive eruption.'

In the meantime there are plenty of other dangers in and around Yellowstone, as was made devastatingly evident on the night of 17 August 1959, at a place called Hebgen Lake just outside the park. At twenty minutes to midnight on that date, Hebgen Lake suffered a catastrophic quake. It was magnitude 7.5, not vast as earthquakes go, but so abrupt and wrenching that it collapsed an entire mountainside. It was the height of the summer season, though fortunately not so many people went to Yellowstone in those days as now. Eighty million tonnes of rock, moving at more than 160 kilometres an hour, just fell off the mountain, travelling with such force and momentum that the leading edge of the landslide ran 120 metres up a mountain on the other side of the valley. Along its path lay part of the Rock Creek Campground. Twenty-eight campers were killed, nineteen of them buried too deep ever to be found again. The devastation was swift but heartbreakingly fickle. Three brothers, sleeping in one tent, were spared. Their parents, sleeping in another tent beside them, were swept away and never seen again.

'A big earthquake – and I mean big – will happen sometime,' Doss told me. 'You can count on that. This is a big fault zone for earthquakes.'

Despite the Hebgen Lake quake and the other known risks, Yellowstone didn't get permanent seismometers until the 1970s.

If you needed a way to appreciate the grandeur and inexorability of geological processes, you could do worse than to consider the Tetons, the sumptuously jagged range that stands just to the south of Yellowstone National Park. Nine million years ago, the Tetons didn't exist. The land around Jackson Hole was just a high grassy plain. But then a 64-kilometre-long fault opened within the Earth and since then, about once every nine hundred years, the Tetons experience a really big earthquake, enough to jerk them another 2 metres higher. It is these repeated jerks over aeons that have raised them to their present majestic heights of 2,000 metres.

That nine hundred years is an average – and a somewhat misleading one. According to Robert B. Smith and Lee J. Siegel in *Windows into the Earth*, a geological history of the region, the last major Teton quake was somewhere between about five thousand and seven thousand years ago. The Tetons, in short, are about the most overdue earthquake zone on the planet.

Hydrothermal explosions are also a significant risk. They can happen any time, pretty much anywhere and without any predictability. 'You know, by design we funnel visitors into thermal basins,' Doss told me after we had watched Old Faithful blow. 'It's what they come to see. Did you know there are more geysers and hot springs at Yellowstone than in all the rest of the world combined?'

'I didn't know that.'

He nodded. 'Ten thousand of them, and nobody knows when a new vent might open.'

We drove to a place called Duck Lake, a body of water a couple of hundred metres across. 'It looks completely innocuous,' he said. 'It's just a big pond. But this big hole didn't used to be here. At some time in the last fifteen thousand years this blew in a really big way. You'd have had several tens of millions of tons of earth and rock and super-heated water blowing out at hypersonic speeds. You can imagine what it would be like if this happened under, say, the parking lot at Old Faithful or one of the visitor centres.' He made an unhappy face.

'Would there be any warning?'

'Probably not. The last significant explosion in the park was at a place called Pork Chop Geyser in 1989. That left a crater about five metres across – not huge by any means, but big enough if you happened to be standing there at the time. Fortunately, nobody was around so nobody was hurt, but that happened without warning. In the very ancient past there have been explosions that have made holes a mile across. And nobody can tell you where or when that might happen again. You just have to hope that you're not standing there when it does.'

Big rockfalls are also a danger. There was a big one at Gardiner Canyon in 1999, but again fortunately no-one was hurt. Late in the afternoon, Doss and I stopped at a place where there was a rock overhang poised above a busy park road. Cracks were clearly visible. 'It could go at any time,' Doss said thoughtfully.

'You're kidding,' I said. There wasn't a moment when there weren't two cars passing beneath it, all filled with, in the most literal sense, happy campers.

'Oh, it's not likely,' he added. 'I'm just saying it *could*. Equally, it could stay like that for decades. There's just no telling. People have to accept that there is risk in coming here. That's all there is to it.'

As we walked back to his vehicle to head back to Mammoth Hot Springs, Doss added: 'But the thing is, most of the time bad things don't happen. Rocks don't fall. Earthquakes don't occur. New vents don't suddenly open up. For all the instability, it's mostly remarkably and amazingly tranquil.'

'Like the Earth itself,' I remarked.

'Precisely,' he agreed.

The risks at Yellowstone apply to park employees as much as to visitors. Doss had got a horrific sense of that in his first week on the job five years earlier. Late one night, three young summer employees were engaging in an illicit activity known as 'hot-potting' – swimming or basking in warm pools. Though the park, for obvious reasons, doesn't publicize it, not all the pools at Yellowstone are dangerously hot. Some are extremely agreeable to lie in, and it was the habit of some of the summer employees to have a dip late at night, even though it was against the rules to do so. Foolishly, the threesome had failed to take a torch, which was extremely dangerous because much of the soil around the warm pools is crusty and thin and one can easily fall through into a scalding vent below. In any case, as they made their way back to their dorm, they came across a stream that they had had to leap over earlier. They backed up a few paces, linked arms and, on the count of three, took a running jump. In fact, it wasn't the stream at all. It was a boiling pool. In the dark they had lost their bearings. None of the three survived.

I thought about this the next morning as I made a brief call, on my way out of the park, at a place called Emerald Pool, in the Upper Geyser Basin. Doss hadn't had time to take me there the day before, but I thought I ought at least to have a look at it, for Emerald Pool is a historic site.

In 1965, a husband and wife team of biologists named Thomas and Louise Brock, while on a summer study trip, had done a crazy thing. They had scooped up some of the yellowy-brown scum that rimmed the pool and examined it for life. To their, and eventually the wider world's, deep surprise, it was full of living microbes. They had found the world's first extremophiles – organisms that could live in water that had previously been assumed to be much too hot or acid or choked with sulphur to bear life. Emerald Pool, remarkably, was all these things, yet at least two types of living thing, *Sulpholobus acidocaldarius* and *Thermophilus aquaticus*, as they became known, found it congenial. It had always been supposed that nothing could survive above temperatures of 50 degrees Celsius, but here were organisms basking in rank, acidic waters nearly twice that hot.

For almost twenty years, one of the Brocks' two new bacteria, *Thermophilus aquaticus*, remained a laboratory curiosity – until a scientist in California named Kary B. Mullis realized that heat-resistant enzymes within it could be used to create a bit of chemical wizardry known as a polymerase chain reaction, which allows scientists to generate lots of DNA from very small amounts – as little as a single molecule in ideal conditions. It's a kind of genetic photocopying, and it became the basis for all subsequent genetic science, from academic studies to police forensic work. It won Mullis the Nobel Prize in chemistry in 1993.

Meanwhile, scientists were finding even hardier microbes, now known as hyperthermophiles, which demand temperatures of 80 degrees Celsius or more. The warmest organism found so far, according to Frances Ashcroft in *Life at the Extremes*, is *Pyrolobus fumarii,* which dwells in the walls of ocean vents where the temperature can reach 113 degrees Celsius. The upper limit for life is thought to be about 120 degrees Celsius, though no-one actually knows. At all events, the Brocks' findings completely changed our perception of the living world. As NASA scientist Jay Bergstralh has put it: 'Wherever we go on Earth – even into what's seemed like the most hostile possible environments for life – as long as there is liquid water and some source of chemical energy we find life.'

Life, it turns out, is infinitely more clever and adaptable than anyone had ever supposed. This is a very good thing for, as we are about to see, we live in a world that doesn't altogether seem to want us here.

V

LIFE ITSELF

The more I examine the universe and study the details of its architecture, the more evidence I find that the universe in some sense must have known we were coming.

Freeman Dyson

16

LONELY PLANET

It isn't easy being an organism. In the whole universe, as far as we yet know, there is only one place, an inconspicuous outpost of the Milky Way called the Earth, that will sustain you, and even it can be pretty grudging.

From the bottom of the deepest ocean trench to the top of the highest mountain, the zone that covers nearly the whole of known life is only around 20 kilometres thick – not much when set against the roominess of the cosmos at large.

For humans it is even worse because we happen to belong to the portion of living things that took the rash but venturesome decision 400 million years ago to crawl out of the seas and become land-based and oxygen-breathing. In consequence, no less than 99.5 per cent of the world's habitable space by volume, according to one estimate, is fundamentally – in practical terms completely – off limits to us.

It isn't simply that we can't breathe in water, but that we couldn't bear the pressures. Because water is about 1,300 times heavier than air, pressures rise swiftly as you descend – by the equivalent of one atmosphere for every 10 metres of depth. On land, if you rose to the top of a 150-metre eminence – Cologne Cathedral or the Washington Monument, say – the change in pressure would be so slight as to be indiscernible. At the same depth under water, however, your veins would collapse and your lungs would compress to the approximate dimensions of a Coke can. Amazingly, people do voluntarily dive to such depths, without breathing apparatus, for the fun of it, in a sport known as free diving. Apparently, the experience of having your internal organs rudely deformed is thought exhilarating (though not, presumably, as exhilarating as having them return to their former dimensions upon resurfacing). To reach such depths, however, divers must be dragged down, and quite briskly, by weights. Without assistance, the deepest anyone has gone and lived to talk about it afterwards is 72 metres – a feat performed by an Italian named Umberto Pelizzari, who in 1992 dived to that depth, lingered for a nanosecond and then shot back to the surface. In terrestrial terms, 72 metres is a good bit shorter than a football pitch. So even in our most exuberant stunts we can hardly claim to be masters of the abyss.

Other organisms do, of course, manage to deal with the pressures at depth, though quite how some of them do so is a mystery. The deepest point in the ocean is the Mariana Trench in the Pacific. There, some 11.3 kilometres down, the pressures rise to over 16,000 pounds per square inch. We have managed just once, briefly, to send humans to that depth in a sturdy diving vessel, yet it is home to colonies of amphipods, a type of crustacean similar to shrimp but transparent, which survive without any protection at all. Most oceans are of course much shallower, but even at the average ocean depth of 4 kilometres the pressure is equivalent to being squashed beneath a stack of fourteen loaded cement trucks.

Nearly everyone, including the authors of some popular books on oceanography, assumes that the human body would crumple under the immense pressures of the deep ocean. In fact, this appears not to be the case. Because we are made largely of water ourselves, and water is 'virtually incompressible', in the words of Frances Ashcroft of Oxford University, 'the body remains at the same pressure as the surrounding water, and is not crushed at depth.' It is the gases inside your body, particularly in the lungs, that cause the trouble. These do compress, though at what point the compression becomes fatal is not known. Until quite recently it was thought that anyone diving to 100 metres or so would die painfully as his or her lungs imploded or chest wall collapsed, but the free divers have repeatedly proved otherwise. It appears, according to Ashcroft, that 'humans may be more like whales and dolphins than had been expected.'

Plenty else can go wrong, however. In the days of diving suits – the sort that were connected to the surface by long hoses – divers sometimes experienced a dreaded phenomenon known as 'the squeeze'. This occurred when the surface pumps failed, leading to a catastrophic loss of pressure in the suit. The air would leave the suit with such violence that the hapless diver would be, all too literally, sucked up into the helmet and hosepipe. When hauled to the surface, 'all that is left in the suit are his bones and some rags of flesh,' the biologist J. B. S. Haldane wrote in 1947, adding for the benefit of doubters, 'This has happened.'

(Incidentally, the original diving helmet, designed in 1823 by an Englishman named Charles Deane, was intended not for diving but for fire fighting. It was called a 'smoke helmet', but, being made of metal, it was hot and cumbersome; as Deane soon discovered, fire-fighters had no particular eagerness to enter burning structures in any form of attire, but most especially not in something that heated up like a kettle and made them clumsy into the bargain. In an attempt to save his investment, Deane tried it under water and found it was ideal for salvage work.)

The real terror of the deep, however, is the bends – not so much because they are unpleasant, though of course they are, as because they are so much more likely. The air we breathe is 80 per cent nitrogen. Put the human body under pressure, and that nitrogen is transformed into tiny bubbles that migrate into the blood and tissues. If the pressure is changed too rapidly – as with a too-quick ascent by a diver – the bubbles trapped within the body will begin to fizz in exactly the manner of a freshly opened bottle of champagne, clogging tiny blood vessels, depriving cells of oxygen and causing pain so excruciating that sufferers are prone to bend double in agony – hence 'the bends'.

The bends have been an occupational hazard for sponge and pearl divers since time immemorial, but didn't attract much attention in the Western world until the nineteenth century, and then it was among people who didn't get wet at all (or at least, not very wet and not generally much above the ankles). They were caisson workers. Caissons were enclosed dry chambers built on river beds to facilitate the construction of bridge piers. They were filled with compressed air, and often when the workers emerged after an extended period of working under this artificial pressure they experienced mild symptoms like tingling or itchy skin. But an unpredictable few felt more insistent pain in the joints and occasionally collapsed in agony, sometimes never to get up again.

It was all most puzzling. Sometimes the workers would go to bed feeling fine, but wake up paralysed. Sometimes they wouldn't wake up at all. Ashcroft relates a story concerning the directors of a new tunnel under the Thames who held a celebratory banquet as the tunnel neared completion. To their consternation their champagne failed to fizz when uncorked in the compressed air of the tunnel. However, when at length they emerged into the fresh air of a London evening, the bubbles sprang instantly to fizziness, memorably enlivening the digestive process.

Apart from avoiding high-pressure environments altogether, only two strategies are reliably successful against the bends. The first is to suffer only a very short exposure to the changes in pressure. That is why the free divers I mentioned earlier can descend to depths of 150 metres without ill effect. They don't stay down long enough for the nitrogen in their system to dissolve into their tissues. The other solution is to ascend by careful stages. This allows the little bubbles of nitrogen to dissipate harmlessly.

A great deal of what we know about surviving at extremes is owed to the extraordinary father and son team of John Scott and J. B. S. Haldane. Even by the demanding standards of British intellectuals, the Haldanes were outstandingly eccentric. The senior Haldane was born in 1860 to an aristocratic Scottish family (his brother was Viscount Haldane), but spent most of his career in comparative modesty as a professor of physiology at Oxford. He was famously absent-minded. Once, after his wife had sent him upstairs to change for a dinner party, he failed to return and was discovered asleep in bed in his pyjamas. When roused, Haldane explained that he had found himself disrobing and assumed it was bedtime. His idea of a holiday was to travel to Cornwall to study hookworm in miners. Aldous Huxley, the novelist grandson of T. H. Huxley, who lived with the Haldanes for a time, parodied him, a touch mercilessly, as the scientist Edward Tantamount in the novel *Point Counter Point*.

Haldane's gift to diving was to work out the rest intervals necessary to manage an ascent from the depths without getting the bends, but his interests ranged across the whole of physiology, from studying altitude sickness in climbers to the problems of heatstroke in desert regions. He had a particular interest in the effects of toxic gases on the human body. To understand more exactly how carbon monoxide leaks killed miners, he methodically poisoned himself, carefully taking and measuring his own blood samples the while. He quit only when he was on the verge of losing all muscle control and his blood saturation level had reached 56 per cent – a level, as Trevor Norton notes in his entertaining history of diving, *Stars Beneath the Sea,* only fractionally removed from nearly certain lethality.

Haldane's son Jack, known to posterity as J.B.S., was a remarkable prodigy who took an interest in his father's work almost from infancy. At the age of three he was overheard demanding peevishly of his father, 'But is it oxyhaemoglobin or carboxyhaemoglobin?' Throughout his youth, the young Haldane helped his father with experiments. By the time he was a teenager, the two often tested gases and gas masks together, taking it in turns to see how long it took them to pass out.

Though J. B. S. Haldane never took a degree in science (he studied classics at Oxford), he became a brilliant scientist in his own right, mostly working for the government at Cambridge. The biologist Peter Medawar, who spent his life around mental Olympians, called him 'the cleverest man I ever knew'. Huxley parodied the younger Haldane too, in his novel *Antic Hay*, but also used his ideas on genetic manipulation of humans as the basis for the plot of *Brave New World*. Among many other achievements, Haldane played a central role in marrying Darwinian principles of evolution to the genetic work of Gregor Mendel to produce what is known to geneticists as the Modern Synthesis.

Perhaps uniquely among human beings, the younger Haldane found the First World War 'a very enjoyable experience' and freely admitted that he 'enjoyed the opportunity of killing people'. He was himself wounded twice. After the war he became a successful popularizer of science and wrote twenty-three books (as well as over four hundred scientific papers). His books are still thoroughly readable and instructive, though not always easy to find. He also became an enthusiastic Marxist. It has been suggested, not altogether cynically, that this was out of a purely contrarian instinct and that if he had been born in the Soviet Union he would have been a passionate monarchist. At all events, most of his articles first appeared in the Communist *Daily Worker*.

Whereas his father's principal interests concerned miners and poisoning, the younger Haldane became obsessed with saving submariners and divers from the unpleasant consequences of their work. With Admiralty funding, he acquired a decompression chamber that he called the 'pressure pot'. This was a metal cylinder into which three people at a time could be sealed and subjected to tests of various types, all painful and nearly all dangerous. Volunteers might be required to sit in ice water while breathing 'aberrant atmosphere', or subjected to rapid changes of pressurization. In one experiment, Haldane himself simulated a dangerously hasty ascent to see what would happen. What happened was that the dental fillings in his teeth exploded. 'Almost every experiment', Norton writes, 'ended with someone having a seizure, bleeding or vomiting.' The chamber was virtually soundproof, so the only way for occupants to signal unhappiness or distress was to tap insistently on the chamber wall or to hold up notes to a small window.

On another occasion, while poisoning himself with elevated levels of oxygen, Haldane had a fit so severe that he crushed several vertebrae. Collapsed lungs were a routine hazard. Perforated eardrums were quite common, too; but, as Haldane reassuringly noted in one of his essays, 'the drum generally heals up; and if a hole remains in it, although one is somewhat deaf, one can blow tobacco smoke out of the ear in question, which is a social accomplishment.'

What was extraordinary about this was not that Haldane

was willing to subject himself to such risk and discomfort in the pursuit of science, but that he had no trouble talking colleagues and loved ones into climbing into the chamber, too. Sent on a simulated descent, his wife once had a fit that lasted thirteen minutes. When at last she stopped bouncing across the floor, she was helped to her feet and sent home to cook dinner. Haldane happily employed whoever happened to be around, including on one memorable occasion a former Prime Minister of Spain, Juan Negrín. Dr Negrín complained afterwards of minor tingling and 'a curious velvety sensation on the lips' but otherwise seems to have escaped unharmed. He may have considered himself very lucky. A similar experiment with oxygen deprivation left Haldane without feeling in his buttocks and lower spine for six years.

Among Haldane's many specific preoccupations was nitrogen intoxication. For reasons that are still poorly understood, at depths beyond about 30 metres nitrogen becomes a powerful intoxicant. Under its influence divers had been known to offer their air hoses to passing fish or to decide to try to have a smoke break. It also produced wild mood swings. In one test, Haldane noted, the subject 'alternated between depression and elation, at one moment begging to be decompressed because he felt "bloody awful" and the next minute laughing and attempting to interfere with his colleague's dexterity test'. In order to measure the rate of deterioration in the subject, a scientist had to go into the chamber with the volunteer to conduct simple mathematical tests. But after a few minutes, as Haldane later recalled, 'the tester was usually as intoxicated as the testee, and often forgot to press the spindle of his stopwatch, or to take proper notes.' The cause of the inebriation is even now a mystery. It is thought that it may be the same thing that causes alcohol intoxication, but as no-one knows for certain what causes *that*, we are none the wiser. At all events, without the greatest care, it is easy to get in trouble once you leave the surface world.

Which brings us back (well, nearly) to our earlier observation that the Earth is not the easiest place to be an organism, even if it is the only place. Of the small portion of the planet's surface that is dry enough to stand on, a surprisingly large amount is too hot or cold or dry or steep or lofty to be of much use to us. Partly, it must be conceded, this is our fault. In terms of adaptability, humans are pretty amazingly useless. Like most animals, we don't much like really hot places, but because we sweat so freely and easily succumb to strokes, we are especially vulnerable. In the worst circumstances – on foot without water in a hot desert – most people will grow delirious and keel over, possibly never to rise again, in no more than seven or eight hours. We are no less helpless in the face of cold. Like all mammals, humans are good at generating heat; but – because we are so nearly hairless – we are not good at keeping it. Even in quite mild weather half the calories you burn go to keep your body warm. Of course, we can counter these frailties to a large extent by employing clothing and shelter, but even so the portions of the Earth on which we are prepared or able to live are modest indeed: just 12 per cent of the total land area, and only 4 per cent of the whole surface if you include the seas.

Yet when you consider conditions elsewhere in the known universe, the wonder is not that we use so little of our planet but that we have managed to find a planet of which we can use even a bit. You have only to look at our own solar system – or, come to that, the Earth at certain periods in its own history – to appreciate that most places are much harsher and much less amenable to life than our mild, blue, watery globe.

So far, space scientists have discovered about seventy planets outside the solar system, out of the ten billion trillion or so that are thought to be out there, so humans can hardly claim to speak with authority on the matter; but it appears that if you wish to have a planet suitable for life, you have to be just awfully lucky, and the more advanced the life, the luckier you have to be. Various observers have identified about two dozen particularly fortunate breaks we have had on the Earth, but this is a flying survey so we'll distil them down to the principal four.

Excellent location. We are, to an almost uncanny degree, the right distance from the right sort of star, one that is big enough to radiate lots of energy, but not so big as to burn itself out swiftly. It is a curiosity of physics that the larger a star is, the more rapidly it burns. Had our sun been ten times as massive, it would have exhausted itself after ten million years instead of ten billion and we wouldn't be here now. We are also fortunate to orbit where we do. Too much nearer, and everything on Earth would have boiled away. Much further away, and everything would have frozen.

In 1978, an astrophysicist named Michael Hart made some calculations and concluded that the Earth would have been uninhabitable had it been just 1 per cent further from or 5 per cent closer to the Sun. That's not much, and in fact it wasn't enough. The figures have since been refined and made a little more generous – 5 per cent nearer and 15 per cent further are thought to be more accurate assessments for our zone of habitability – but that is still a narrow belt.*

To appreciate just how narrow, you have only to look at Venus. Venus is only 25 million miles closer to the Sun than

*The discovery of extremophiles in the boiling mudpots of Yellowstone and of similar organisms elsewhere made scientists realize that actually life of a type could range much further than that – even perhaps beneath the icy skin of Pluto. What we are talking about here are the conditions that would produce reasonably complex surface creatures.

we are. The Sun's warmth reaches it just two minutes before it touches us. In size and composition, Venus is very like the Earth, but the small difference in orbital distance made all the difference to how it turned out. It appears that during the early years of the solar system Venus was only slightly warmer than the Earth and probably had oceans. But those few degrees of extra warmth meant that Venus could not hold onto its surface water, with disastrous consequences for its climate. As its water evaporated, the hydrogen atoms escaped into space and the oxygen atoms combined with carbon to form a dense atmosphere of the greenhouse gas carbon dioxide. Venus became stifling. Although people of my age will recall a time when astronomers hoped that Venus might harbour life beneath its padded clouds, possibly even a kind of tropical verdure, we now know that it is much too fierce an environment for any kind of life that we can reasonably conceive of. Its surface temperature is a roasting 470 degrees Celsius, which is hot enough to melt lead, and the atmospheric pressure at the surface is ninety times that of Earth, more than any human body could withstand. We lack the technology to make suits or even spaceships that would allow us to visit. Our knowledge of Venus's surface is based on distant radar imagery and some startled squawks from an unmanned Soviet probe that was dropped hopefully into the clouds in 1972 and functioned for barely an hour before permanently shutting down.

So that's what happens when you move two light minutes closer to the Sun. Travel further out and the problem becomes not heat but cold, as Mars frigidly attests. It, too, was once a much more congenial place, but couldn't retain a usable atmosphere and turned into a frozen waste.

But just being the right distance from the Sun cannot be the whole story, for otherwise the Moon would be forested and fair, which patently it is not. For that you need to have:

The right kind of planet. I don't imagine even many geophysicists, when asked to count their blessings, would include living on a planet with a molten interior, but it's a pretty near certainty that without all that magma swirling around beneath us we wouldn't be here now. Apart from much else, our lively interior created the outpourings of gas that helped to build an atmosphere and provided us with the magnetic field that shields us from cosmic radiation. It also gave us plate tectonics, which continually renews and rumples the surface. If the Earth were perfectly smooth, it would be covered everywhere with water to a depth of 4 kilometres. There might be life in that lonesome ocean, but there certainly wouldn't be football.

In addition to having a beneficial interior, we also have the right elements in the correct proportions. In the most literal way, we are made of the right stuff. This is so crucial to our well-being that we are going to discuss it more fully in a minute, but first we need to consider the two remaining factors, beginning with another one that is often overlooked:

We're a twin planet. Not many of us normally think of the Moon as a companion planet, but that is, in effect, what it is. Most moons are tiny in relation to their master planet. The Martian satellites of Phobos and Deimos, for instance, are only about 10 kilometres in diameter. Our Moon, however, is more than a quarter the diameter of the Earth, which makes ours the only planet in the solar system with a sizeable moon in comparison to itself (except Pluto, which doesn't really count because Pluto is itself so small) – and what a difference that makes to us.

Without the Moon's steadying influence, the Earth would wobble like a dying top, with goodness knows what consequences for climate and weather. The Moon's steady gravitational influence keeps the Earth spinning at the right speed and angle to provide the sort of stability necessary for the long and successful development of life. This won't go on for ever. The Moon is slipping from our grasp at a rate of about 4 centimetres a year. In another two billion years it will have receded so far that it won't keep us steady and we will have to come up with some other solution, but in the meantime you should think of it as much more than just a pleasant feature in the night sky.

For a long time, astronomers assumed that either the Moon and the Earth formed together or that the Earth captured the Moon as it drifted by. We now believe, as you will recall from an earlier chapter, that about 4.4 billion years ago a Mars-sized object slammed into Earth, blowing out enough material to create the Moon from the debris. This was obviously a very good thing for us – but especially so as it happened such a long time ago. If it had happened in 1896 or last Wednesday, clearly we wouldn't be nearly so pleased about it. Which brings us to our fourth and in many ways most crucial consideration:

Timing. The universe is an amazingly fickle and eventful place and our existence within it is a wonder. If a long and unimaginably complex sequence of events stretching back 4.6 billion years or so hadn't played out in a particular manner at particular times – if, to take just one obvious instance, the dinosaurs hadn't been wiped out by a meteor when they were – you might well be a few centimetres long, with whiskers and a tail, and reading this in a burrow.

We don't really know, because we have nothing else to which we can compare our own existence, but it seems evident that if you wish to end up as a moderately advanced, thinking society, you need to be at the right end of a very long chain of outcomes involving reasonable periods of stability interspersed with just the right amount of stress and challenge (ice ages appear to be especially helpful in this regard) and marked by a total absence of real cataclysm. As we shall see in the pages that remain to us, we are very lucky to find ourselves in that position.

And on that note, let us now turn briefly to the elements that made us.

There are ninety-two naturally occurring elements on the Earth, plus a further twenty or so that have been created in labs, but some of these we can immediately put to one side – as, in fact, chemists themselves tend to do. Not a few of our earthly chemicals are surprisingly little known. Astatine, for instance, is practically unstudied. It has a name and a place on the periodic table (next door to Marie Curie's polonium), but almost nothing else. The problem isn't scientific indifference, but rarity. There just isn't much astatine out there. The most elusive element of all, however, appears to be francium, which is so rare that it is thought that our entire planet may contain, at any given moment, fewer than twenty francium atoms. Altogether, only about thirty of the naturally occurring elements are widespread on Earth, and barely half a dozen are of central importance to life.

As you might expect, oxygen is our most abundant element, accounting for just under 50 per cent of the Earth's crust, but after that the relative abundances are often surprising. Who would guess, for instance, that silicon is the second most common element on the Earth, or that titanium is tenth? Abundance has little to do with their familiarity or utility to us. Many of the more obscure elements are actually more common than the better-known ones. There is more cerium on the Earth than copper, more neodymium and lanthanum than cobalt or nitrogen. Tin barely makes it into the top fifty, eclipsed by such relative obscurities as praseodymium, samarium, gadolinium and dysprosium.

Abundance also has little to do with ease of detection.

Aluminium is the fourth most common element on Earth, accounting for nearly a tenth of everything that's underneath your feet, but its existence wasn't even suspected until it was discovered in the nineteenth century by Humphry Davy, and for a long time after that it was treated as rare and precious. Congress nearly put a shiny lining of aluminium foil atop the Washington Monument to show what a classy and prosperous nation we had become, and the French imperial family in the same period discarded the state silver dinner service and replaced it with an aluminium one. The fashion was cutting edge even if the knives weren't.

Nor does abundance necessarily relate to importance. Carbon is only the fifteenth most common element, accounting for a very modest 0.048 per cent of Earth's crust, but we would be lost without it. What sets the carbon atom apart is that it is shamelessly promiscuous. It is the party animal of the atomic world, latching on to many other atoms (including itself) and holding tight, forming molecular conga lines of hearty robustness – the very trick of nature necessary to build proteins and DNA. As Paul Davies has written: 'If it wasn't for carbon, life as we know it would be impossible. Probably any sort of life would be impossible.' Yet carbon is not all that plentiful even in us who so vitally depend on it. Of every 200 atoms in your body, 126 are hydrogen, 51 are oxygen, and just 19 are carbon.*

Other elements are critical not for creating life but for sustaining it. We need iron to manufacture haemoglobin, and without it we would die. Cobalt is necessary for the creation of vitamin B_{12}. Potassium and a very little sodium are literally good for your nerves. Molybdenum, manganese and vanadium help to keep your enzymes purring. Zinc – bless it – oxidizes alcohol.

---

*Of the remaining four, three are nitrogen and the remaining atom is divided among all the other elements.

We have evolved to utilize or tolerate these things – we could hardly be here otherwise – but even then we live within narrow ranges of acceptance. Selenium is vital to all of us, but take in just a little too much and it will be the last thing you ever do. The degree to which organisms require or tolerate certain elements is a relic of their evolution. Sheep and cattle now graze side by side, but actually have very different mineral requirements. Modern cattle need quite a lot of copper because they evolved in parts of Europe and Africa where copper was abundant. Sheep, on the other hand, evolved in copper-poor areas of Asia Minor. As a rule, and not surprisingly, our tolerance for elements is directly proportionate to their abundance in the Earth's crust. We have evolved to expect, and in some cases actually need, the tiny amounts of rare elements that accumulate in the flesh or fibre that we eat. But step up the doses, in some cases by only a tiny amount, and we can soon cross a threshold. Much of this is only imperfectly understood. No-one knows, for example, whether a tiny amount of arsenic is necessary for our well-being or not. Some authorities say it is; some not. All that is certain is that too much of it will kill you.

The properties of the elements can become more curious still when they are combined. Oxygen and hydrogen, for instance, are two of the most combustion-friendly elements around, but put them together and they make incombustible water.* Odder still in combination are sodium, one of the most unstable of all elements, and chlorine, one of the most toxic. Drop a small lump of pure sodium into ordinary water

*Oxygen itself is not combustible; it merely facilitates the combustion of other things. This is just as well, for if oxygen were combustible, each time you lit a match all the air around you would burst into flame. Hydrogen gas, on the other hand, *is* extremely combustible, as the dirigible *Hindenburg* demonstrated on 6 May 1937, in Lakehurst, New Jersey, when the hydrogen that provided its lift burst explosively into flame, killing thirty-six people.

and it will explode with enough force to kill. Chlorine is even more notoriously hazardous. Though useful in small concentrations for killing micro-organisms (it's chlorine you smell in bleach), in larger volumes it is lethal. Chlorine was the element of choice for many of the poison gases of the First World War. And, as many a sore-eyed swimmer will attest, even in exceedingly dilute form the human body doesn't appreciate it. Yet put these two nasty elements together and what do you get? Sodium chloride – common table salt.

By and large, if an element doesn't naturally find its way into our systems – if it isn't soluble in water, say – we tend to be intolerant of it. Lead poisons us because we were never exposed to it until we began to fashion it into food vessels and pipes for plumbing. (Not incidentally, lead's symbol is Pb for the Latin *plumbum*, the source word for our modern *plumbing*.) The Romans also flavoured their wine with lead, which may be part of the reason they are not the force they used to be. As we have seen elsewhere, our own performance with lead (not to mention mercury, cadmium and all the other industrial pollutants with which we routinely dose ourselves) does not leave us a great deal of room for smirking. When elements don't occur naturally on Earth, we have evolved no tolerance for them and so they tend to be extremely toxic to us, as with plutonium. Our tolerance for plutonium is zero: there is no level at which it is not going to make you want to lie down.

I have brought you a long way to make a small point: a big part of the reason that Earth seems so miraculously accommodating is that we evolved to suit its conditions. What we marvel at is not that it is suitable to life but that it is suitable to *our* life – and hardly surprising really. It may be that many of the things that make it so splendid to us – well-proportioned Sun, doting Moon, sociable carbon, more molten magma than you can shake a stick at and all the rest

– seem splendid simply because they are what we were born to count on. No-one can altogether say.

Other worlds may harbour beings thankful for their silvery lakes of mercury and drifting clouds of ammonia. They may be delighted that their planet doesn't shake them silly with its grinding plates or spew messy gobs of lava over the landscape, but rather exists in a permanent non-tectonic tranquillity. Any visitors to the Earth from afar would almost certainly, at the very least, be bemused to find us living in an atmosphere composed of nitrogen, a gas sulkily disinclined to react with anything, and oxygen, which is so partial to combustion that we must place fire stations throughout our cities to protect ourselves from its livelier effects. But even if our visitors were oxygen-breathing bipeds with shopping malls and a fondness for action movies, it is unlikely that they would find the Earth ideal. We couldn't even give them lunch because all our foods contain traces of manganese, selenium, zinc and other elemental particles at least some of which would be poisonous to them. To them the Earth might not seem a wondrously congenial place at all.

The physicist Richard Feynman used to make a joke about *a posteriori* conclusions – reasoning from known facts back to possible causes. 'You know, the most amazing thing happened to me tonight,' he would say. 'I saw a car with the licence plate ARW 357. Can you imagine? Of all the millions of licence plates in the state, what was the chance that I would see that particular one tonight? Amazing!' His point, of course, is that it is easy to make any banal situation seem extraordinary if you treat it as fateful.

So it is possible that the events and conditions that led to the rise of life on the Earth are not quite as extraordinary as we like to think. Still, they were extraordinary enough, and one thing is certain: they will have to do until we find some better.

17

INTO THE TROPOSPHERE

Thank goodness for the atmosphere. It keeps us warm. Without it, Earth would be a lifeless ball of ice with an average temperature of minus 50 degrees Celsius. In addition, the atmosphere absorbs or deflects incoming swarms of cosmic rays, charged particles, ultraviolet rays and the like. Altogether, the gaseous padding of the atmosphere is equivalent to a 4.5-metre thickness of protective concrete, and without it these invisible visitors from space would slice through us like tiny daggers. Even raindrops would pound us senseless if it weren't for the atmosphere's slowing drag.

The most striking thing about our atmosphere is that there isn't very much of it. It extends upwards for about 190 kilometres, which might seem reasonably bounteous when viewed from ground level, but if you shrank the Earth to the size of a standard desktop globe it would only be about the thickness of a couple of coats of varnish.

For scientific convenience, the atmosphere is divided into four unequal layers: troposphere, stratosphere, mesosphere and ionosphere (now often called the thermosphere). The troposphere is the part that's dear to us. It alone contains enough warmth and oxygen to allow us to function, though even it swiftly becomes uncongenial to life as you climb up through it. From ground level to its highest point, the troposphere (or 'turning sphere') is about 16 kilometres thick at the equator and no more than 10 or 11 kilometres high in the temperate latitudes where most of us live. Eighty per cent of the atmosphere's mass, virtually all the water and thus virtually all the weather are contained within this thin and wispy layer. There really isn't much between you and oblivion.

Beyond the troposphere is the stratosphere. When you see the top of a storm cloud flattening out into the classic anvil shape, you are looking at the boundary between the troposphere and the stratosphere. This invisible ceiling is known as the tropopause and was discovered in 1902 by a Frenchman in a balloon, Léon-Philippe Teisserenc de Bort. *Pause* in this sense doesn't mean to stop momentarily but to cease altogether; it's from the same Greek root as *menopause*. Even at the troposphere's greatest extent, the tropopause is not very distant. A fast lift of the sort used in modern skyscrapers would get you there in about twenty minutes, though you would be well advised not to make the trip. Such a rapid ascent without pressurization would, at the very least, result in severe cerebral and pulmonary oedemas, a dangerous excess of fluids in the body's tissues. When the doors opened at the viewing platform, anyone inside would almost certainly be dead or dying. Even a more measured ascent would be accompanied by a great deal of discomfort. The temperature 10 kilometres up can be minus 57 degrees Celsius and you would need, or at least very much appreciate, supplementary oxygen.

After you have left the troposphere the temperature soon warms up again, to about 4 degrees Celsius, thanks to the absorptive effects of ozone (something else de Bort discovered on his daring 1902 ascent). It then plunges to as low as minus 90 degrees Celsius in the mesosphere before skyrocketing to 1,500 degrees Celsius or more in the aptly named but very erratic thermosphere, where temperatures can vary by over 500 degrees from day to night – though it must be said that 'temperature' at such a height becomes a somewhat notional concept. Temperature is really just a measure of the activity of molecules. At sea level, air molecules are so thick that one molecule can move only the tiniest distance – about eight-millionths of a centimetre, to be precise – before banging into another. Because trillions of molecules are constantly colliding, a lot of heat gets exchanged. But at the height of the thermosphere, at 80 kilometres or more, the air is so thin that any two molecules will be miles apart and hardly ever come into contact. So although each molecule is very warm, there are few interactions between them and thus little heat transference. This is good news for satellites and spaceships, because if the exchange of heat were more efficient any manmade object orbiting at that level would burst into flame.

Even so, spaceships have to take care in the outer atmosphere, particularly on return trips to Earth, as the space shuttle *Columbia* demonstrated all too tragically in February 2003. Although the atmosphere is very thin, if a craft comes in at too steep an angle – more than about 6 degrees – or too swiftly it can strike enough molecules to generate drag of an exceedingly combustible nature. Conversely, if an incoming vehicle hit the thermosphere at too shallow an angle, it could well bounce back into space, like a pebble skipped across water.

But you needn't venture to the edge of the atmosphere to be reminded of what hopelessly ground-hugging beings we are. As anyone who has spent time in a lofty city will know, you don't have to rise too many hundreds of metres from sea level before your body begins to protest. Even experienced mountaineers, with the benefits of fitness, training and bottled oxygen, quickly become vulnerable at height to confusion, nausea, exhaustion, frostbite, hypothermia, migraine, loss of appetite and a great many other stumbling dysfunctions. In a hundred emphatic ways the human body reminds its owner that it wasn't designed to operate so far above sea level.

'Even under the most favorable circumstances,' the climber Peter Habeler has written of conditions atop Everest, 'every step at that altitude demands a colossal effort of will. You must force yourself to make every movement, reach for every handhold. You are perpetually threatened by a leaden, deadly fatigue.' In *The Other Side of Everest*, the British mountaineer and film-maker Matt Dickinson records how Howard Somervell, on a 1924 British expedition up Everest, 'found himself choking to death after a piece of infected flesh came loose and blocked his windpipe'. With a supreme effort Somervell managed to cough up the obstruction. It turned out to be 'the entire mucous lining of his larynx'.

Bodily distress is notorious above 7,500 metres – the area known to climbers as the Death Zone – but many people become severely debilitated, even dangerously ill, at heights of no more than 4,500 metres or so. Susceptibility has little to do with fitness. Grannies sometimes caper about in lofty situations while their fitter offspring are reduced to helpless, groaning heaps until conveyed to lower altitudes.

The absolute limit of human tolerance for continuous living appears to be about 5,500 metres, but even people conditioned to living at altitude could not tolerate such heights for long. Frances Ashcroft, in *Life at the Extremes*, notes that there are Andean sulphur mines at 5,800 metres, but that the miners prefer to descend 460 metres each evening, and climb back up the following day, rather than live continuously at that elevation. People who habitually live at altitude have often spent thousands of years developing disproportionately large chests and lungs, and increasing their density of oxygen-bearing red blood cells by almost a third, though there are limits to how much thickening with red cells the blood supply can stand before it becomes too thick to flow smoothly. Moreover, above 5,500 metres even the most well-adapted women cannot provide a growing foetus with enough oxygen to bring it to its full term.

In the 1780s, when people began to make experimental balloon ascents in Europe, something that surprised them was how chilly it got as they rose. The temperature drops about 1.6 degrees Celsius with every 1,000 metres you climb. Logic would seem to indicate that the closer you get to a source of heat, the warmer you should feel. Part of the explanation is that you are not really getting nearer the Sun in any meaningful sense. The Sun is 93 million miles away. To move a few hundred metres closer to it is like taking one step closer to a bushfire in Australia and expecting to smell smoke when you are standing in Ohio. The answer again takes us back to the question of the density of molecules in the atmosphere. Sunlight energizes atoms. It increases the rate at which they jiggle and jounce, and in their enlivened state they crash into one another, releasing heat. When you feel the sun warm on your back on a summer's day, it's really excited atoms you feel. The higher you climb, the fewer molecules there are, and so the fewer collisions between them. Air is deceptive stuff. Even at sea level, we tend to think of the air as being ethereal and all but weightless. In fact, it has plenty of bulk, and that bulk often exerts itself. As a marine scientist named Wyville Thomson wrote more than a century ago: 'We sometimes find when we get up in the morning, by a rise of an inch in the barometer, that nearly half a ton has been quietly piled upon us during the night, but we experience no inconvenience, rather a feeling of exhilaration and buoyancy, since it requires a little less exertion to move our bodies in the denser medium.' The reason you don't feel crushed under that extra half-ton of pressure is the same reason your body would not be crushed deep beneath the sea: it is made mostly of incompressible fluids, which push back, equalizing the pressures within and without.

But get air in motion, as with a hurricane or even a stiff breeze, and you will quickly be reminded that it has very considerable mass. Altogether there are about 5,200 million million tonnes of air around us – 25 million tonnes for every square mile of the planet – a not inconsequential volume. When you get millions of tonnes of atmosphere rushing past at 50 or 60 kilometres an hour, it's hardly a surprise that tree-limbs snap and roof tiles go flying. As Anthony Smith notes, a typical weather front may consist of 750 million tonnes of cold air pinned beneath a billion tonnes of warmer air. Hardly a wonder that the result is at times meteorologically exciting.

Certainly there is no shortage of energy in the world above our heads. One thunderstorm, it has been calculated, can contain an amount of energy equivalent to four days' use of electricity for the whole United States. In the right conditions, storm clouds can rise to heights of 10 to 15 kilometres and contain updrafts and downdrafts of over 150 kilometres an hour. These are often side by side, which is why pilots don't want to fly through them. In all the internal turmoil, particles within the cloud pick up electrical charges. For reasons not entirely understood, the lighter particles tend to become positively charged and to be wafted by air currents to the top of the cloud. The heavier particles linger at the base, accumulating negative charges. These negatively charged particles have a powerful urge to rush to the positively charged Earth and good luck to anything that gets in their way. A bolt of lightning travels at 435,000 kilometres an hour and can heat the air around it to a decidedly crisp 28,000 degrees Celsius, several times hotter than the surface of the sun. At any one moment 1,800 thunderstorms are in progress around the globe – some 40,000 a day. Day and night across the planet, every second about a hundred lightning bolts hit the ground. The sky is a lively place.

Much of our knowledge of what goes on up there is surprisingly recent. Jet streams, usually located about 9,000–10,000 metres up, can bowl along at up to nearly 300 kilometres an hour and vastly influence weather systems over whole continents, yet their existence wasn't suspected until pilots began to fly into them during the Second World War. Even now a great deal of atmospheric phenomena is barely understood. A form of wave motion popularly known as clear-air turbulence occasionally enlivens aeroplane flights. About twenty such incidents a year are serious enough to need reporting. They are not associated with cloud structures or anything else that can be detected visually or by radar. They are just pockets of startling turbulence in the middle of tranquil skies. In a typical incident, a plane en route from Singapore to Sydney was flying over central Australia in calm conditions when it suddenly fell 90 metres – enough to fling unsecured people against the ceiling. Twelve people were injured, one seriously. No-one knows what causes such disruptive cells of air.

The process that moves air around in the atmosphere is the same process that drives the internal engine of the planet, namely convection. Moist, warm air from the equatorial regions rises until it hits the barrier of the tropopause and spreads out. As it travels away from the equator and cools, it sinks. When it hits bottom, some of the sinking air looks for an area of low pressure to fill and heads back for the equator, completing the circuit.

At the equator the convection process is generally stable and the weather predictably fair, but in temperate zones the patterns are far more seasonal, localized and random, which results in an endless battle between systems of high-pressure and low-pressure air. Low-pressure systems are created by rising air, which conveys water molecules into the sky, forming clouds and eventually rain. Warm air can hold more moisture than cool air, which is why tropical and summer storms tend to be the heaviest. Thus low areas tend to be associated with cloud and rain, and highs generally spell sunshine and fair weather. When two such systems meet, it often becomes manifest in the clouds. For instance, stratus clouds – those unlovable, featureless sprawls that give us our overcast skies – happen when moisture-bearing updrafts lack the oomph to break through a level of more stable air above, and instead spread out, like smoke hitting a ceiling. Indeed, if you watch a smoker sometime, you can get a very good idea of how things work by watching how smoke rises from a cigarette in a still room. At first, it goes straight up (this is called a laminar flow if you need to impress anyone) and then it spreads out in a diffused, wavy layer. The greatest supercomputer in the world, taking measurements in the most carefully controlled environment, cannot accurately predict what forms these ripplings will take, so you can imagine the difficulties that confront meteorologists when they try to forecast such motions in a spinning, windy, large-scale world.

What we do know is that because heat from the Sun is unevenly distributed, differences in air pressure arise on the planet. Air can't abide this, so it rushes around trying to equalize things everywhere. Wind is simply the air's way of trying to keep things in balance. Air always flows from areas of high pressure to areas of low pressure (as you would expect; think of anything with air under pressure – a balloon or an air tank or an airplane with a missing window – and think how insistently that pressured air wants to go somewhere else), and the greater the discrepancy in pressures, the faster the wind blows.

Incidentally, wind speeds, like most things that accumulate, grow exponentially, so a wind blowing at 300 kilometres an hour is not simply ten times stronger than a wind blowing at 30 kilometres an hour, but a hundred times stronger – and hence that much more destructive. Introduce several million tonnes of air to this accelerator effect and the result can be exceedingly energetic. A tropical hurricane can release in twenty-four hours as much energy as a rich, medium-sized nation like Britain or France uses in a year.

The impulse of the atmosphere to seek equilibrium was first suspected by Edmond Halley – the man who was everywhere – and elaborated upon in the eighteenth century by his fellow Briton George Hadley, who saw that rising and falling columns of air tended to produce 'cells' (known ever since as 'Hadley cells'). Though a lawyer by profession, Hadley had a keen interest in the weather (he was, after all, English) and also suggested a link between his cells, the Earth's spin and the apparent deflections of air that give us our trade winds. However, it was an engineering professor at the École Polytechnique in Paris, Gustave-Gaspard de Coriolis, who worked out the details of these interactions in 1835, and thus we call it the Coriolis effect. (Coriolis's other distinction at the school was to introduce water coolers, which are still known there as Corios, apparently.) The Earth revolves at a brisk 1,675 kilometres an hour at the equator, though as you move towards the poles the speed slopes off considerably, to about 900 kilometres an hour in London or Paris, for instance. The reason for this is self-evident when you think about it. If you are on the equator the spinning Earth has to carry you quite a distance – about 40,000 kilometres – to get you back to the same spot, whereas if you stand beside the North Pole you may need to travel only a few metres to complete a revolution; yet in both cases it takes twenty-four hours to get you back to where you began. Therefore, it follows that the closer you get to the equator the faster you must be spinning.

The Coriolis effect explains why anything moving

through the air in a straight line laterally to the Earth's spin will, given enough distance, seem to curve to the right in the northern hemisphere and to the left in the southern as the Earth revolves beneath it. The standard way to envision this is to imagine yourself at the centre of a large carousel and tossing a ball to someone positioned on the edge. By the time the ball gets to the perimeter, the target person has moved on and the ball passes behind him. From his perspective, it looks as if it has curved away from him. That is the Coriolis effect and it is what gives weather systems their curl and sends hurricanes spinning off like tops. The Coriolis effect is also why naval guns firing artillery shells have to adjust to left or right; a shell fired 15 miles would otherwise deviate by about 100 yards and plop harmlessly into the sea.

Considering the practical and psychological importance of the weather to nearly everyone, meteorology didn't really get going as a science until shortly before the beginning of the nineteenth century (though the term *meteorology* itself had been around since 1626, when it was coined by a T. Granger in a book of logic).

Part of the problem was that successful meteorology requires the precise measurements of temperatures, and thermometers for a long time proved more difficult to make than you might expect. An accurate reading was dependent on getting a very even bore in a glass tube, and that wasn't easy to do. The first person to solve the problem was Daniel Gabriel Fahrenheit, a Dutch maker of instruments, who produced an accurate thermometer in 1717. However, for reasons unknown he calibrated the instrument in a way that put freezing at 32 degrees and boiling at 212 degrees. From the outset this numeric eccentricity bothered some people and in 1742 Anders Celsius, a Swedish astronomer, came up with a competing scale. In proof of the proposition that inventors seldom get matters entirely right, Celsius made boiling point zero and freezing point 100 on his scale, but that was soon reversed.

The person most frequently identified as the father of modern meteorology was an English pharmacist named Luke Howard, who came to prominence at the beginning of the nineteenth century. Howard is chiefly remembered now for giving cloud types their names in 1803. Although he was an active and respected member of the Linnaean Society and employed Linnaean principles in his new scheme, Howard chose the rather more obscure Askesian Society as the forum in which to announce his new scheme of classification. (The Askesian Society, you may just recall from an earlier chapter, was the body whose members were unusually devoted to the pleasures of nitrous oxide, so we can only hope they treated Howard's presentation with the sober attention it deserved. It is a point on which Howard scholars are curiously silent.)

Howard divided clouds into three groups: stratus for the layered clouds, cumulus for the fluffy ones (the word means heaped in Latin) and cirrus (meaning curled) for the high, thin feathery formations that generally presage colder weather. To these he subsequently added a fourth term, nimbus (from the Latin for cloud), for a rain cloud. The beauty of Howard's system was that the basic components could be freely recombined to describe every shape and size of passing cloud – stratocumulus, cirrostratus, cumulonimbus, and so on. It was an immediate hit, and not just in England. Goethe was so taken with the system that he dedicated four poems to Howard.

Howard's system has been much added to over the years, so much so that the encyclopedic if little-read *International Cloud Atlas* runs to two volumes, but interestingly virtually all the post-Howard cloud types – mammatus, pileus, nebulosis, spissatus, floccus and mediocris are a sampling –

have never caught on with anyone outside meteorology and not terribly much within it, I'm told. Incidentally, the first, much thinner edition of that atlas, produced in 1896, divided clouds into ten basic types, of which the plumpest and most cushiony-looking was number nine, cumulonimbus.* That seems to have been the source of the expression 'to be on cloud nine'.

For all the heft and fury of the occasional anvil-headed storm cloud, the average cloud is actually a benign and surprisingly insubstantial thing. A fluffy summer cumulus several hundred metres to a side may contain no more than 100–150 litres of water – 'about enough to fill a bathtub', as James Trefil has noted. You can get some sense of the immaterial quality of clouds by strolling through fog – which is, after all, nothing more than a cloud that lacks the will to fly. To quote Trefil again: 'If you walk 100 yards through a typical fog, you will come into contact with only about half a cubic inch of water – not enough to give you a decent drink.' In consequence clouds are not great reservoirs of water. Only about 0.035 per cent of the Earth's fresh water is floating around above us at any moment.

Depending on where it falls, the prognosis for a water molecule varies widely. If it lands in fertile soil it will be soaked up by plants or re-evaporated directly within hours or days. If it finds its way down to the groundwater, however, it may not see sunlight again for many years –

*If you have ever been struck by how beautifully crisp and well defined the edges of cumulus clouds tend to be, while other clouds are more blurry, the explanation is that there is a pronounced boundary between the moist interior of a cumulus cloud and the dry air beyond it. Any water molecule that strays beyond the edge of the cloud is immediately zapped by the dry air beyond, allowing the cloud to keep its fine edge. Much higher cirrus clouds are composed of ice and the zone between the edge of the cloud and the air beyond it not so clearly delineated, which is why they tend to be blurry at the edges.

thousands if it gets really deep. When you look at a lake, you are looking at a collection of molecules that have been there on average for about a decade. In the ocean the residence time is thought to be more like a hundred years. Altogether, about 60 per cent of water molecules in a rainfall are returned to the atmosphere within a day or two. Once evaporated, they spend no more than a week or so – Drury says twelve days – in the sky before falling again as rain.

Evaporation is a swift process, as you can easily gauge by the fate of a puddle on a summer's day. Even something as large as the Mediterranean would dry out in a thousand years if it were not continually replenished. Such an event occurred a little under six million years ago and provoked what is known to science as the Messinian Salinity Crisis. What happened was that continental movement closed the Strait of Gibraltar. As the Mediterranean dried, its evaporated contents fell as fresh-water rain into other seas, mildly diluting their saltiness – indeed, making them just dilute enough to freeze over larger areas than normal. The enlarged area of ice bounced back more of the Sun's heat and pushed Earth into an ice age. So, at least, the theory goes.

What is certainly true, as far as we can tell, is that a little change in the Earth's dynamics can have repercussions beyond our imagining. Such an event, as we shall see a little further on, may even have created us.

The real powerhouse of the planet's surface behaviour are the oceans. Indeed, meteorologists increasingly treat oceans and atmosphere as a single system, which is why we must give them a little of our attention here. Water is marvellous at holding and transporting heat – unimaginably vast quantities of it. Every day, the Gulf Stream carries an amount of heat to Europe equivalent to the world's output of coal for ten years, which is why Britain and Ireland have such mild winters compared with Canada and Russia. But water also warms slowly, which is why lakes and swimming pools are cold even on the hottest days. For that reason there tends to be a lag in the official, astronomical start of a season and the actual feeling that that season has started. So spring may officially start in the northern hemisphere in March, but it doesn't feel like it in most places until April at the very earliest.

The oceans are not one uniform mass of water. Their differences in temperature, salinity, depth, density and so on have huge effects on how they move heat around, which in turn affects climate. The Atlantic, for instance, is saltier than the Pacific, and a good thing too. The saltier a water is the denser it is, and dense water sinks. Without its extra burden of salt, the Atlantic currents would proceed up to the Arctic, warming the North Pole, but depriving Europe of all that kindly warmth. The main agent of heat transfer on Earth is what is known as thermohaline circulation, which originates in slow, deep currents far below the surface – a process first detected by the scientist-adventurer Count von Rumford in 1797.* What happens is that surface waters, as they get to the vicinity of Europe, grow dense and sink to great depths and begin a slow trip back to the southern hemisphere. When they reach Antarctica, they are caught up in the Antarctic Circumpolar Current, where they are driven onward into the Pacific. The process is very slow – it can take fifteen hundred years for water to travel from the North

*The term means a number of things to different people, it appears. In November 2002, Carl Wunsch of MIT published a report in *Science*, 'What Is the Thermohaline Circulation?', in which he noted that the expression has been used in leading journals to signify at least seven different phenomena (circulation at the abyssal level, circulation driven by differences in density or buoyancy, 'meridional overturning circulation of mass' and so on) – though all are to do with ocean circulations and the transfer of heat, the cautiously vague and embracing sense in which I have employed it here.

Atlantic to the mid-Pacific – but the volumes of heat and water they move are very considerable and the influence on the climate is enormous.

(As for the question of how anyone could possibly figure out how long it takes a drop of water to get from one ocean to another, the answer is that scientists can measure compounds in the water like chlorofluorocarbons and work out how long it has been since they were last in the air. By comparing a lot of measurements from different depths and locations, they can reasonably chart the water's movement.)

Thermohaline circulation not only moves heat around, but also helps to stir up nutrients as the currents rise and fall, making greater volumes of the ocean habitable for fish and other marine creatures. Unfortunately, it appears the circulation may also be very sensitive to change. According to computer simulations, even a modest dilution of the ocean's salt content – from increased melting of the Greenland ice sheet, for instance – could disrupt the cycle disastrously.

The seas do one other great favour for us. They soak up tremendous volumes of carbon and provide a means for it to be safely locked away. One of the oddities of our solar system is that the Sun burns about 25 per cent more brightly now than when the solar system was young. This should have resulted in a much warmer Earth. Indeed, as the English geologist Aubrey Manning has put it, 'This colossal change should have had an absolutely catastrophic effect on the Earth and yet it appears that our world has hardly been affected.'

So what keeps the planet stable and cool? Life does. Trillions upon trillions of tiny marine organisms that most of us have never heard of – foraminiferans and coccoliths and calcareous algae – capture atmospheric carbon, in the form of carbon dioxide, when it falls as rain and use it (in combination with other things) to make their tiny shells. By locking the carbon up in their shells, they keep it from being re-evaporated into the atmosphere where it would build up dangerously as a greenhouse gas. Eventually all the tiny foraminiferans and coccoliths and so on die and fall to the bottom of the sea, where they are compressed into limestone. It is remarkable, when you behold an extraordinary natural feature like the White Cliffs of Dover in England, to reflect that it is made up almost entirely of tiny deceased marine organisms, but even more remarkable when you realize how much carbon they cumulatively sequester. A six-inch cube of Dover chalk will contain well over a thousand litres of compressed carbon dioxide that would otherwise be doing us no good at all. Altogether there is about twenty thousand times as much carbon locked away in the Earth's rocks as in the atmosphere. Eventually much of that limestone will end up feeding volcanoes and the carbon will return to the atmosphere and fall to the Earth in rain, which is why the whole is called the long-term carbon cycle. The process takes a very long time – about half a million years for a typical carbon atom – but in the absence of any other disturbance it works remarkably well at keeping the climate stable.

Unfortunately, human beings have a careless predilection for disrupting this cycle by putting lots of extra carbon into the atmosphere whether the foraminiferans are ready for it or not. Since 1850, it has been estimated, we have lofted about 100 billion tonnes of extra carbon into the air, a total that increases by about 7 billion tonnes each year. Overall, that's not actually all that much. Nature – mostly through the belchings of volcanoes and the decay of plants – sends about 200 billion tonnes of carbon dioxide into the atmosphere each year, nearly thirty times as much as we do with our cars and factories. But you have only to look at the haze that hangs over our cities or the Grand Canyon or even, sometimes, the White Cliffs of Dover to see what a difference our contribution makes.

We know from samples of very old ice that the 'natural' level of carbon dioxide in the atmosphere – that is, before we started inflating it with industrial activity – is about 280 parts per million. By 1958, when people in lab coats started to pay attention to it, it had risen to 315 parts per million. Today it is over 360 parts per million and rising by roughly one-quarter of 1 per cent a year. By the end of the twenty-first century it is forecast to rise to about 560 parts per million.

So far, the Earth's oceans and forests (which also pack away a lot of carbon) have managed to save us from ourselves, but, as Peter Cox of the British Meteorological Office puts it: 'There is a critical threshold where the natural biosphere stops buffering us from the effects of our emissions and actually starts to amplify them.' The fear is that there would be a very rapid increase in the Earth's warming. Unable to adapt, many trees and other plants would die, releasing their stores of carbon and adding to the problem. Such cycles have occasionally happened in the distant past even without a human contribution. The good news is that even here, nature is quite wonderful. It is almost certain that eventually the carbon cycle would reassert itself and return the Earth to a situation of stability and happiness. The last time this happened, it took a mere sixty thousand years.

18

THE BOUNDING MAIN

Imagine trying to live in a world dominated by dihydrogen oxide, a compound that has no taste or smell and is so variable in its properties that it is generally benign but at other times swiftly lethal. Depending on its state, it can scald you or freeze you. In the presence of certain organic molecules it can form carbonic acids so nasty that they can strip the leaves from trees and eat the faces off statuary. In bulk, when agitated, it can strike with a fury that no human edifice could withstand. Even for those who have learned to live with it, it is an often murderous substance. We call it water.

Water is everywhere. A potato is 80 per cent water, a cow 74 per cent, a bacterium 75 per cent. A tomato, at 95 per cent, is little *but* water. Even humans are 65 per cent water, making us more liquid than solid by a margin of almost two to one. Water is strange stuff. It is formless and transparent, and yet we long to be beside it. It has no taste and yet we love the taste of it. We will travel great distances and pay small fortunes to see it in sunshine. And even though we know it is dangerous and drowns tens of thousands of people every year, we can't wait to frolic in it.

Because water is so ubiquitous we tend to overlook what an extraordinary substance it is. Almost nothing about it can be used to make reliable predictions about the properties of other liquids, and vice versa. If you knew nothing of water and based your assumptions on the behaviour of compounds most chemically akin to it – hydrogen selenide or hydrogen sulphide, notably – you would expect it to boil at minus 93 degrees Celsius and to be a gas at room temperature.

Most liquids when chilled contract by about 10 per cent. Water does too, but only down to a point. Once it is within whispering distance of freezing, it begins – perversely, beguilingly, extremely improbably – to expand. By the time it is solid, it is almost a tenth more voluminous than it was before. Because it expands, ice floats on water – 'an utterly bizarre property', according to John Gribbin. If it lacked this splendid waywardness, ice would sink, and lakes and oceans would freeze from the bottom up. Without surface ice to hold heat in, the water's warmth would radiate away, leaving it even chillier and creating yet more ice. Soon even the oceans would freeze and almost certainly stay that way for a very long time, probably for ever – hardly the conditions to nurture life. Thankfully for us, water seems unaware of the rules of chemistry or laws of physics.

Everyone knows that water's chemical formula is H_2O, which means that it consists of one largish oxygen atom with two smaller hydrogen atoms attached to it. The hydrogen atoms cling fiercely to their oxygen host, but also make casual bonds with other water molecules. The nature of a water molecule means that it engages in a kind of dance with other water molecules, briefly pairing and then moving on, like the ever-changing partners in a quadrille, to use Robert Kunzig's nice phrase. A glass of water may not appear terribly lively, but every molecule in it is changing partners billions of times a second. That's why water molecules stick together to form bodies like puddles and lakes, but not so tightly that they can't be easily separated as when, for instance, you dive into a pool of them. At any given moment only 15 per cent of them are actually touching.

In one sense the bond is very strong – it is why water molecules can flow uphill when siphoned and why water droplets on a car bonnet show such a singular determination to bead with their partners. It is also why water has surface tension. The molecules at the surface are attracted more powerfully to the like molecules beneath and beside them than to the air molecules above. This creates a sort of membrane strong enough to support insects and skipping stones. It is what gives the sting to a belly-flop.

I hardly need point out that we would be lost without it. Deprived of water, the human body rapidly falls apart. Within days, the lips vanish 'as if amputated, the gums blacken, the nose withers to half its length, and the skin so contracts around the eyes as to prevent blinking', according to one account. Water is so vital to us that it is easy to overlook that all but the smallest fraction of the water on Earth is poisonous to us – deadly poisonous – because of the salts within it.

We need salt to live, but only in very small amounts, and sea water contains way more – about seventy times more – salt than we can safely metabolize. A typical litre of sea water will contain only about 2.5 teaspoons of common salt – the kind we sprinkle on food – but much larger amounts of other elements, compounds and other dissolved solids, which are collectively known as salts. The proportions of these salts and minerals in our tissues are uncannily similar to those in sea water – we sweat and cry sea water, as Margulis and Sagan have put it – but curiously we cannot tolerate them as an input. Take a lot of salt into your body and your metabolism very quickly goes into crisis. From every cell, water molecules rush off like so many volunteer firemen to try to dilute and carry off the sudden intake of salt. This leaves the cells dangerously short of the water they need to carry out their normal functions. They become, in a word, dehydrated. In extreme situations, dehydration will lead to seizures, unconsciousness and brain damage. Meanwhile, the overworked blood cells carry the salt to the kidneys, which eventually become overwhelmed and shut down. Without functioning kidneys you die. That is why we don't drink sea water.

There are 1.3 billion cubic kilometres of water on Earth and that is all we're ever going to get. The system is closed: practically speaking, nothing can be added or subtracted. The water you drink has been around doing its job since the Earth was young. By 3.8 billion years ago, the oceans had (at least more or less) achieved their present volumes.

The water realm is known as the hydrosphere and it is overwhelmingly oceanic. Ninety-seven per cent of all the water on Earth is in the seas, the greater part of it in the Pacific, which is bigger than all the land masses put together. Altogether the Pacific holds just over half of all the ocean water (51.6 per cent); the Atlantic has 23.6 per cent and the Indian Ocean 21.2 per cent, leaving just 3.6 per cent to be accounted for by all the other seas. The average depth of the ocean is 3.86 kilometres, with the Pacific on average about 300 metres deeper than the Atlantic and Indian Oceans. Sixty per cent of the planet's surface is ocean more than 1.6 kilometres deep. As Philip Ball notes, we would better call our planet not Earth but Water.

Of the 3 per cent of Earth's water that is fresh, most exists as ice sheets. Only the tiniest amount – 0.036 per cent – is found in lakes, rivers and reservoirs, and an even smaller part – just 0.001 per cent – exists in clouds or as vapour. Nearly 90 per cent of the planet's ice is in Antarctica and most of the rest is in Greenland. Go to the South Pole and you will be standing on over 2 miles of ice, at the North Pole just 15 feet of it. Antarctica alone has 6 million cubic miles of ice – enough to raise the oceans by a height of 200 feet if it all melted. But if all the water in the atmosphere fell as rain, evenly everywhere, the oceans would deepen by only a couple of centimetres.

Sea level, incidentally, is an almost entirely notional concept. Seas are not level at all. Tides, winds, the Coriolis force and other effects alter water levels considerably from one ocean to another and even within oceans. The Pacific is about a foot and a half higher along its western edge – a consequence of the centrifugal force created by the Earth's spin. Just as when you pull on a tub of water the water tends to flow towards the other end, as if reluctant to come with you, so the eastward spin of Earth piles water up against the ocean's western margins.

Considering the age-old importance of the seas to us, it is striking how long it took the world to take a scientific interest in them. Until well into the nineteenth century most of what was known about the oceans was based on what washed ashore or came up in fishing nets, and nearly all that was written was based more on anecdote and supposition than on physical evidence. In the 1830s, the British naturalist Edward Forbes surveyed ocean beds throughout the Atlantic and Mediterranean and declared that there was no life at all in the seas below 600 metres. It seemed a reasonable assumption. There was no light at that depth, so no plant life, and the pressures of water at such depths were known to be extreme. So it came as something of a surprise when, in 1860, one of the first transatlantic telegraph cables was hauled up for repairs from more than 3 kilometres down and found to be thickly encrusted with corals, clams and other living detritus.

The first really organized investigation of the seas didn't come until 1872, when a joint expedition set up by the British Museum, the Royal Society and the British government set forth from Portsmouth on a former warship called HMS *Challenger*. For three and a half years they sailed the world, sampling waters, netting fish and hauling a dredge through sediments. It was evidently dreary work. Out of a complement of 240 scientists and crew, one in four jumped ship and eight more died or went mad – 'driven to distraction by the mind-numbing routine of years of dredging', in the words of the historian Samantha Weinberg. But they sailed across almost 70,000 nautical miles of sea, collected over 4,700 new species of marine organisms, gathered enough information to create a fifty-volume report (which took nineteen years to put together), and gave the world the name of a new scientific discipline: *oceanography*. They also discovered, by means of depth measurements, that there appeared to be submerged mountains in mid-Atlantic, prompting some excited observers to speculate that they had found the lost continent of Atlantis.

Because the institutional world mostly ignored the seas, it fell to devoted – and very occasional – amateurs to tell us what was down there. Modern deep-water exploration begins with Charles William Beebe and Otis Barton in 1930. Although they were equal partners, the more colourful Beebe has always received far more written attention. Born in 1877 into a well-to-do family in New York City, Beebe studied zoology at Columbia University, then took a job as a birdkeeper at the New York Zoological Society. Tiring of that, he decided to adopt the life of an adventurer and for the next quarter-century travelled extensively through Asia and South America with a succession of attractive female assistants whose jobs were inventively described as 'historian and technicist' or 'assistant in fish problems'. He supported these endeavours with a succession of popular books with titles like *Edge of the Jungle* and *Jungle Days*, though he also produced some respectable books on wildlife and ornithology.

In the mid-1920s, on a trip to the Galápagos Islands, he discovered 'the delights of dangling', as he described deep-sea diving. Soon afterwards he teamed up with Barton, who came from an even wealthier family, had also attended Columbia and also longed for adventure. Although Beebe nearly always gets the credit, it was in fact Barton who designed the first bathysphere (from the Greek word for 'deep') and funded the $12,000 cost of its construction. It was a tiny and necessarily robust chamber, made of cast iron 1.5 inches thick and with two small portholes containing quartz blocks 3 inches thick. It held two men, but only if they were prepared to become extremely well acquainted. Even by the standards of the age, the technology was unsophisticated. The sphere had no manoeuvrability – it simply hung on the end of a long cable – and only the most primitive breathing system: to neutralize their own carbon dioxide they set out open cans of soda lime, and to absorb moisture they opened a small tub of calcium chloride, over which they sometimes waved palm fronds to encourage chemical reactions.

But the nameless little bathysphere did the job it was intended to do. On the first dive, in June 1930 in the Bahamas, Barton and Beebe set a world record by descending to 183 metres. By 1934, they had pushed the record to over 900 metres, where it would stay until after the Second World War. Barton was confident the device was safe to a depth of about 1,400 metres, though the strain on every bolt and rivet was audibly evident with every fathom they descended. At any depth, it was brave and risky work. At 900 metres, their little porthole was subjected to 19 tons of pressure per square inch. Should they pass the structure's limits of tolerance, death at such a depth would have been instantaneous, as Beebe never failed to observe in his many books, articles and radio broadcasts. Their main concern, however, was that the shipboard winch, straining to hold onto a metal ball and two tons of steel cable, would snap and send the two men plunging to the sea floor. In such an event, nothing could have saved them.

The one thing their descents didn't produce was a great deal of worthwhile science. Although they encountered many creatures that had not been seen before, the limits of visibility and the fact that neither of the intrepid aquanauts was a trained oceanographer meant they often weren't able to describe their findings in the kind of detail that real scientists craved. The sphere didn't carry an external light, merely a 250-watt bulb they could hold up to the window, but the water below 150 metres was practically impenetrable anyway, and they were peering into it through three inches of quartz, so anything they hoped to view would have to be nearly as interested in them as they were in it. About all they could report, in consequence, was that there were a lot of strange things down there. On one dive in 1934, Beebe was startled to spy a giant serpent 'more than twenty feet long and very wide'. It passed too swiftly to be more than a shadow. Whatever it was, nothing like it has been seen by anyone since. Because of such vagueness, their reports were generally ignored by academics.

After their record-breaking descent of 1934, Beebe lost interest in diving and moved on to other adventures, but Barton persevered. To his credit, Beebe always told anyone who asked that Barton was the real brains behind the enterprise, but Barton seemed unable to step from the shadows. He, too, wrote thrilling accounts of their underwater adventures and even starred in a Hollywood movie called *Titans of the Deep*, featuring a bathysphere and many exciting and largely fictionalized encounters with aggressive giant squid and the like. He even advertised Camel cigarettes ('They don't give me jittery nerves'). In 1948 he increased the depth record by 50 per cent, with a dive to 1,370 metres in the Pacific Ocean near California, but the world seemed determined to overlook him. One newspaper reviewer of *Titans of the Deep* actually thought the star of the film was Beebe. Nowadays, Barton is lucky to get a mention.

At all events, he was about to be comprehensively eclipsed by a father and son team from Switzerland, Auguste and Jacques Piccard, who were designing a new type of probe called a bathyscaphe (meaning 'deep boat'). Christened *Trieste*, after the Italian city in which it was built, the new device manoeuvred independently, though it did little more than just go up and down. On one of its early dives, in early 1954, it descended to below 4,000 metres, nearly three times Barton's record-breaking dive of six years earlier. But deep-sea dives required a great deal of costly support and the Piccards were gradually going broke.

In 1958, they did a deal with the US Navy which gave the Navy ownership but left them in control. Now flush with funds, the Piccards rebuilt the vessel, giving it walls nearly 13 centimetres thick and shrinking the windows to just 5 centimetres in diameter – little more than peepholes. But it was now strong enough to withstand truly enormous pressures, and in January 1960 Jacques Piccard and Lt Don Walsh of the US Navy sank slowly to the bottom of the ocean's deepest canyon, the Mariana Trench, some 400 kilometres off Guam in the western Pacific (and discovered, not incidentally, by Harry Hess with his fathometer). It took just under four hours to fall 10,918 metres, or almost 7 miles. Although the pressure at that depth was nearly 17,000 pounds per square inch, they noticed with surprise that they disturbed a bottom-dwelling flatfish just as they touched down. They had no facilities for taking photographs, so there is no visual record of the event.

After just twenty minutes at the world's deepest point, they returned to the surface. It was the only occasion in which human beings have gone so deep.

Forty years later, the question that naturally occurs is: why has no-one gone back since? To begin with, further dives were vigorously opposed by Vice Admiral Hyman G. Rickover, a man with a lively temperament, forceful views and, most pertinently, control of the departmental chequebook. He thought underwater exploration a waste of resources and pointed out that the Navy was not a research institute. The nation, moreover, was about to become fully preoccupied with space travel and the quest to send a man to the Moon, which made deep sea investigations seem unimportant and rather old-fashioned. But the decisive consideration is that the *Trieste* descent didn't actually achieve much. As a navy official explained years later: 'We didn't learn a hell of a lot from it, other than that we could do it. Why do it again?' It was, in short, a long way to go to find a flatfish, and expensive too. Repeating the exercise today, it has been estimated, would cost at least $100 million.

When underwater researchers realized that the Navy had no intention of pursuing a promised exploration programme, there was a pained outcry. Partly to placate its critics, the Navy provided funding for a more advanced submersible, to be operated by the Woods Hole Oceanographic Institution of Massachusetts. Called *Alvin*, in somewhat contracted honour of the oceanographer Allyn C. Vine, it would be a fully manoeuvrable mini-submarine, though it wouldn't go anywhere near as deep as *Trieste*. There was just one problem: the designers couldn't find anyone willing to build it. According to William J. Broad in *The Universe Below*: 'No big company like General Dynamics, which made submarines for the Navy, wanted to take on a project disparaged by both the Bureau of Ships and Admiral Rickover, the gods of naval patronage.' Eventually, not to say improbably, *Alvin* was constructed by General Mills, the food company, at a factory where it made the machines to produce breakfast cereals.

As for what else was down there, people really had very little idea. Well into the 1950s, the best maps available to oceanographers were overwhelmingly based on a little detail from scattered surveys going back to 1929 grafted onto, essentially, an ocean of guesswork. The US Navy had excellent charts with which to guide submarines through canyons and around guyots, but it didn't wish such information to fall into Soviet hands, so it kept its knowledge classified. Academics therefore had to make do with sketchy and antiquated surveys or rely on hopeful surmise. Even today our knowledge of the ocean floors remains remarkably low resolution. If you look at the Moon with a standard backyard telescope you will see substantial craters – Fracastorious, Blancanus, Zach, Planck and many others familiar to any lunar scientist – that would be unknown if they were on our own ocean floors. We have better maps of Mars than we do of our own seabeds.

At the surface level, investigative techniques have also been a trifle ad hoc. In 1994, 34,000 ice hockey gloves were swept overboard from a Korean cargo ship during a storm in the Pacific. The gloves washed up all over, from Vancouver to Vietnam, helping oceanographers to trace currents more accurately than they ever had before.

Today *Alvin* is nearly forty years old, but it remains the world's premier research vessel. There are still no submersibles that can go anywhere near the depth of the Mariana Trench and only five, including *Alvin*, that can reach the depths of the 'abyssal plain' – the deep ocean floor – which covers more than half the planet's surface. A typical submersible costs about $25,000 a day to operate, so they are hardly dropped into the water on a whim, still less put to sea in the hope that they will randomly stumble on something of interest. It's rather as if our first-hand experience of the surface world were based on the work of five guys exploring on garden tractors after dark. According to Robert Kunzig, humans may have scrutinized 'perhaps a millionth or a billionth of the sea's darkness. Maybe less. Maybe much less.'

But oceanographers are nothing if not industrious and they have made several important discoveries with their limited resources – including, in 1977, one of the most important and startling biological discoveries of the twentieth century. In that year *Alvin* found teeming colonies of large organisms living on and around deep-sea vents off the Galápagos Islands – tube worms over 3 metres long, clams 30 centimetres wide, shrimps and mussels in profusion, wriggling spaghetti worms. They all owed their existence to vast colonies of bacteria that were deriving *their* energy and sustenance from hydrogen sulphides – compounds profoundly toxic to surface creatures – that were pouring steadily from the vents. It was a world independent of sunlight, oxygen or anything else normally associated with life. This was a living system based not on photosynthesis but on chemosynthesis, an arrangement that biologists would have dismissed as preposterous had anyone been imaginative enough to suggest it.

Huge amounts of heat and energy are released by these vents. Two dozen of them together will produce as much energy as a large power station and the range of temperatures around them is enormous. The temperature at the point of outflow can be as much as 400 degrees Celsius, while a couple of metres away the water may be only two or three degrees above freezing. A type of worm called alvinellids were found living right on the margins, with the water temperature 78 degrees Celsius warmer at their heads than at their tails. Before this it had been thought that no complex organisms could survive in water warmer than about 54 degrees Celsius, and here was one that was surviving warmer temperatures than that *and* extreme cold to boot. The discovery transformed our understanding of the requirements for life.

It also answered one of the great puzzles of oceanography – something that many of us didn't realize *was* a puzzle – namely, why the oceans don't grow saltier with time. At the risk of stating the obvious, there is a lot of salt in the sea –

enough to bury every bit of land on the planet to a depth of about 150 metres. It had been known for centuries that rivers carry minerals to the sea and that these minerals combine with ions in the ocean water to form salts. So far no problem. But what was puzzling was that the salinity levels of the sea were stable. Millions of gallons of fresh water evaporate from the ocean daily, leaving all their salts behind, so logically the seas ought to grow more salty with the passing years, but they don't. Something takes an amount of salt out of the water equivalent to the amount being put in. For a very long time, no-one could figure out what could be responsible for this.

Alvin's discovery of the deep-sea vents provided the answer. Geophysicists realized that the vents were acting much like the filters in a fish tank. As water is taken down into the Earth's crust, salts are stripped from it, and eventually clean water is blown out again through the chimney stacks. The process is not swift – it can take up to ten million years to clean an ocean – but if you are not in a hurry it is marvellously efficient.

Perhaps nothing speaks more clearly of our psychological remoteness from the ocean depths than that the main expressed goal for oceanographers during International Geophysical Year, 1957/8, was to study 'the use of ocean depths for the dumping of radioactive wastes'. This wasn't a secret assignment, you understand, but a proud public boast. In fact, though it wasn't much publicized, by 1957/8 the dumping of radioactive wastes had already been going on, with a certain appalling vigour, for over a decade. Since 1946, the United States had been ferrying 55-gallon drums of radioactive gunk out to the Fallarone Islands, some 50 kilometres off the California coast near San Francisco, where it simply threw them overboard.

It was all quite extraordinarily sloppy. Most of the drums

were exactly the sort you see rusting behind petrol stations or standing outside factories, with no protective linings of any type. When they failed to sink, which was usually, navy gunners riddled them with bullets to let water in (and, of course, plutonium, uranium and strontium out). Before this dumping was halted in the 1990s, the United States had dumped many hundreds of thousands of drums into about fifty ocean sites – almost fifty thousand of them in the Fallarones alone. But the United States was by no means alone. Among the other enthusiastic dumpers were Russia, China, Japan, New Zealand, and nearly all the nations of Europe.

And what effect might all this have had on life beneath the seas? Well, little, we hope, but we actually have no idea. We are astoundingly, sumptuously, radiantly ignorant of life beneath the seas. Even the most substantial ocean creatures are often remarkably little known to us – including the most mighty of them all, the great blue whale, a creature of such leviathan proportions that (to quote David Attenborough) its 'tongue weighs as much as an elephant, its heart is the size of a car and some of its blood vessels are so wide that you could swim down them'. It is the most gargantuan beast the Earth has yet produced, bigger even than the most cumbrous dinosaurs. Yet the lives of blue whales are largely a mystery to us. Much of the time we have no idea where they are – where they go to breed, for instance, or what routes they follow to get there. What little we know of them comes almost entirely from eavesdropping on their songs, but even these are a mystery. Blue whales will sometimes break off a song, then pick it up again at exactly the same spot six months later. Sometimes they strike up with a new song, which no member can have heard before but which each already knows. How they do this and why are not remotely understood. And these are animals that must routinely come to the surface to breathe.

For animals that need never surface, obscurity can be even more tantalizing. Consider our knowledge of the fabled giant squid. Though nothing on the scale of the blue whale, it is a decidedly substantial animal, with eyes the size of soccer balls and trailing tentacles that can reach lengths of 18 metres. It weighs nearly a tonne and is Earth's largest invertebrate. If you dumped one in a small swimming pool, there wouldn't be much room for anything else. Yet no scientist – no person, as far as we know – has ever seen a giant squid alive. Zoologists have devoted careers to trying to capture, or just glimpse, living giant squid and have always failed. They are known mostly from being washed up on beaches – particularly, for unknown reasons, the beaches of the South Island of New Zealand. They must exist in large numbers because they form a central part of the sperm whale's diet, and sperm whales take a lot of feeding.*

According to one estimate, there could be as many as 30 million species of animals living in the sea, most still undiscovered. The first hint of how truly abundant life is in the deep seas didn't come until as recently as the 1960s with the invention of the epibenthic sled – a dredging device that captures organisms not just on and near the sea floor but also buried in the sediments beneath. In a single one-hour trawl along the continental shelf, at a depth of about 1.5 kilometres, Woods Hole oceanographers Howard Sandler and Robert Hessler netted over twenty-five thousand creatures – worms, starfish, sea cucumbers and the like – representing 365 species. Even at a depth of nearly 5 kilometres, they found some 3,700 creatures representing

*The indigestible parts of giant squid, in particular their beaks, accumulate in sperm whales' stomachs into the substance known as ambergris, which is used as a fixative in perfumes. The next time you spray on Chanel Number 5 (assuming you do), you may wish to reflect that you are dousing yourself in distillate of unseen sea monster.

almost two hundred species of organism. But the dredge could capture only those things that were too slow or stupid to get out of the way. In the late 1960s a marine biologist named John Isaacs had the idea of lowering a camera with bait attached to it, and found still more, in particular dense swarms of writhing hagfish, a primitive eel-like creature, as well as darting shoals of grenadier fish. Where a good food source is suddenly available – for instance, when a whale dies and sinks to the bottom – as many as 390 species of marine creature have been found dining off it. Intriguingly, many of these creatures were found to have come from vents up to 1,600 kilometres away. These included such types as mussels and clams, which are hardly known as great travellers. It is now thought that the larvae of certain organisms may drift through the water until, by some unknown chemical means, they detect that they have arrived at a food opportunity and fall onto it.

So why, if the seas are so vast, do we so easily overtax them? Well, to begin with, the world's seas are not uniformly bounteous. Altogether less than a tenth of the ocean is considered naturally productive. Most aquatic species like to be in shallow waters, where there are warmth and light and an abundance of organic matter to prime the food chain. Coral reefs, for instance, constitute well under 1 per cent of the ocean's space but are home to about 25 per cent of its fish.

Elsewhere, the oceans aren't nearly so rich. Take Australia. With 36,735 kilometres of coastline and over 23 million square kilometres of territorial waters, it has more sea lapping its shores than any other country, yet, as Tim Flannery notes, it doesn't even make it into the top fifty among fishing nations. Indeed, Australia is a large net importer of seafood. This is because much of Australia's water is, like much of Australia itself, essentially desert. (A notable exception is the Great Barrier Reef off Queensland, which is sumptuously fecund.) Because the soil is poor, it produces practically no nutrients in its run-offs.

Even where life thrives, it is often extremely sensitive to disturbance. In the 1970s, fishermen from Australia and, to a lesser extent, New Zealand discovered shoals of a little-known fish living at a depth of about 800 metres on their continental shelves. They were known as orange roughy, they were delicious and they existed in huge numbers. In no time at all, fishing fleets were hauling in 40,000 tonnes of roughy a year. Then marine biologists made some alarming discoveries. Roughy are extremely long-lived and slow-maturing. Some may be 150 years old; any roughy you have eaten may well have been born when Victoria was Queen. Roughy have adopted this exceedingly unhurried lifestyle because the waters they live in are so resource-poor. In such waters, some fish spawn just once in a lifetime. Clearly these are populations that cannot stand a great deal of disturbance. Unfortunately, by the time this was realized the stocks had been severely depleted. Even with good management it will be decades before the populations recover, if they ever do.

Elsewhere, however, the misuse of the oceans has been more wanton than inadvertent. Many fishermen 'fin' sharks – that is, slice their fins off, then dump them back into the water to die. In 1998, shark fins sold in the Far East for over $110 a kilo, and a bowl of shark-fin soup retailed in Tokyo for $100. The World Wildlife Fund estimated in 1994 that the number of sharks killed each year was between 40 million and 70 million.

As of 1995, some 37,000 industrial-sized fishing ships, plus about a million smaller boats, were between them taking twice as many fish from the sea as they had just twenty-five years earlier. Trawlers are sometimes now as big as cruise ships and haul behind them nets big enough to hold a dozen jumbo jets. Some even use spotter planes to locate shoals of fish from the air.

It is estimated that about a quarter of every fishing net hauled up contains 'by-catch' – fish that can't be landed because they are too small or of the wrong type or caught in the wrong season. As one observer told *The Economist*: 'We're still in the Dark Ages. We just drop a net down and see what comes up.' Perhaps as much as 22 million tonnes of such unwanted fish are dumped back in the sea each year, mostly in the form of corpses. For every kilogram of shrimp harvested, about four kilograms of fish and other marine creatures are destroyed.

Large areas of the North Sea floor are dragged clean by beam trawlers as many as seven times a year, a degree of disturbance that no ecosystem can withstand. At least two-thirds of species in the North Sea, by many estimates, are being overfished. Across the Atlantic things are no better. Halibut once abounded in such numbers off New England that individual boats could land 20,000 pounds of it in a day. Now halibut is all but extinct off the northeast coast of America.

Nothing, however, compares with the fate of cod. In the late fifteenth century, the explorer John Cabot found cod in incredible numbers on North America's eastern banks – shallow areas of water popular with bottom-feeding fish like cod. The fish existed in such numbers, an astonished Cabot reported, that sailors scooped them up in baskets. Some of the banks were vast. Georges Banks off Massachusetts is bigger than the state it abuts. The Grand Banks off Newfoundland is bigger still, and for centuries was always dense with cod. They were thought to be inexhaustible. Of course they were anything but.

By 1960, the number of spawning cod in the north Atlantic had fallen to an estimated 1.6 million tonnes. By 1990 this had sunk to 22,000 tonnes. In commercial terms, the cod were extinct. 'Fishermen', wrote Mark Kurlansky in his fascinating history, *Cod*, 'had caught them all.' The cod may have lost the western Atlantic for ever. In 1992, cod fishing was stopped altogether on the Grand Banks, but as of autumn 2002, according to a report in *Nature*, stocks had still not staged a comeback. Kurlansky notes that the fish of fish fillets or fish fingers was originally cod, but then was replaced by haddock, then by redfish and lately by Pacific pollock. These days, he notes drily, 'fish' is 'whatever is left'.

Much the same can be said of many other seafoods. In the New England fisheries off Rhode Island, it was once routine to haul in lobsters weighing 9 kilograms. Sometimes they reached over 13 kilos. Left unmolested, lobsters can live for decades – as much as 70 years, it is thought – and they never stop growing. Nowadays few lobsters weigh more than 1 kilogram on capture. 'Biologists', according to the *New York Times*, 'estimate that 90 per cent of lobsters are caught within a year after they reach the legal minimum size at about age six.' Despite declining catches, New England fishermen continue to receive state and federal tax incentives that encourage them – in some cases all but compel them – to acquire bigger boats and to harvest the seas more intensively. Today the fishermen of Massachusetts are reduced to fishing the hideous hagfish, for which there is a slight market in the Far East, but even their numbers are now falling.

We are remarkably ignorant of the dynamics that rule life in the sea. While marine life is poorer than it ought to be in areas that have been overfished, in some naturally impoverished waters there is far more life than there ought to be. The southern oceans around Antarctica produce only about 3 per cent of the world's phytoplankton – far too little, it would seem, to support a complex ecosystem, and yet they do. Crab-eater seals are not a species of animal that most of us have heard of, but they may actually be the second most numerous large species of animal on Earth, after humans. As many as 15 million of them may live on the pack ice around Antarctica. There are also perhaps two million Weddel seals, at least half a million Emperor penguins, and maybe as many as four million Adelie penguins. The food chain is thus hopelessly top-heavy, but somehow it works. Remarkably, no-one knows how.

All this is a very roundabout way of making the point that we know very little about Earth's biggest system. But then, as we shall see in the pages remaining to us, once you start talking about life, there is a great deal we don't know – not least, how it got going in the first place.

19

THE RISE OF LIFE

In 1953 Stanley Miller, a graduate student at the University of Chicago, took two flasks – one containing a little water to represent a primeval ocean, the other holding a mixture of methane, ammonia and hydrogen sulphide gases to represent the Earth's early atmosphere – connected them with rubber tubes and introduced some electrical sparks as a stand-in for lightning. After a few days, the water in the flasks had turned green and yellow in a hearty broth of amino acids, fatty acids, sugars and other organic compounds. 'If God didn't do it this way,' observed Miller's delighted supervisor, the Nobel laureate Harold Urey, 'He missed a good bet.'

Press reports of the time made it sound as if about all that was needed now was for somebody to give the flasks a good shake and life would crawl out. As time has shown, it wasn't nearly so simple. Despite half a century of further study, we are no nearer to synthesizing life today than we were in 1953 – and much further away from thinking we can. Scientists are now pretty certain that the early atmosphere was nothing like as primed for development as Miller and Urey's gaseous stew, but rather was a much less reactive blend of nitrogen and carbon dioxide. Repeating Miller's experiments with these more challenging inputs has so far produced only one fairly primitive amino acid. At all events, creating amino acids is not really the problem. The problem is proteins.

Proteins are what you get when you string amino acids together, and we need a lot of them. No-one really knows, but there may be as many as a million types of protein in the human body, and each one is a little miracle. By all the laws of probability proteins shouldn't exist. To make a protein you need to assemble amino acids (which I am obliged by long tradition to refer to here as 'the building blocks of life') in a particular order, in much the same way that you assemble letters in a particular order to spell a word. The problem is that words in the amino-acid alphabet are often exceedingly long. To spell 'collagen', the name of a common type of protein, you need to arrange eight letters in the right order. To *make* collagen, you need to arrange 1,055 amino acids in precisely the right sequence. But – and here's an obvious but crucial point – *you don't* make it. It makes itself, spontaneously, without direction, and this is where the unlikelihoods come in.

The chances of a 1,055-sequence molecule like collagen spontaneously self-assembling are, frankly, nil. It just isn't going to happen. To grasp what a long shot its existence is, visualize a standard Las Vegas slot machine but broadened greatly – to about 27 metres, to be precise – to accommodate 1,055 spinning wheels instead of the usual three or four, and with twenty symbols on each wheel (one for each common amino acid).* How long would you have to pull

*There are actually twenty-two naturally occurring amino acids known on Earth, and more may await discovery, but only twenty of them are necessary to produce us and most other living things. The twenty-second, called pyrrolysine, was discovered in 2002 by researchers at Ohio State University and is found only in a single type of Archaean (a basic form of life that we will discuss a little further on in the story) called *Methanosarcina barkeri*.

the handle before all 1,055 symbols came up in the right order? Effectively, for ever. Even if you reduced the number of spinning wheels to 200, which is actually a more typical number of amino acids for a protein, the odds against all 200 coming up in a prescribed sequence are 1 in 10^{260} (that is a 1 followed by 260 zeros). That in itself is a larger number than all the atoms in the universe.

Proteins, in short, are complex entities. Haemoglobin is only 146 amino acids long, a runt by protein standards, yet even it offers 10^{190} possible amino-acid combinations, which is why it took the Cambridge University chemist Max Perutz twenty-three years – a career, more or less – to unravel it. For random events to produce even a single protein would seem a stunning improbability – like a whirlwind spinning through a junkyard and leaving behind a fully assembled jumbo jet, in the colourful simile of the astronomer Fred Hoyle.

Yet we are talking about several hundred thousand types of protein, perhaps a million, each unique and each, as far as we know, vital to the maintenance of a sound and happy you. And it goes on from there. To be of use, a protein must not only assemble amino acids in the right sequence, it must then engage in a kind of chemical origami and fold itself into a very specific shape. Even having achieved this structural complexity, a protein is no good to you if it can't reproduce itself, and proteins can't. For this you need DNA. DNA is a whiz at replicating – it can make a copy of itself in seconds – but can do virtually nothing else. So we have a paradoxical situation. Proteins can't exist without DNA and DNA has no purpose without proteins. Are we to assume, then, that they arose simultaneously with the purpose of supporting each other? If so: wow.

And there is more still. DNA, proteins and the other components of life couldn't prosper without some sort of membrane to contain them. No atom or molecule has ever achieved life independently. Pluck any atom from your body and it is no more alive than is a grain of sand. It is only when they come together within the nurturing refuge of a cell that these diverse materials can take part in the amazing dance that we call life. Without the cell, they are nothing more than interesting chemicals. But without the chemicals, the cell has no purpose. As Davies puts it, 'If everything needs everything else, how did the community of molecules ever arise in the first place?' It is rather as if all the ingredients in your kitchen somehow got together and baked themselves into a cake – but a cake that could moreover divide when necessary to produce *more* cakes. It is little wonder that we call it the miracle of life. It is also little wonder that we have barely begun to understand it.

So what accounts for all this wondrous complexity? Well, one possibility is that perhaps it isn't quite – not *quite* – so wondrous as at first it seems. Take those amazingly improbable proteins. The wonder we see in their assembly comes in assuming that they arrived on the scene fully formed. But what if the protein chains didn't assemble all at once? What if, in the great slot machine of creation, some of the wheels could be held, as a gambler might hold a number of promising cherries? What if, in other words, proteins didn't suddenly burst into being, but *evolved*?

Imagine if you took all the components that make up a human being – carbon, hydrogen, oxygen and so on – and put them in a container with some water, gave it a vigorous stir and out stepped a completed person. That would be amazing. Well, that's essentially what Hoyle and others (including many ardent creationists) argue when they suggest that proteins spontaneously formed all at once. They didn't – they can't have. As Richard Dawkins argues in *The Blind Watchmaker*, there must have been some kind of cumulative selection process that allowed amino acids to assemble in chunks. Perhaps two or three amino acids linked up for some simple purpose and then after a time bumped into some other similar small cluster and in so doing 'discovered' some additional improvement.

Chemical reactions of the sort associated with life are actually something of a commonplace. It may be beyond us to cook them up in a lab, à la Stanley Miller and Harold Urey, but the universe does it readily enough. Lots of molecules in nature get together to form long chains called polymers. Sugars constantly assemble to form starches. Crystals can do a number of lifelike things – replicate, respond to environmental stimuli, take on a patterned complexity. They've never achieved life itself, of course, but they demonstrate repeatedly that complexity is a natural, spontaneous, entirely reliable event. There may or may not be a great deal of life in the universe at large, but there is no shortage of ordered self-assembly, in everything from the transfixing symmetry of snowflakes to the comely rings of Saturn.

So powerful is this natural impulse to assemble that many scientists now believe that life may be more inevitable than we think – that it is, in the words of the Belgian biochemist and Nobel laureate Christian de Duve, 'an obligatory manifestation of matter, bound to arise wherever conditions are appropriate'. De Duve thought it likely that such conditions would be encountered perhaps a million times in every galaxy.

Certainly there is nothing terribly exotic in the chemicals that animate us. If you wished to create another living object, whether a goldfish or a head of lettuce or a human being, you would need really only four principal elements, carbon, hydrogen, oxygen and nitrogen, plus small amounts of a few others, principally sulphur, phosphorus, calcium and iron. Put these together in three dozen or so combinations to form some sugars, acids and other basic compounds and you can build anything that lives. As Dawkins notes: 'There is nothing special about the substances from which living things are made. Living things are collections of molecules, like everything else.'

The bottom line is that life is amazing and gratifying, perhaps even miraculous, but hardly impossible – as we repeatedly attest with our own modest existences. To be sure, many of the fine details of life's beginnings remain pretty imponderable. Every scenario you have ever read concerning the conditions necessary for life involves water – from the 'warm little pond' where Darwin supposed life began to the bubbling sea vents that are now the most popular candidates for life's beginnings – but all this overlooks the fact that to turn monomers into polymers (which is to say, to begin to create proteins) involves a type of reaction known to biology as 'dehydration linkages'. As one leading biology text puts it, with perhaps just a tiny hint of discomfort, 'Researchers agree that such reactions would not have been energetically favorable in the primitive sea, or indeed in any aqueous medium, because of the mass action law.' It is a little like putting sugar in a glass of water and having it become a cube. It shouldn't happen, but somehow in nature it does. The actual chemistry of all this is a little arcane for our purposes here, but it is enough to know that if you make monomers wet they don't turn into polymers – except when creating life on the Earth. How and why it happens then and not otherwise is one of biology's great unanswered questions.

One of the biggest surprises in the earth sciences in recent decades was discovering just how early in Earth's history life arose. Well into the 1950s, it was thought that life was less than six hundred million years old. By the 1970s, a few adventurous souls felt that maybe it went back 2.5 billion years. But the present date of 3.85 billion years is stunningly early. The Earth's surface didn't become solid until about 3.9 billion years ago.

'We can only infer from this rapidity that it is not "difficult" for life of bacterial grade to evolve on planets with appropriate conditions,' Stephen Jay Gould observed in the *New York Times* in 1996. Or as he put it elsewhere, it is hard to avoid the conclusion that 'life, arising as soon as it could, was chemically destined to be.'

Life emerged so swiftly, in fact, that some authorities think it must have had help – perhaps a good deal of help. The idea that earthly life might have arrived from space has a surprisingly long and even occasionally distinguished history. The great Lord Kelvin himself raised the possibility as long ago as 1871 at a meeting of the British Association for the Advancement of Science, when he suggested that 'the germs of life might have been brought to the earth by some meteorite.' But it remained little more than a fringe notion until one Sunday in September 1969 when tens of thousands of Australians were startled by a series of sonic booms and the sight of a fireball streaking from east to west across the sky. The fireball made a strange crackling sound as it passed and left behind a smell that some likened to methylated spirits and others described as just awful.

The fireball exploded above Murchison, a town of six hundred people in the Goulburn Valley north of Melbourne, and came raining down in chunks, some weighing over 5 kilograms. Fortunately, no-one was hurt. The meteorite was of a rare type known as a carbonaceous chondrite, and the townspeople helpfully collected and brought in some 90 kilograms of it. The timing could hardly have been better. Less than two months earlier, the Apollo 11 astronauts had returned to Earth with a bag full of lunar rocks, so labs throughout the world were geared up – indeed, clamouring – for rocks of extraterrestrial origin.

The Murchison meteorite was found to be 4.5 billion years old, and it was studded with amino acids – seventy-four types in all, eight of which are involved in the formation of earthly proteins. In late 2001, more than thirty years after it crashed, a team at the Ames Research Center in California announced that the Murchison rock also contained complex strings of sugars called polyols, which had not been found off the Earth before.

A few other carbonaceous chondrites have strayed into the Earth's path since 1969 – one that landed near Tagish Lake in Canada's Yukon in January 2000 was seen over large parts of North America – and they have likewise confirmed that the universe is actually rich in organic compounds. Halley's comet, it is now thought, is about 25 per cent organic molecules. Get enough of those crashing into a suitable place – Earth, for instance – and you have the basic elements you need for life.

There are two problems with notions of panspermia, as extraterrestrial theories are known. The first is that it doesn't answer any questions about how life arose, but merely moves responsibility for it elsewhere. The other is that panspermia tends sometimes to excite even the most respectable adherents to levels of speculation that can be safely called imprudent. Francis Crick, co-discoverer of the structure of DNA, and his colleague Leslie Orgel have suggested that Earth was 'deliberately seeded with life by intelligent aliens', an idea that Gribbin calls 'at the very fringe of scientific respectability' – or, put another way, a notion that would be considered wildly lunatic were it voiced by anyone other than a Nobel laureate. Fred Hoyle and his colleague Chandra Wickramasinghe further eroded enthusiasm for panspermia by suggesting, as noted in Chapter 3, that outer space brought us not only life but also many diseases such as flu and bubonic plague, ideas that were easily disproved by biochemists.

Whatever prompted life to begin, it happened just once. That is the most extraordinary fact in biology, perhaps the most extraordinary fact we know. Everything that has ever lived, plant or animal, dates its beginnings from the same primordial twitch. At some point in an unimaginably distant past some little bag of chemicals fidgeted to life. It absorbed some nutrients, gently pulsed, had a brief existence. This much may have happened before, perhaps many times. But this ancestral packet did something additional and extraordinary: it cleaved itself and produced an heir. A tiny bundle of genetic material passed from one living entity to another, and has never stopped moving since. It was the moment of creation for us all. Biologists sometimes call it the Big Birth.

'Wherever you go in the world, whatever animal, plant, bug or blob you look at, if it is alive, it will use the same dictionary and know the same code. All life is one,' says Matt Ridley. We are all the result of a single genetic trick handed down from generation to generation over nearly four billion years, to such an extent that you can take a fragment of human genetic instruction and patch it into a faulty yeast cell and the yeast cell will put it to work as if it were its own. In a very real sense, it *is* its own.

The dawn of life – or something very like it – sits on a shelf in the office of a friendly isotope geochemist named Victoria Bennett in the Earth Sciences building of the Australian National University in Canberra. An American, Ms Bennett came to the ANU from California on a two-year contract in 1989 and has been there ever since. When I visited her, in late 2001, she handed me a modestly hefty hunk of rock composed of thin alternating stripes of white quartz and a grey-green material called clinopyroxene. The rock came from Akilia Island in Greenland, where unusually ancient rocks were found in 1997. The rocks are 3.85 billion years old and represent the oldest marine sediments ever found.

'We can't be certain that what you are holding once contained living organisms because you'd have to pulverize it to find out,' Bennett told me. 'But it comes from the same deposit where the oldest life was excavated, so it *probably* had life in it.' Nor would you find actual fossilized microbes, however carefully you searched. Any simple organisms, alas, would have been baked away by the processes that turned ocean mud to stone. Instead, what we would see if we crunched up the rock and examined it microscopically would be the chemical residues that the organisms left behind – carbon isotopes and a type of phosphate called apatite, which together provide strong evidence that the rock once contained colonies of living things. 'We can only guess what the organism might have looked like,' Bennett said. 'It was probably about as basic as life can get – but it was life nonetheless. It lived. It propagated.'

And eventually it led to us.

If you are into very old rocks, and Ms Bennett indubitably is, the ANU has long been a prime place to be. This is largely thanks to the ingenuity of a man named Bill Compston, who is now retired but in the 1970s built the world's first Sensitive High Resolution Ion Micro Probe – or SHRIMP, as it is more affectionately known from its initial letters. This is a machine that measures the decay rate of uranium in tiny minerals called zircons. Zircons appear in most rocks apart from basalts and are extremely durable, surviving every natural process but subduction. Most of the Earth's crust has been slipped back into the interior at some point, but just occasionally – in Western Australia and Greenland, for example – geologists have found outcrops of rocks that have remained always at the surface. Compston's machine allowed such rocks to be dated with unparalleled precision. The prototype SHRIMP was built and machined in the Earth Sciences Department's own workshops, and looked like something that had been built from spare parts on a budget, but it worked great. On its first formal test, in 1982, it dated the oldest thing ever found

– a 4.3-billion-year-old rock from Western Australia.

'It caused quite a stir at the time,' Bennett told me, 'to find something so important so quickly with brand-new technology.'

She took me down the hall to see the current model, SHRIMP II. It was a big, heavy piece of stainless-steel apparatus, perhaps 3.5 metres long and 1.5 metres high, and as solidly built as a deep-sea probe. At a console in front of it, keeping an eye on ever-changing strings of figures on a screen, was a man named Bob from Canterbury University in New Zealand. He had been there since 4 a.m., he told me. It was just after 9 a.m. and Bob had the machine until noon. SHRIMP II runs twenty-four hours a day; there are that many rocks to date. Ask a pair of geochemists how something like this works, and they will start talking about isotopic abundances and ionization levels with an enthusiasm that is more endearing than fathomable. The upshot of it, however, was that the machine, by bombarding a sample of rock with streams of charged atoms, is able to detect subtle differences in the amounts of lead and uranium in the zircon samples, by which means the age of rocks can be accurately adduced. Bob told me that it takes about seventeen minutes to read one zircon and it is necessary to read dozens from each rock to make the data reliable. In practice, the process seemed to involve about the same level of scattered activity, and about as much stimulation, as a trip to a launderette. Bob seemed very happy, however; but then, people from New Zealand very generally do.

The Earth Sciences compound was an odd combination of things – part office, part lab, part machine shed. 'We used to build everything here,' she said. 'We even had our own glassblower, but he's retired. But we still have two full-time rock crushers.' She caught my look of mild surprise. 'We get through a *lot* of rocks. And they have to be very carefully prepared. You have to make sure there is no contamination from previous samples – no dust or anything. It's quite a meticulous process.' She showed me the rock-crushing machines, which were indeed pristine, though the rock crushers had apparently gone for coffee. Beside the machines were large boxes containing rocks of all shapes and sizes. They do indeed get through a lot of rocks at the ANU.

Back in Bennett's office after our tour, I noticed hanging on her wall a poster giving an artist's colourfully imaginative interpretation of the Earth as it might have looked 3.5 billion years ago, just when life was getting going, in the ancient period known to earth science as the Archaean. The poster showed an alien landscape of huge, very active volcanoes, and a steamy, copper-coloured sea beneath a harsh red sky. Stromatolites, a kind of bacterial rock, filled the shallows in the foreground. It didn't look like a very promising place to create and nurture life. I asked her if the painting was accurate.

'Well, one school of thought says it was actually cool then because the sun was much weaker.' (I later learned that biologists, when they are feeling jocose, refer to this as 'the Chinese restaurant problem' – because we had a dim sun.) 'Without an atmosphere ultraviolet rays from the sun, even from a weak sun, would have tended to break apart any incipient bonds made by molecules. And yet right there' – she tapped the stromatolites – 'you have organisms almost at the surface. It's a puzzle.'

'So we don't know what the world was like back then?'

'Mmmm,' she agreed thoughtfully.

'Either way it doesn't seem very conducive to life.'

She nodded amiably. 'But there must have been something that suited life. Otherwise we wouldn't be here.'

It certainly wouldn't have suited us. If you were to step from a time machine into that ancient Archaean world, you would very swiftly scamper back inside, for there was no more oxygen to breathe on the Earth back then than there is on Mars today. It was also full of noxious vapours from hydrochloric and sulphuric acids powerful enough to eat through clothing and blister skin. Nor would it have provided the clean and glowing vistas depicted in the poster in Victoria Bennett's office. The chemical stew that was the atmosphere then would have allowed little sunlight to reach the Earth's surface. What little you could see would be illumined only briefly by bright and frequent lightning flashes. In short, it was the Earth, but an Earth we wouldn't recognize as our own.

Anniversaries were few and far between in the Archaean world. For two billion years bacterial organisms were the only forms of life. They lived, they reproduced, they swarmed, but they didn't show any particular inclination to move on to another, more challenging level of existence. At some point in the first billion years of life, cyanobacteria, or blue-green algae, learned to tap into a freely available resource – the hydrogen that exists in spectacular abundance in water. They absorbed water molecules, supped on the hydrogen and released the oxygen as waste, and in so doing invented photosynthesis. As Margulis and Sagan note, photosynthesis is 'undoubtedly the most important single metabolic innovation in the history of life on the planet' – and it was invented not by plants but by bacteria.

As cyanobacteria proliferated the world began to fill with O_2, to the consternation of those organisms that found it poisonous – which in those days was all of them. In an anaerobic (or non-oxygen-using) world, oxygen is extremely poisonous. Our white blood cells actually use oxygen to kill invading bacteria. That oxygen is fundamentally toxic often comes as a surprise to those of us who find it so convivial to our well-being, but that is only because we have evolved to exploit it. To other things it is a terror. It is what turns butter rancid and makes iron rust. Even we can tolerate it only up to a point. The oxygen level in our cells is only about a tenth the level found in the atmosphere.

The new oxygen-using organisms had two advantages. Oxygen was a more efficient way to produce energy, and it vanquished competitor organisms. Some retreated into the oozy, anaerobic world of bogs and lake bottoms. Others did likewise but then later (much later) migrated to the digestive tracts of beings like you and me. Quite a number of these primeval entities are alive inside your body right now, helping to digest your food, but abhorring even the tiniest hint of O_2. Untold number of others failed to adapt and died.

The cyanobacteria were a runaway success. At first, the extra oxygen they produced didn't accumulate in the atmosphere, but combined with iron to form ferric oxides, which sank to the bottom of primitive seas. For millions of years, the world literally rusted – a phenomenon vividly recorded in the banded iron deposits that provide so much of the world's iron ore today. For many tens of millions of years not a great deal more than this happened. If you went back to that early Proterozoic world you wouldn't find many signs of promise for the Earth's future life. Perhaps here and there in sheltered pools you'd encounter a film of living scum or a coating of glossy greens and browns on shoreline rocks, but otherwise life remained invisible.

But about 3.5 billion years ago something more emphatic became apparent. Wherever the seas were shallow, visible structures began to appear. As they went through their chemical routines, the cyanobacteria became very slightly tacky, and that tackiness trapped micro-particles of dust and sand, which became bound together to form slightly weird but solid structures – the stromatolites that featured in the shallows of the poster on Victoria Bennett's office wall. Stromatolites came in various shapes and sizes. Sometimes they looked like enormous cauliflowers, sometimes like fluffy mattresses (stromatolite comes from the Greek for mattress); sometimes they came in the form of columns, rising tens of metres above the surface of the water – on occasion as high as 100 metres. In all their manifestations, they were a kind of living rock, and they represented the world's first co-operative venture, with some varieties of primitive organism living just at the surface and others living just underneath, each taking advantage of conditions created by the other. The world had its first ecosystem.

For many years, scientists knew about stromatolites from fossil formations, but in 1961 they got a real surprise with the discovery of a community of living stromatolites at Shark Bay on the remote northwest coast of Australia. This was most unexpected – so unexpected, in fact, that it was some years before scientists realized quite what they had found. Today, however, Shark Bay is a tourist attraction – or at least as much of a tourist attraction as a place hundreds of miles from anywhere much and dozens of miles from anywhere at all can ever be. Boardwalks have been built out into the bay so that visitors can stroll over the water to get a good look at the stromatolites, quietly respiring just beneath the surface. They are lustreless and grey and look, as I recorded in an earlier book, like very large cow-pats. But it is a curiously giddying moment to find yourself staring at living remnants of the Earth as it was 3.5 billion years ago. As Richard Fortey has put it: 'This is truly time travelling, and if the world were attuned to its real wonders this sight would be as well-known as the pyramids of Giza.' Although you'd never guess it, these dull rocks swarm with life, with an estimated (well, obviously estimated) three billion individual organisms on every square yard of rock. Sometimes when you look carefully you can see tiny strings of bubbles rising to the surface as they give up their oxygen. In two billion years such tiny exertions raised the level of oxygen in the Earth's atmosphere to 20 per cent, preparing the way for the next, more complex chapter in life's history.

It has been suggested that the cyanobacteria at Shark Bay are perhaps the most slowly evolving organisms on Earth, and certainly now they are among the rarest. Having prepared the way for more complex life forms, they were then grazed out of existence nearly everywhere by the very organisms whose existence they had made possible. (They exist at Shark Bay because the waters are too saline for the creatures that would normally feast on them.)

One reason life took so long to grow complex was that the world had to wait until the simpler organisms had oxygenated the atmosphere sufficiently. 'Animals could not summon up the energy to work,' as Fortey has put it. It took about two billion years, roughly 40 per cent of Earth's history, for oxygen levels to reach more or less modern levels of concentration in the atmosphere. But once the stage was set, and apparently quite suddenly, an entirely new type of cell arose – one containing a nucleus and other little bodies collectively called *organelles* (from a Greek word meaning 'little tools'). The process is thought to have started when some blundering or adventuresome bacterium either invaded or was captured by some other bacterium and it turned out that this suited them both. The captive bacterium became, it is thought, a mitochondrion. This mitochondrial invasion (or endosymbiotic event, as biologists like to term it) made complex life possible. (In plants a similar invasion produced chloroplasts, which enable plants to photosynthesize.)

Mitochondria manipulate oxygen in a way that liberates energy from foodstuffs. Without this niftily facilitating trick, life on Earth today would be nothing more than a sludge of simple microbes. Mitochondria are very tiny – you could pack a billion into the space occupied by a grain of sand – but also very hungry. Almost every nutriment you absorb goes to feeding them.

We couldn't live for two minutes without them, yet even after a billion years mitochondria behave as if they think things might not work out between us. They maintain their own DNA, RNA and ribosomes. They reproduce at a different time from their host cells. They look like bacteria, divide like bacteria and sometimes respond to antibiotics in the way bacteria do. They don't even speak the same genetic language as the cell in which they live. In short, they keep their bags packed. It is like having a stranger in your house, but one who has been there for a billion years.

The new type of cells are known as eukaryotes (meaning 'truly nucleated'), as contrasted with the old type, which are known as prokaryotes ('pre-nucleated'), and they seem to have arrived suddenly in the fossil record. The oldest eukaryotes yet known, called *Grypania*, were discovered in iron sediments in Michigan in 1992. Such fossils have been found just once and then no more are known for 500 million years.

Earth had taken its first step towards becoming a truly interesting planet. Compared with the new eukaryotes the old prokaryotes were little more than 'bags of chemicals', to borrow from the British geologist Stephen Drury. Eukaryotes were bigger – eventually as much as ten thousand times bigger – than their simpler cousins, and could carry as much as a thousand times more DNA. Gradually, thanks to these breakthroughs, life became complex and created two types of organism – those that expel oxygen (like plants) and those that take it in (like you and me).

Single-celled eukaryotes were once called *protozoa* ('pre-animals'), but that term is increasingly disdained. Today the common term for them is *protists*. Compared with the bacteria that had gone before, these new protists were wonders of design and sophistication. The simple amoeba, just one cell big and without any ambitions but to exist, contains 400 million bits of genetic information in its DNA

– enough, as Carl Sagan noted, to fill 80 books of 500 pages.

Eventually the eukaryotes learned an even more singular trick. It took a long time – a billion years or so – but it was a good one when they mastered it. They learned to form together into complex multicellular beings. Thanks to this innovation, big, complicated, visible entities like us were possible. Planet Earth was ready to move on to its next ambitious phase.

But before we get too excited about that, it is worth remembering that the world, as we are about to see, still belongs to the very small.

20

SMALL WORLD

It's probably not a good idea to take too personal an interest in your microbes. Louis Pasteur, the great French chemist and bacteriologist, became so preoccupied with his that he took to peering critically at every dish placed before him with a magnifying glass, a habit that presumably did not win him many repeat invitations to dinner.

In fact, there is no point in trying to hide from your bacteria, for they are on and around you always, in numbers you can't conceive of. If you are in good health and averagely diligent about hygiene, you will have a herd of about one trillion bacteria grazing on your fleshy plains – about a hundred thousand of them on every square centimetre of skin. They are there to dine off the ten billion or so flakes of skin you shed every day, plus all the tasty oils and fortifying minerals that seep out from every pore and fissure. You are for them the ultimate buffet, with the convenience of warmth and constant mobility thrown in. By way of thanks, they give you B.O.

And those are just the bacteria that inhabit your skin. There are trillions more tucked away in your gut and nasal passages, clinging to your hair and eyelashes, swimming over the surface of your eyes, drilling through the enamel of your teeth. Your digestive system alone is host to more than a hundred trillion microbes, of at least four hundred types. Some deal with sugars, some with starches, some attack other bacteria. A surprising number, like the ubiquitous intestinal spirochetes, have no detectable function at all. They just seem to like to be with you. Every human body consists of about ten quadrillion cells, but is host to about a hundred quadrillion bacterial cells. They are, in short, a big part of us. From the bacteria's point of view, of course, we are a rather small part of them.

Because we humans are big and clever enough to produce and use antibiotics and disinfectants, it is easy to convince ourselves that we have banished bacteria to the fringes of existence. Don't you believe it. Bacteria may not build cities or have interesting social lives, but they will be here when the Sun explodes. This is their planet, and we are on it only because they allow us to be.

Bacteria, never forget, got along for billions of years without us. We couldn't survive a day without them. They process our wastes and make them usable again; without their diligent munching nothing would rot. They purify our water and keep our soils productive. Bacteria synthesize vitamins in our gut, convert the things we eat into useful sugars and polysaccharides, and go to war on alien microbes that slip down our gullet.

We depend totally on bacteria to pluck nitrogen from the air and convert it into useful nucleotides and amino acids for us. It is a prodigious and gratifying feat. As Margulis and Sagan note, to do the same thing industrially (as when making fertilizers) manufacturers must heat the source materials to 500 degrees Celsius and squeeze them to 300 times normal pressures. Bacteria do the same thing all the time without fuss, and thank goodness, for no larger organism could survive without the nitrogen they pass on. Above all, microbes continue to provide us with the air we breathe and to keep the atmosphere stable. Microbes, including the modern versions of cyanobacteria, supply the greater part of the planet's breathable oxygen. Algae and other tiny organisms bubbling away in the sea blow out about 150 billion kilograms of the stuff every year.

And they are amazingly prolific. The more frantic among them can yield a new generation in less than ten minutes; *Clostridium perfringens*, the disagreeable little organism that causes gangrene, can reproduce in nine minutes and then begin at once to split again. At such a rate, a single bacterium could theoretically produce more offspring in two days than there are protons in the universe. 'Given an adequate supply of nutrients, a single bacterial cell can generate 280,000 billion individuals in a single day,' according to the Belgian biochemist and Nobel laureate Christian de Duve. In the same period, a human cell can just about manage a single division.

About once every million divisions, they produce a mutant. Usually this is bad luck for the mutant – for an organism, change is always risky – but just occasionally the new bacterium is endowed with some accidental advantage, such as the ability to elude or shrug off an attack of antibiotics. With this ability to evolve rapidly goes another, even scarier advantage. Bacteria share information. Any bacterium can take pieces of genetic coding from any other. Essentially, as Margulis and Sagan put it, all bacteria swim in a single gene pool. Any adaptive change that occurs in one area of the bacterial universe can spread to any other. It's rather as if a human could go to an insect to get the necessary genetic coding to sprout wings or walk on ceilings. It means that from a genetic point of view bacteria have become a single superorganism – tiny, dispersed, but invincible.

They will live and thrive on almost anything you spill, dribble or shake loose. Just give them a little moisture – as when you run a damp cloth over a counter – and they will bloom as if created from nothing. They will eat wood, the glue in wallpaper, the metals in hardened paint. Scientists in Australia found microbes known as *Thiobacillus concretivorans* which lived in – indeed, could not live without – concentrations of sulphuric acid strong enough to dissolve metal. A species called *Micrococcus radiophilus* was found living happily in the waste tanks of nuclear reactors, gorging itself on plutonium and whatever else was there. Some bacteria break down chemical materials from which, as far as we can tell, they gain no benefit at all.

They have been found living in boiling mud pots and lakes of caustic soda, deep inside rocks, at the bottom of the sea, in hidden pools of icy water in the McMurdo Dry Valleys of Antarctica, and 11 kilometres down in the Pacific Ocean where pressures are more than a thousand times greater than at the surface, or equivalent to being squashed beneath fifty jumbo jets. Some of them seem to be practically indestructible. *Deinococcus radiodurans* is, according to *The Economist*, 'almost immune to radioactivity'. Blast its DNA with radiation and the pieces immediately re-form 'like the scuttling limbs of an undead creature from a horror movie'.

Perhaps the most extraordinary survival yet found was that of a *Streptococcus* bacterium that was recovered from the sealed lens of a camera that had stood on the Moon for two years. In short, there are few environments in which bacteria aren't prepared to live. 'They are finding now that when they push probes into ocean vents so hot that the probes actually start to melt, there are bacteria even there', Victoria Bennett told me.

In the 1920s two scientists at the University of Chicago, Edson Bastin and Frank Greer, announced that they had isolated from oil wells strains of bacteria that had been living at depths of 600 metres. The notion was dismissed as fundamentally preposterous – there was nothing to live *on* at 600 metres – and for fifty years it was assumed that their samples had been contaminated with surface microbes. We now know that there are a lot of microbes living deep within the Earth, many of which have nothing at all to do with the conventionally organic world. They eat rocks or, rather, the stuff that's in rocks – iron, sulphur, manganese and so on. And they breathe odd things too – iron, chromium, cobalt, even uranium. Such processes may be instrumental in concentrating gold, copper and other precious metals, and possibly deposits of oil and natural gas. It has even been suggested that their tireless nibblings created the Earth's crust.

Some scientists now think that there could be as much as 100 trillion tonnes of bacteria living beneath our feet in what are known as subsurface lithoautotrophic microbial ecosystems – SLiME for short. Thomas Gold of Cornell University has estimated that if you took all the bacteria out of the Earth's interior and dumped them on the surface, they would cover the planet to a depth of 15 metres – the height of a four-storey building. If the estimates are correct, there could be more life under the Earth than on top of it.

At depth, microbes shrink in size and become extremely sluggish. The liveliest of them may divide no more than once a century, some no more than perhaps once in five hundred years. As *The Economist* has put it: 'The key to long life, it seems, is not to do too much.' When things are really tough, bacteria are prepared to shut down all systems and wait for better times. In 1997 scientists successfully activated some anthrax spores that had lain dormant for eighty years in a museum display in Trondheim, Norway. Other microorganisms have leaped back to life after being released from a 118-year-old can of meat and a 166-year-old bottle of beer. In 1996, scientists at the Russian Academy of Science claimed to have revived bacteria frozen in Siberian permafrost for three million years. But the record claim for durability so far is one made by Russell Vreeland and colleagues at West Chester University in Pennsylvania in 2000, when they announced that they had resuscitated 250-million-year-old bacteria called *Bacillus permians* that had been trapped in salt deposits 600 metres underground in Carlsbad, New Mexico. If so, this microbe is older than the continents.

The report met with some understandable dubiousness. Many biochemists maintained that over such a span the microbe's components would have become uselessly degraded unless the bacterium roused itself from time to time. However, if the bacterium did stir occasionally, there was no plausible internal source of energy that could have lasted so long. The more doubtful scientists suggested that the sample might have been contaminated, if not during its retrieval then perhaps while still buried. In 2001, a team from Tel Aviv University argued that *B. permians* was almost identical to a strain of modern bacteria, *Bacillus marismortui*, found in the Dead Sea. Only two of its genetic sequences differed, and then only slightly.

'Are we to believe', the Israeli researchers wrote, 'that in 250 million years *B. permians* has accumulated the same amount of genetic differences that could be achieved in just 3–7 days in the laboratory?' In reply, Vreeland suggested that 'bacteria evolve faster in the lab than they do in the wild.'

Maybe.

It is a remarkable fact that well into the space age, most school textbooks divided the world of the living into just two categories – plant and animal. Micro-organisms hardly featured. Amoebas and similar single-celled organisms were treated as proto-animals and algae as proto-plants. Bacteria were usually lumped in with plants, too, even though everyone knew they didn't belong there. As far back as the late nineteenth century the German naturalist Ernst Haeckel had suggested that bacteria deserved to be placed in a separate kingdom, which he called Monera, but the idea didn't begin to catch on among biologists until the 1960s, and then only among some of them. (I note that my trusty *American Heritage* desk dictionary from 1969 doesn't recognize the term.)

Many organisms in the visible world were also poorly served by the traditional division. Fungi, the group that includes mushrooms, moulds, mildews, yeasts and puffballs, were nearly always treated as botanical objects, though in fact almost nothing about them – how they reproduce and respire, how they build themselves – matches anything in the plant world. Structurally, they have more in common with animals in that they build their cells from chitin, a material that gives them their distinctive texture. The same substance is used to make the shells of insects and the claws of mammals, though it isn't nearly so tasty in a stag beetle as in a Portobello mushroom. Above all, unlike all plants, fungi don't photosynthesize, so they have no chlorophyll and thus are not green. Instead they grow directly on their food source, which can be almost anything. Fungi will eat the sulphur off a concrete wall or the decaying matter between your toes – two things no plant will do. Almost the only plant-like quality they have is that they root.

Even less comfortably susceptible to categorization was the peculiar group of organisms formally called myxomycetes but more commonly known as slime moulds. The name no doubt has much to do with their obscurity. An appellation that sounded a little more dynamic – 'ambulant self-activating protoplasm', say – and less like the stuff you find when you reach deep into a clogged drain would almost certainly have earned these extraordinary entities a more immediate share of the attention they deserve, for slime moulds are, make no mistake, among the most interesting organisms in nature. When times are good, they exist as one-celled individuals, much like amoebas. But when conditions grow tough, they crawl to a central gathering place and become, almost miraculously, a slug. The slug is not a thing of beauty and it doesn't go terribly far – usually just from the bottom of a pile of leaf litter to the top, where it is in a slightly more exposed position – but for millions of years this may well have been the niftiest trick in the universe.

And it doesn't stop there. Having hauled itself up to a more favourable locale, the slime mould transforms itself yet again, taking on the form of a plant. By some curious orderly process the cells reconfigure, like the members of a tiny marching band, to make a stalk atop of which forms a bulb known as a fruiting body. Inside the fruiting body are millions of spores which, at the appropriate moment, are released to the wind to blow away to become single-celled organisms that can start the process again.

For years, slime moulds were claimed as protozoa by zoologists and as fungi by mycologists, though most people could see they didn't really belong anywhere. When genetic testing arrived, people in lab coats were surprised to find that slime moulds were so distinctive and peculiar that they weren't directly related to anything else in nature, and sometimes not even to each other.

In 1969, in an attempt to bring some order to the growing inadequacies of classification, an ecologist from Cornell named R. H. Whittaker unveiled in the journal *Science* a proposal to divide life into five principal branches – kingdoms, as they are known – called Animalia, Plantae, Fungi, Protista and Monera. Protista was a modification of an earlier term, *Protoctista*, which had been suggested a century earlier by a Scottish biologist named John Hogg, and was meant to describe any organisms that were neither plant nor animal.

Though Whittaker's new scheme was a great improvement, Protista remained ill defined. Some taxonomists reserved the term for large unicellular organisms – the eukaryotes – but others treated it as the kind of odd-sock drawer of biology, putting into it anything that didn't fit anywhere else. It included (depending on which text you consulted) slime moulds, amoebas, even seaweed, among much else. By one calculation it contained as many as two hundred thousand different species of organism all told. That's a lot of odd socks.

Ironically, just as Whittaker's five-kingdom classification was beginning to find its way into textbooks, an unassuming academic at the University of Illinois was groping his way towards a discovery that would challenge everything. His name was Carl Woese (rhymes with rose) and since the mid-1960s – or about as early as it was possible to do so – he had been quietly studying genetic sequences in bacteria. In the early days, this was an exceedingly painstaking process. Work on a single bacterium could easily consume a year. At that time, according to Woese, only about five hundred species of bacteria were known, which is fewer than the number of species you have in your mouth. Today the number is about ten times that, though that is still far short of the 26,900 species of algae, 70,000 of fungi, and 30,800 of amoebas and related organisms whose biographies fill the annals of biology.

It isn't simple indifference that keeps the total low. Bacteria can be exasperatingly difficult to isolate and study. Only about 1 per cent will grow in culture. Considering how wildly adaptable they are in nature, it is an odd fact that the one place they seem not to wish to live is a petri dish. Plop them on a bed of agar and pamper them as you will, and most will just lie there, declining every inducement to bloom. Any bacterium that thrives in a lab is by definition exceptional, and yet these were, almost exclusively, the organisms studied by microbiologists. It was, said Woese, 'like learning about animals from visiting zoos'.

Genes, however, allowed Woese to approach microorganisms from another angle. As he worked, Woese realized that there were more fundamental divisions in the microbial world than anyone suspected. A lot of little organisms that looked like bacteria and behaved like bacteria were actually something else altogether – something that had branched off from bacteria a long time ago. Woese called these organisms archaebacteria, later shortened to archaea.

It has to be said that the attributes that distinguish archaea from bacteria are not the sort that would quicken the pulse of any but a biologist. They are mostly differences in their lipids and an absence of something called peptidoglycan. But in practice they make a world of difference. Archaea are more different from bacteria than you and I are from a crab or spider. Singlehandedly, Woese had discovered an unsuspected division of life, so fundamental that it stood above the level of kingdom at the apogee of the Universal Tree of Life, as it is rather reverentially known.

In 1976 he startled the world – or at least the little bit of it that was paying attention – by redrawing the Tree of Life to incorporate not five main divisions, but twenty-three. These he grouped under three new principal categories – Bacteria, Archaea and Eukarya (sometimes spelled Eucarya) – which he called domains. The new arrangement was as follows:

- *Bacteria*: cyanobacteria, purple bacteria, gram-positive bacteria, green non-sulphur bacteria, flavobacteria and thermotogales
- *Archaea*: halophilic archaeans, methanosarcina, methanobacterium, methanoncoccus, thermoceler, thermoproteus and pyrodictium
- *Eukarya*: diplomads, microsporidia, trichomonads, flagellates, entameba, slime moulds, ciliates, plants, fungi and animals

Woese's new divisions did not take the biological world by storm. Some dismissed his system as much too heavily weighted towards the microbial. Many just ignored it. Woese, according to Frances Ashcroft, 'felt bitterly disappointed'. But slowly his new scheme began to catch on among microbiologists. Botanists and zoologists were much slower to appreciate its virtues. It's not hard to see why. In Woese's model, the worlds of botany and zoology are relegated to a few twigs on the outermost branch of the Eukaryan limb. Everything else belongs to unicellular beings.

'These folks were brought up to classify in terms of gross morphological similarities and differences,' Woese told an interviewer in 1996. 'The idea of doing so in terms of molecular sequence is a bit hard for many of them to swallow.' In short, if they couldn't see a difference with their own eyes, they didn't like it. And so they persisted with the more conventional five-kingdom division – an arrangement that Woese called 'not very useful' in his milder moments and 'positively misleading' much of the rest of the time. 'Biology, like physics before it,' Woese wrote, 'has moved to a level where the objects of interest and their interactions often cannot be perceived through direct observation.'

In 1998 the great and ancient Harvard zoologist Ernst Mayr (who then was in his ninety-fourth year and at the time of my writing is nearing one hundred and still going strong) stirred the pot further by declaring that there should be just two prime divisions of life – 'empires' he called them. In a paper published in the *Proceedings of the National Academy of Sciences*, Mayr said that Woese's findings were interesting but ultimately misguided, noting that 'Woese was not trained as a biologist and quite naturally does not have an extensive familiarity with the principles of classification,' which is perhaps as close as one distinguished scientist can come to saying of another that he doesn't know what he is talking about.

The specifics of Mayr's criticisms are highly technical – they involve issues of meiotic sexuality, Hennigian cladification and controversial interpretations of the genome of *Methanobacterium thermoautrophicum*, among rather a lot else – but essentially he argues that Woese's arrangement unbalances the Tree of Life. The bacterial realm, Mayr notes, consists of no more than a few thousand species while the archaean has a mere 175 named specimens, with perhaps a few thousand more to be found – 'but hardly more than that'. By contrast, the eukaryotic realm – that is, the complicated organisms with nucleated cells, like us – numbers already in the millions of species. For the sake of 'the principle of balance', Mayr argues for combining the simple bacterial organisms in a single category, Prokaryota, while placing the more complex and 'highly evolved' remainder in the empire Eukaryota, which would stand alongside as an equal. Put another way, he argues for keeping things much as they were before. This division between simple cells and complex cells 'is where the great break is in the living world'.

If Woese's new arrangement teaches us anything it is that life really is various and that most of that variety is small, unicellular and unfamiliar. It is a natural human impulse to think of evolution as a long chain of improvements, of a never-ending advance towards largeness and complexity – in a word, towards us. We flatter ourselves. Most of the real diversity in evolution has been small-scale. We large things are just flukes – an interesting side branch. Of the twenty-three main divisions of life, only three – plants, animals and fungi – are large enough to be seen by the human eye, and even they contain species that are microscopic. Indeed, according to Woese, if you totalled up all the biomass of the planet – every living thing, plants included – microbes would account for at least 80 per cent of all there is, perhaps more. The world belongs to the very small – and it has done for a very long time.

* * *

So why, you are bound to ask at some point in your life, do microbes so often want to hurt us? What possible satisfaction could there be to a microbe in having us grow feverish or chilled, or disfigured with sores, or above all deceased? A dead host, after all, is hardly going to provide long-term hospitality.

To begin with, it is worth remembering that most micro-organisms are neutral or even beneficial to human well-being. The most rampantly infectious organism on Earth, a bacterium called Wolbachia, doesn't hurt humans at all – or, come to that, any other vertebrates – but if you are a shrimp or worm or fruit fly, it can make you wish you had never been born. Altogether, only about one microbe in a thousand is a pathogen for humans, according to the *National Geographic* – though, knowing what some of them can do, we could be forgiven for thinking that that is quite enough. Even if most of them are benign, microbes are still the number three killer in the Western world – and even many that don't kill us make us deeply rue their existence.

Making a host unwell has certain benefits for the microbe. The symptoms of an illness often help to spread the disease. Vomiting, sneezing and diarrhoea are excellent methods of getting out of one host and into position for boarding another. The most effective strategy of all is to enlist the help of a mobile third party. Infectious organisms love mosquitoes because the mosquito's sting delivers them directly into a bloodstream where they can get straight to work before the victim's defence mechanisms can figure out what's hit them. This is why so many grade A diseases – malaria, yellow fever, dengue fever, encephalitis and a hundred or so other less celebrated but often rapacious maladies – begin with a mosquito bite. It is a fortunate fluke for us that HIV, the AIDS agent, isn't among them – at least not yet. Any HIV the mosquito sucks up on its travels is dissolved by the mosquito's own metabolism. When the day comes that the virus mutates its way around this, we may be in real trouble.

It is a mistake, however, to consider the matter too carefully from the position of logic because micro-organisms clearly are not calculating entities. They don't care what they do to you any more than you care what distress you cause when you slaughter them by the millions with a soapy shower or a swipe of deodorant. The only time your continuing well-being is of consequence to a pathogen is when it kills you too well. If they eliminate you before they can move on, then they may well die out themselves. History, Jared Diamond notes, is full of diseases that 'once caused terrifying epidemics and then disappeared as mysteriously as they had come'. He cites the robust but mercifully transient English sweating sickness, which raged from 1485 to 1552, killing tens of thousands as it went, before burning itself out. Too much efficiency is not a good thing for any infectious organism.

A great deal of sickness arises not because of what the organism has done to you but because of what your body is trying to do to the organism. In its quest to rid the body of pathogens, the immune system sometimes destroys cells or damages critical tissues, so often when you are unwell what you are feeling is not the pathogens but your own immune responses. Anyway, getting sick is a sensible response to infection. Sick people retire to their beds and thus are less of a threat to the wider community.

Because there are so many things out there with the potential to hurt you, your body holds lots of different varieties of defensive white blood cells – some ten million types in all, each designed to identify and destroy a particular sort of invader. It would be impossibly inefficient to maintain ten million separate standing armies, so each variety of white blood cell keeps only a few scouts on active duty.

When an infectious agent – what's known as an antigen – invades, relevant scouts identify the attacker and put out a call for reinforcements of the right type. While your body is manufacturing these forces, you are likely to feel wretched. The onset of recovery begins when the troops finally swing into action.

White cells are merciless and will hunt down and kill every last pathogen they can find. To avoid extinction, attackers have evolved two elemental strategies. Either they strike quickly and move on to a new host, as with common infectious illnesses like flu, or they disguise themselves so that the white cells fail to spot them, as with HIV, the virus responsible for AIDS, which can sit harmlessly and unnoticed in the nuclei of cells for years before springing into action.

One of the odder aspects of infection is that microbes that normally do no harm at all sometimes get into the wrong parts of the body and 'go kind of crazy', in the words of Dr Bryan Marsh, an infectious diseases specialist at Dartmouth–Hitchcock Medical Center in Lebanon, New Hampshire. 'It happens all the time with car accidents when people suffer internal injuries. Microbes that are normally benign in the gut get into other parts of the body – the bloodstream, for instance – and cause terrible havoc.'

The scariest, most out-of-control bacterial disorder of the moment is a disease called necrotizing fasciitis in which bacteria essentially eat the victim from the inside out, devouring internal tissue and leaving behind a pulpy, noxious residue. Patients often come in with comparatively mild complaints – a skin rash and fever, typically – but then dramatically deteriorate. When they are opened up it is often found that they are simply being consumed. The only treatment is what is known as 'radical excisional surgery' – cutting out every bit of infected area. Seventy per cent of victims die; many of the rest are left terribly disfigured. The source of the infection is a mundane family of bacteria called Group A Streptococcus, which normally do no more than cause strep throat. Very occasionally, for reasons unknown, some of these bacteria get through the lining of the throat and into the body proper, where they wreak the most devastating havoc. They are completely resistant to antibiotics. About a thousand cases a year occur in the United States and no-one can say that it won't get worse.

Precisely the same thing happens with meningitis. At least 10 per cent of young adults, and perhaps 30 per cent of teenagers, carry the deadly meningococcal bacterium, but it lives quite harmlessly in the throat. Just occasionally – in about one young person in a hundred thousand – it gets into the bloodstream and makes them very ill indeed. In the worst cases, death can come in twelve hours. That's shockingly quick. 'You can have a person who's in perfect health at breakfast and dead by evening,' says Marsh.

We would have much more success with bacteria if we weren't so profligate with our best weapon against them: antibiotics. Remarkably, by one estimate some 70 per cent of the antibiotics used in the developed world are given to farm animals, often routinely in stock feed, simply to promote growth or as a precaution against infection. Such applications give bacteria every opportunity to evolve a resistance to them. It is an opportunity that they have enthusiastically seized.

In 1952, penicillin was fully effective against all strains of staphylococcus bacteria, to such an extent that by the early 1960s the US surgeon-general, William Stewart, felt confident enough to declare: 'The time has come to close the book on infectious diseases. We have basically wiped out infection in the United States.' Even as he spoke, however, some 90 per cent of those strains were in the process of developing immunity to penicillin. Soon one of these new strains, called methicillin-resistant staphylococcus aureus, began to show up in hospitals. Only one type of antibiotic, vanomycin, remained effective against it, but in 1997 a hospital in Tokyo reported the appearance of a strain that could resist even that. Within months it had spread to six other Japanese hospitals. All over, the microbes are beginning to win the war again: in US hospitals alone, some fourteen thousand people a year die from infections they pick up there. As James Surowiecki noted in a *New Yorker* article, given a choice between developing antibiotics that people will take every day for two weeks and antidepressants that people will take every day for ever, drug companies not surprisingly opt for the latter. Although a few antibiotics have been toughened up a bit, the pharmaceutical industry hasn't given us an entirely new antibiotic since the 1970s.

Our carelessness is all the more alarming since the discovery that many other ailments may be bacterial in origin. The process of discovery began in 1983 when Barry Marshall, a doctor in Perth, Western Australia, found that many stomach cancers and most stomach ulcers are caused by a bacterium called *Helicobacter pylori*. Even though his findings were easily tested, the notion was so radical that more than a decade would pass before they were generally accepted. America's National Institutes of Health, for instance, didn't officially endorse the idea until 1994. 'Hundreds, even thousands of people must have died from ulcers who wouldn't have,' Marshall told a reporter from *Forbes* in 1999.

Since then, further research has shown that there is or may well be a bacterial component in all kinds of other disorders – heart disease, asthma, arthritis, multiple sclerosis, several types of mental disorders, many cancers, even, it has been suggested (in *Science* no less), obesity. The day may not be far off when we desperately require an effective antibiotic and haven't got one to call on.

It may come as a slight comfort to know that bacteria can

themselves get sick. They are sometimes infected by bacteriophages (or simply phages), a type of virus. A virus is a strange and unlovely entity – 'a piece of nucleic acid surrounded by bad news' in the memorable phrase of the Nobel laureate Peter Medawar. Smaller and simpler than bacteria, viruses aren't themselves alive. In isolation they are inert and harmless. But introduce them into a suitable host and they burst into busyness – into life. About five thousand types of virus are known, and between them they afflict us with many hundreds of diseases, ranging from the flu and common cold to those that are most invidious to human well-being: smallpox, rabies, yellow fever, Ebola, polio and AIDS.

Viruses prosper by hijacking the genetic material of a living cell, and using it to produce more virus. They reproduce in a fanatical manner, then burst out in search of more cells to invade. Not being living organisms themselves, they can afford to be very simple. Many, including HIV, have ten genes or fewer, whereas even the simplest bacteria require several thousand. They are also very tiny, much too small to be seen with a conventional microscope. It wasn't until 1943 and the invention of the electron microscope that science got its first look at them. But they can do immense damage. Smallpox in the twentieth century alone killed an estimated 300 million people.

They also have an unnerving capacity to burst upon the world in some new and startling form and then to vanish again as quickly as they came. In 1916, in one such case, people in Europe and America began to come down with a strange sleeping sickness, which became known as encephalitis lethargica. Victims would go to sleep and not wake up. They could be roused without great difficulty to take food or go to the lavatory, and would answer questions sensibly – they knew who and where they were – though their manner was always apathetic. However, the moment they were permitted to rest, they would at once sink back into deepest slumber and remain in that state for as long as they were left. Some went on in this manner for months before dying. A very few survived and regained consciousness but not their former liveliness. They existed in a state of profound apathy, 'like extinct volcanoes', in the words of one doctor. In ten years the disease killed some five million people and then quietly went away. It didn't get much lasting attention because in the meantime an even worse epidemic – indeed, the worst in history – swept across the world.

It is sometimes called the Great Swine Flu epidemic and sometimes the Great Spanish Flu epidemic, but in either case it was ferocious. The First World War killed 21 million people in four years; swine flu did the same in its first four months. Almost 80 per cent of American casualties in the First World War came not from enemy fire, but from flu. In some units the mortality rate was as high as 80 per cent.

Swine flu arose as a normal, non-lethal flu in the spring of 1918, but somehow, over the following months – no-one knows how or where – it mutated into something more severe. A fifth of victims suffered only mild symptoms, but the rest became gravely ill and many died. Some succumbed within hours; others held on for a few days.

In the United States, the first deaths were recorded among sailors in Boston in late August 1918, but the epidemic quickly spread to all parts of the country. Schools closed, public entertainments were shut down, people everywhere wore masks. It did little good. Between autumn 1918 and spring the following year, 548,452 people died of the flu in America. The toll in Britain was 220,000, with similar numbers in France and Germany. No-one knows the global toll, as records in the third world were often poor, but it was not less than twenty million and probably more like fifty million. Some estimates have put the global total as high as a hundred million.

In an attempt to devise a vaccine, medical authorities conducted experiments on volunteers at a military prison on Deer Island in Boston Harbor. The prisoners were promised pardons if they survived a battery of tests. These tests were rigorous to say the least. First, the subjects were injected with infected lung tissue taken from the dead and then sprayed in the eyes, nose and mouth with infectious aerosols. If they still failed to succumb, they had their throats swabbed with discharges taken straight from the sick and dying. If all else failed, they were required to sit open-mouthed while a gravely ill victim was sat up slightly and made to cough into their faces.

Out of – somewhat amazingly – three hundred men who volunteered, the doctors chose sixty-two for the tests. None contracted the flu – not one. The only person who did grow ill was the ward doctor, who swiftly died. The probable explanation for this is that the epidemic had passed through the prison a few weeks earlier and the volunteers, all of whom had survived that visitation, had a natural immunity.

Much about the 1918 flu epidemic is understood poorly or not at all. One mystery is how it erupted suddenly, all over, in places separated by oceans, mountain ranges and other earthly impediments. A virus can survive for no more than a few hours outside a host body, so how could it appear in Madrid, Bombay and Philadelphia all in the same week?

The probable answer is that it was incubated and spread by people who had only slight symptoms or none at all. Even in normal outbreaks, about 10 per cent of people in any given population have the flu but are unaware of it because they experience no ill effects. And because they remain in circulation they tend to be the great spreaders of the disease.

That would account for the 1918 outbreak's widespread distribution, but it still doesn't explain how it managed to lie low for several months before erupting so explosively at more or less the same time all over. Even more mysterious is that it was most devastating to people in the prime of life. Flu normally is hardest on infants and the elderly, but in the 1918 outbreak deaths were overwhelmingly among people in their twenties and thirties. Older people may have benefited from resistance gained from an earlier exposure to the same strain, but why the very young were similarly spared is unknown. The greatest mystery of all is why the 1918 flu was so ferociously deadly when most flus are not. We still have no idea.

From time to time certain strains of virus return. A disagreeable Russian virus known as H1N1 caused severe outbreaks over wide areas in 1933, then again in the 1950s and yet again in the 1970s. Where it went in the meantime each time is uncertain. One suggestion is that viruses hide out unnoticed in populations of wild animals before trying their hand at a new generation of humans. No-one can rule out the possibility that the great swine flu epidemic might once again rear its head.

And if it doesn't, others well might. New and frightening viruses crop up all the time. Ebola, Lassa and Marburg fevers all have tended to flare up and die down again, but no-one can say that they aren't quietly mutating away somewhere, or simply awaiting the right opportunity to burst forth in a catastrophic manner. It is now apparent that AIDS has been among us much longer than anyone originally suspected. Researchers at the Manchester Royal Infirmary discovered that a sailor who had died of mysterious, untreatable causes in 1959 in fact had AIDS. Yet, for whatever reasons, the disease remained generally quiescent for another twenty years.

The miracle is that other such diseases haven't gone rampant. Lassa fever, which wasn't first detected until 1969, in West Africa, is extremely virulent and little understood. In 1969, a doctor at a Yale University lab in New Haven,

Connecticut, who was studying Lassa fever came down with it. He survived, but, more alarmingly, a technician in a nearby lab, with no direct exposure, also contracted the disease and died.

Happily the outbreak stopped there, but we can't count on always being so fortunate. Our lifestyles invite epidemics. Air travel makes it possible to spread infectious agents across the planet with amazing ease. An Ebola virus could begin the day in, say, Benin, and finish it in New York or Hamburg or Nairobi, or all three. It means also that medical authorities increasingly need to be acquainted with pretty much every malady that exists everywhere, but of course they are not. In 1990, a Nigerian living in Chicago was exposed to Lassa fever on a visit to his homeland, but didn't develop symptoms until he had returned to the United States. He died in a Chicago hospital without diagnosis and without anyone taking any special precautions in treating him, unaware that he had one of the most lethal and infectious diseases on the planet. Miraculously, no-one else was infected. We may not be so lucky next time.

And on that sobering note, it's time to return to the world of the visibly living.

21

LIFE GOES ON

It isn't easy to become a fossil. The fate of nearly all living organisms – over 99.9 per cent of them – is to compost down to nothingness. When your spark is gone, every molecule you own will be nibbled off you or sluiced away to be put to use in some other system. That's just the way it is. Even if you make it into the small pool of organisms, the less than 0.1 per cent, that don't get devoured, the chances of being fossilized are very small.

In order to become a fossil, several things must happen. First, you must die in the right place. Only about 15 per cent of rocks can preserve fossils, so it's no good keeling over on a future site of granite. In practical terms the deceased must become buried in sediment where it can leave an impression, like a leaf in wet mud, or decompose without exposure to oxygen, permitting the molecules in its bones and hard parts (and very occasionally softer parts) to be replaced by dissolved minerals, creating a petrified copy of the original. Then, as the sediments in which the fossil lies are carelessly pressed and folded and pushed about by Earth's processes, the fossil must somehow maintain an identifiable shape. Finally, but above all, after tens of millions or perhaps hundreds of millions of years hidden away, it must be found and recognized as something worth keeping.

Only about one bone in a billion, it is thought, ever becomes fossilized. If that is so, it means that the complete fossil legacy of all the Americans alive today – that's 270 million people with 206 bones each – will only be about fifty bones, one-quarter of a complete skeleton. That's not to say, of course, that any of these bones will ever actually be found. Bearing in mind that they can be buried anywhere within an area of slightly over 9.3 million square kilometres, little of which will ever be turned over, much less examined, it would be something of a miracle if they were. Fossils are in every sense vanishingly rare. Most of what has lived on Earth has left behind no record at all. It has been estimated that less than one species in ten thousand has made it into the fossil record. That in itself is a stunningly infinitesimal proportion. However, if you accept the common estimate that the Earth has produced thirty billion species of creature in its time, and Richard Leakey and Roger Lewin's statement (in *The Sixth Extinction*) that there are 250,000 species of creature in the fossil record, that reduces the proportion to just one in 120,000. Either way, what we possess is the merest sampling of all the life that the Earth has spawned.

Moreover, the record we do have is hopelessly skewed. Most land animals, of course, don't die in sediments. They drop in the open and are eaten or left to rot or weather down to nothing. The fossil record, consequently, is almost absurdly biased in favour of marine creatures. About 95 per cent of all the fossils we possess are of animals that once lived under water, mostly in shallow seas.

I mention all this to explain why on a grey day in February I went to the Natural History Museum in London to meet a cheerful, vaguely rumpled, very likeable palaeontologist named Richard Fortey.

Fortey knows an awful lot about an awful lot. He is the

author of a wry, splendid book called *Life: An Unauthorised Biography*, which covers the whole pageant of animate creation. But his first love is a type of marine creature called trilobites, which once teemed in Ordovician seas but haven't existed for a long time except in fossilized form. All trilobites shared a basic body plan of three parts, or lobes – head, tail, thorax – from which comes the name. Fortey found his first when he was a boy clambering over rocks at St David's Bay in Wales. He was hooked for life.

He took me to a gallery of tall metal cupboards. Each cupboard was filled with shallow drawers, and each drawer was filled with stony trilobites – twenty thousand specimens in all.

'It seems like a big number,' he agreed, 'but you have to remember that millions upon millions of trilobites lived for millions upon millions of years in ancient seas, so twenty thousand isn't a huge number. And most of these are only partial specimens. Finding a complete trilobite fossil is still a big moment for a palaeontologist.'

Trilobites first appeared – fully formed, seemingly from nowhere – about 540 million years ago, near the start of the great outburst of complex life popularly known as the Cambrian explosion, and then vanished, along with a great deal else, in the great and still mysterious Permian extinction three million or so centuries later. As with all extinct creatures, there is a natural temptation to regard them as failures, but in fact they were among the most successful animals ever to live. They reigned for 300 million years – twice the span of dinosaurs, which were themselves among history's great survivors. Humans, Fortey points out, have survived so far for one-half of 1 per cent as long.

With so much time at their disposal, the trilobites proliferated prodigiously. Most remained small, about the size of modern beetles, but some grew to be as big as platters. Altogether they formed at least five thousand genera and sixty thousand species – though more turn up all the time. Fortey had recently been at a conference in South America where he was approached by an academic from a small provincial university in Argentina. 'She had a box that was full of interesting things – trilobites that had never been seen before in South America, or indeed anywhere, and a great deal else. She had no research facilities to study them and no funds to look for more. Huge parts of the world are still unexplored.'

'In terms of trilobites?'

'No, in terms of everything.'

Throughout the nineteenth century, trilobites were almost the only known forms of early complex life, and for that reason were assiduously collected and studied. The big mystery about them was their sudden appearance. Even now, as Fortey says, it can be startling to go to the right formation of rocks and to work your way upwards through the aeons, finding no visible life at all, and then suddenly 'a whole *Profallotaspis* or *Elenellus* as big as a crab will pop into your waiting hands.' These were creatures with limbs, gills, nervous systems, probing antennae, 'a brain of sorts', in Fortey's words, and the strangest eyes ever seen. Made of calcite rods, the same stuff that forms limestone, they constituted the earliest visual systems known. More than this, the earliest trilobites didn't consist of just one venturesome species but dozens, and didn't appear in one or two locations but all over. Many thinking people in the nineteenth century saw this as proof of God's handiwork and refutation of Darwin's evolutionary ideals. If evolution proceeded slowly, they asked, then how did he account for this sudden appearance of complex, fully formed creatures? The fact is, he couldn't.

And so matters seemed destined to remain for ever until

one day in 1909, three months shy of the fiftieth anniversary of the publication of Darwin's *On the Origin of Species*, when a palaeontologist named Charles Doolittle Walcott made an extraordinary find in the Canadian Rockies.

Walcott was born in 1850 and grew up near Utica, New York, in a family of modest means, which became more modest still with the sudden death of his father when Charles was an infant. As a boy Walcott discovered that he had a knack for finding fossils, particularly trilobites, and built up a collection of sufficient distinction that it was bought by Louis Agassiz for his museum at Harvard for a small fortune – about £45,000 in today's money. Although he had barely a high-school education and was self-taught in the sciences, Walcott became a leading authority on trilobites and was the first person to establish that they were arthropods, the group that includes modern insects and crustaceans.

In 1879 Walcott took a job as a field researcher with the newly formed United States Geological Survey and served with such distinction that within fifteen years he had risen to be its head. In 1907 he was appointed secretary of the Smithsonian Institution, where he remained until his death in 1927. Despite his administrative obligations, he continued to do fieldwork and to write prolifically. 'His books fill a library shelf,' according to Fortey. Not incidentally, he was also a founding director of the National Advisory Committee for Aeronautics, which eventually became the National Aeronautics and Space Agency, or NASA, and thus can rightly be considered the grandfather of the space age.

But what he is remembered for now is an astute but lucky find in British Columbia, high above the little town of Field, in the late summer of 1909. The customary version of the story is that Walcott, accompanied by his wife, was riding on a mountain trail when his wife's horse slipped on loose stones. Dismounting to assist her, Walcott discovered that the horse had turned a slab of shale that contained fossil crustaceans of an especially ancient and unusual type. Snow was falling – winter comes early to the Canadian Rockies – so they didn't linger, but the next year at the first opportunity Walcott returned to the spot. Tracing the presumed route of the rocks' slide, he climbed 750 feet to near the mountain's summit. There, 8,000 feet above sea level, he found a shale outcrop, about the length of a city block, containing an unrivalled array of fossils from soon after the moment when complex life burst forth in dazzling profusion – the famous Cambrian explosion. Walcott had found, in effect, the holy grail of palaeontology. The outcrop became known as the Burgess Shale, from the name of the ridge on which it was found, and for a long time it provided 'our sole vista upon the inception of modern life in all its fullness', as the late Stephen Jay Gould recorded in his popular book *Wonderful Life*.

Gould, ever scrupulous, discovered from reading Walcott's diaries that the story of the Burgess Shale's discovery appears to have been somewhat embroidered – Walcott makes no mention of a slipping horse or falling snow – but there is no disputing that it was an extraordinary find.

It is almost impossible for us, whose time on Earth is limited to a breezy few decades, to appreciate how remote in time from us the Cambrian outburst was. If you could fly backwards into the past at the rate of one year per second, it would take you about half an hour to reach the time of Christ, and a little over three weeks to get back to the beginnings of human life. But it would take you twenty years to reach the dawn of the Cambrian period. It was, in other words, an extremely long time ago and the world was a very different place.

For one thing, 500 million years ago and more when the Burgess Shale was formed it wasn't at the top of a mountain but at the foot of one. Specifically, it was in a shallow ocean basin at the bottom of a steep cliff. The seas of that time teemed with life, but normally the animals left no record because they were soft-bodied and decayed upon dying. At Burgess, however, the cliff collapsed and the creatures below, entombed in a mudslide, were pressed like flowers in a book, their features preserved in wondrous detail.

In annual summer trips from 1910 to 1925 (by which time he was seventy-five years old), Walcott excavated tens of thousands of specimens (Gould says eighty thousand; the normally unimpeachable fact checkers of *National Geographic* say sixty thousand), which he brought back to Washington for further study. In both sheer numbers and diversity the collection was unparalleled. Some of the Burgess fossils had shells; many others did not. Some of the creatures were sighted, others blind. The variety was enormous, consisting of 140 species, by one count. 'The Burgess Shale included a range of disparity in anatomical designs never again equaled, and not matched today by all the creatures in the world's oceans,' Gould wrote.

Unfortunately, according to Gould, Walcott failed to discern the significance of what he had found. 'Snatching defeat from the jaws of victory,' Gould wrote in another work, *Eight Little Piggies*, 'Walcott then proceeded to misinterpret these magnificent fossils in the deepest possible way.' He placed them into modern groups, making them ancestral to today's worms, jellyfish and other creatures, and thus failed to appreciate their distinctness. 'Under such an interpretation,' Gould sighed, 'life began in primordial simplicity and moved inexorably, predictably onward to more and better.'

Walcott died in 1927 and the Burgess fossils were largely forgotten. For nearly half a century they stayed shut away in drawers in the American Museum of Natural History in Washington, seldom consulted and never questioned. Then in 1973 a graduate student from Cambridge University named Simon Conway Morris paid a visit to the collection. He was astonished by what he found. The fossils were far more varied and magnificent than Walcott had indicated in his writings. In taxonomy the category that describes the basic body plans of organisms is the phylum, and here, Conway Morris concluded, were drawer after drawer of such anatomical singularities – all amazingly and unaccountably unrecognized by the man who had found them.

With his supervisor, Harry Whittington, and fellow graduate student Derek Briggs, Conway Morris spent the next several years making a systematic revision of the entire collection, and cranking out one exciting monograph after another as discovery piled upon discovery. Many of the creatures employed body plans that were not simply unlike anything seen before or since, but were *bizarrely* different. One, *Opabinia*, had five eyes and a nozzle-like snout with claws on the end. Another, a disc-shaped being called *Peytoia*, looked almost comically like a circular pineapple slice. A third had evidently tottered about on rows of stilt-like legs, and was so odd that they named it *Hallucigenia*. There was so much unrecognized novelty in the collection that at one point upon opening a new drawer Conway Morris famously was heard to mutter, 'Oh fuck, not another phylum.'

The English team's revisions showed that the Cambrian had been a time of unparalleled innovation and experimentation in body designs. For almost four billion years life had dawdled along without any detectable ambitions in the direction of complexity, and then suddenly, in the space of just five or ten million years, it had created all the basic body designs still in use today. Name a creature, from a nematode worm to Cameron Diaz, and they all use architecture first created in the Cambrian party.

What was most surprising, however, was that there were

so many body designs that had failed to make the cut, so to speak, and left no descendants. Altogether, according to Gould, at least fifteen and perhaps as many as twenty of the Burgess animals belonged to no recognized phylum. (The number soon grew in some popular accounts to as many as a hundred – far more than the Cambridge scientists ever actually claimed.) 'The history of life,' wrote Gould, 'is a story of massive removal followed by differentiation within a few surviving stocks, not the conventional tale of steadily increasing excellence, complexity, and diversity.' Evolutionary success, it appeared, was a lottery.

One creature that *did* manage to slip through, a small wormlike being called *Pikaia gracilens*, was found to have a primitive spinal column, making it the earliest known ancestor of all later vertebrates, including us. *Pikaia* were by no means abundant among the Burgess fossils, so goodness knows how close they may have come to extinction. Gould, in a famous quotation, leaves no doubt that he sees our lineal success as a fortunate fluke: 'Wind back the tape of life to the early days of the Burgess Shale; let it play again from an identical starting point, and the chance becomes vanishingly small that anything like human intelligence would grace the replay.'

Gould's *Wonderful Life* was published in 1989 to general critical acclaim and was a great commercial success. What wasn't generally known was that many scientists didn't agree with Gould's conclusions at all, and that it was all soon to get very ugly. In the context of the Cambrian, 'explosion' would soon have more to do with modern tempers than ancient physiological facts.

In fact, we now know, complex organisms existed at least a hundred million years before the Cambrian. We should have known a whole lot sooner. Nearly forty years after Walcott made his discovery in Canada, on the other side of the planet in Australia a young geologist named Reginald Sprigg found something even older and in its way just as remarkable.

In 1946 Sprigg, a young assistant government geologist for the state of South Australia, was sent to make a survey of abandoned mines in the Ediacaran Hills of the Flinders Range, an expanse of baking outback some 500 kilometres north of Adelaide. The idea was to see if there were any old mines that might be profitably reworked using newer technologies, so he wasn't studying surface rocks at all, still less fossils. But one day, while eating his lunch, Sprigg idly overturned a hunk of sandstone and was surprised – to put it mildly – to see that the rock's surface was covered in delicate fossils, rather like the impressions leaves make in mud. These rocks predated the Cambrian explosion. He was looking at the dawn of visible life.

Sprigg submitted a paper to *Nature*, but it was turned down. He read it instead at the next annual meeting of the Australian and New Zealand Association for the Advancement of Science, but it failed to find favour with the association's head, who said the Ediacaran imprints were merely 'fortuitous inorganic markings' – patterns made by wind or rain or tides, but not living beings. His hopes not yet entirely crushed, Sprigg travelled to London and presented his findings to the 1948 International Geological Congress, but failed to excite either interest or belief. Finally, for want of a better outlet, he published his findings in the *Transactions of the Royal Society of South Australia*. Then he quit his government job and took up oil exploration.

Nine years later, in 1957, a schoolboy named John Mason, while walking through Charnwood Forest in the English Midlands, found a rock with a strange fossil in it, similar to a modern sea pen and exactly like some of the specimens Sprigg had found and been trying to tell everyone about ever since. The schoolboy handed it in to a palaeontologist at the University of Leicester, who identified it at once as Precambrian. Young Mason got his picture in the papers and was treated as a precocious hero; he still is in many books. The specimen was named in his honour *Charnia masoni*.

Today some of Sprigg's original Ediacaran specimens, along with many of the other fifteen hundred that have been found throughout the Flinders Range since that time, can be seen in a glass case in an upstairs room of the stout and lovely South Australian Museum in Adelaide, but they don't attract a great deal of attention. The delicately etched patterns are rather faint and not terribly arresting to the untrained eye. They are mostly small and disc-shaped, with occasional, vague trailing ribbons. Fortey has described them as 'soft-bodied oddities'.

There is still very little agreement about what these things were or how they lived. They had, as far as can be told, no mouth or anus with which to take in and discharge digestive materials, and no internal organs with which to process them along the way. 'In life,' Fortey says, 'most of them probably simply lay upon the surface of the sandy sediment, like soft, structureless and inanimate flatfish.' At their liveliest, they were no more complex than jellyfish. All the Ediacaran creatures were diploblastic, meaning they were built from two layers of tissue. With the exception of jellyfish, all animals today are triploblastic.

Some experts think they weren't animals at all, but more like plants or fungi. The distinctions between plant and animal are not always clear even now. The modern sponge spends its life fixed to a single spot and has no eyes or brain or beating heart, and yet is an animal. 'When we go back to the Precambrian the differences between plants and animals were probably even less clear,' says Fortey. 'There isn't any rule that says you have to be demonstrably one or the other.'

Nor is it agreed that the Ediacaran organisms are in any way ancestral to anything alive today (except possibly some jellyfish). Many authorities see them as a kind of failed experiment, a stab at complexity that didn't take, possibly because the sluggish Ediacaran organisms were devoured or outcompeted by the lither and more sophisticated animals of the Cambrian period.

'There is nothing closely similar alive today,' Fortey has written. 'They are difficult to interpret as any kind of ancestors of what was to follow.'

The feeling was that ultimately they weren't terribly important to the development of life on Earth. Many authorities believe that there was a mass extermination at the Precambrian–Cambrian boundary and that all the Ediacaran creatures (except the uncertain jellyfish) failed to move on to the next phase. The real business of complex life, in other words, started with the Cambrian explosion. That's how Gould saw it, in any case.

As for the revisions of the Burgess Shale fossils, almost at once people began to question the interpretations and, in particular, Gould's interpretation of the interpretations. 'From the first there were a number of scientists who doubted the account that Steve Gould had presented, however much they admired the manner of its delivery,' Fortey wrote in *Life*. That is putting it mildly.

'If only Stephen Gould could think as clearly as he writes!' barked the Oxford academic Richard Dawkins in the opening line of a review (in the *Sunday Telegraph*) of *Wonderful Life*. Dawkins acknowledged that the book was 'unputdownable' and a 'literary tour-de-force', but accused Gould of engaging in a 'grandiloquent and near-disingenuous' misrepresentation of the facts by suggesting that the Burgess revisions had stunned the palaeontological community. 'The view that he is attacking – that evolution marches inexorably towards a pinnacle such as man – has not been believed for 50 years,' Dawkins fumed.

That was a subtlety lost on many general reviewers. One, writing in the *New York Times Book Review*, cheerfully suggested that as a result of Gould's book scientists 'have been throwing out some preconceptions that they had not examined for generations. They are, reluctantly or enthusiastically, accepting the idea that humans are as much an accident of nature as a product of orderly development.'

But the real heat directed at Gould arose from the belief that many of his conclusions were simply mistaken or carelessly inflated. Writing in the journal *Evolution*, Dawkins attacked Gould's assertions that 'evolution in the Cambrian was a different *kind* of process from today' and expressed exasperation at Gould's repeated suggestions that 'the Cambrian was a period of evolutionary "experiment", evolutionary "trial and error", evolutionary "false starts" . . . It was the fertile time when all the great "fundamental body plans" were invented. Nowadays, evolution just tinkers with old body plans. Back in the Cambrian, new phyla and new classes arose. Nowadays we only get new species!'

Noting how often this idea – that there are no new body plans – is picked up, Dawkins says: 'It is as though a gardener looked at an oak tree and remarked, wonderingly: "Isn't it strange that no major new boughs have appeared on this tree for many years? These days, all the new growth appears to be at the twig level."'

'It was a strange time,' Fortey says now, 'especially when you reflected that this was all about something that happened five hundred million years ago, but feelings really did run quite high. I joked in one of my books that I felt as if I ought to put a safety helmet on before writing about the Cambrian period, but it did actually feel a bit like that.'

Strangest of all was the response of one of the heroes of

Wonderful Life, Simon Conway Morris, who startled many in the palaeontological community by rounding abruptly on Gould in a book of his own, *The Crucible of Creation*. 'I have never encountered such spleen in a book by a professional,' Fortey wrote later. 'The casual reader of *The Crucible of Creation*, unaware of the history, would never gather that the author's views had once been close to (if not actually shared with) Gould's.'

When I asked Fortey about it, he said: 'Well, it was very strange, quite shocking really, because Gould's portrayal of him had been so flattering. I could only assume that Simon was embarrassed. You know, science changes but books are permanent, and I suppose he regretted being so irremediably associated with views that he no longer altogether held. There was all that stuff about "Oh fuck, another phylum" and I expect he regretted being famous for that. You'd never know from reading Simon's book that his views had once been nearly identical to Gould's.'

What happened was that the early Cambrian fossils began to undergo a period of critical reappraisal. Fortey and Derek Briggs – one of the other principals in Gould's book – used a method known as cladistics to compare the various Burgess fossils. In simple terms, cladistics consists of organizing organisms on the basis of shared features. Fortey gives as an example the idea of comparing a shrew and an elephant. If you considered the elephant's large size and striking trunk you might conclude that it could have little in common with a tiny, sniffling shrew. But if you compared both of them with a lizard, you would see that the elephant and shrew were in fact built to much the same plan. In essence, what Fortey is saying is that Gould saw elephants and shrews where he and Briggs saw mammals. The Burgess creatures, they believed, weren't as strange and various as they appeared at first sight. 'They were often no stranger than trilobites,' Fortey says now. 'It is just that we have had a century or so to get used to trilobites. Familiarity, you know, breeds familiarity.'

This wasn't, I should note, because of sloppiness or inattention. Interpreting the forms and relationships of ancient animals on the basis of often distorted and fragmentary evidence is clearly a tricky business. Edward O. Wilson has noted that if you took selected species of modern insects and presented them as Burgess-style fossils nobody would ever guess that they were all from the same phylum, so different are their body plans. Also instrumental in helping revisions were the discoveries of two further early Cambrian sites, one in Greenland and one in China, plus more scattered finds, which among them yielded many additional and often better specimens.

The upshot is that the Burgess fossils were found to be not so different after all. *Hallucigenia*, it turned out, had been reconstructed upside down. Its stilt-like legs were actually spikes along its back. *Peytoia*, the weird creature that looked like a pineapple slice, was found to be not a distinct creature but merely part of a larger animal called *Anomalocaris*. Many of the Burgess specimens have now been assigned to living phyla – just where Walcott put them in the first place. *Hallucigenia* and some others are thought to be related to *Onychophora*, a group of caterpillar-like animals. Others have been reclassified as precursors of the modern annelids. In fact, says Fortey, 'there are relatively few Cambrian designs that are wholly novel. More often they turn out to be just interesting elaborations of well-established designs.' As he wrote in *Life*: 'None was as strange as a present day barnacle, nor as grotesque as a queen termite.'

So the Burgess Shale specimens weren't so spectacular after all. This made them, as Fortey has written, 'no less interesting, or odd, just more explicable'. Their weird body plans were just a kind of youthful exuberance – the evolutionary equivalent, as it were, of spiked hair and tongue studs. Eventually the forms settled into a staid and stable middle age.

But that still left the enduring question of where all these animals had come from – how they had suddenly appeared from nowhere.

Alas, it turns out the Cambrian explosion may not have been quite so explosive as all that. The Cambrian animals, it is now thought, were probably there all along, but were just too small to see. Once again it was trilobites that provided the clue – in particular, that seemingly mystifying appearance of different types of trilobite in widely scattered locations around the globe, all at more or less the same time.

On the face of it, the sudden appearance of lots of fully formed but varied creatures would seem to enhance the miraculousness of the Cambrian outburst, but in fact it did the opposite. It is one thing to have one well-formed creature like a trilobite burst forth in isolation – that really is a wonder – but to have many of them, all distinct but clearly related, turning up simultaneously in the fossil record in places as far apart as China and New York, clearly suggests that we are missing a big part of their history. There could be no stronger evidence that they simply had to have a forebear – some grandfather species that started the line in a much earlier past.

And the reason we haven't found these earlier species, it is now thought, is that they were too tiny to be preserved. Says Fortey: 'It isn't necessary to be big to be a perfectly functioning, complex organism. The sea swarms with tiny arthropods today that have left no fossil record.' He cites the little copepod, which numbers in the trillions in modern seas and clusters in shoals large enough to turn vast areas of the ocean black, and yet our total knowledge of its ancestry is a single specimen found in the body of an ancient fossilized fish.

'The Cambrian explosion, if that's the word for it,

probably was more an increase in size than a sudden appearance of new body types,' Fortey says. 'And it could have happened quite swiftly, so in that sense I suppose it was an explosion.' The idea is that, just as mammals bided their time for a hundred million years until the dinosaurs cleared off and then seemingly burst forth in profusion all over the planet, so too perhaps the arthropods and other triploblasts waited in semi-microscopic anonymity for the dominant Ediacaran organisms to have their day. Says Fortey: 'We know that mammals increased in size quite dramatically after the dinosaurs went – though when I say quite abruptly I of course mean it in a geological sense. We're still talking millions of years.'

Incidentally, Reginald Sprigg did eventually get a measure of overdue credit. One of the main early genera, *Spriggina*, was named in his honour, as were several species, and the whole became known as the Ediacaran fauna after the hills through which he had searched. By this time, however, Sprigg's fossil-hunting days were long over. After leaving geology he founded a successful oil company and eventually retired to an estate in his beloved Flinders Range where he created a wildlife reserve. He died in 1994 a rich man.

22

GOODBYE TO ALL THAT

When you consider it from a human perspective, and clearly it would be difficult for us to do otherwise, life is an odd thing. It couldn't wait to get going, but then, having got going, it seemed in very little hurry to move on.

Consider the lichen. Lichens are just about the hardiest visible organisms on Earth, but among the least ambitious. They will grow happily enough in a sunny churchyard, but they particularly thrive in environments where no other organism would go – on blowy mountaintops and Arctic wastes, wherever there is little but rock and rain and cold, and almost no competition. In areas of Antarctica where virtually nothing else will grow, you can find vast expanses of lichen – 400 types of them – adhering devotedly to every wind-whipped rock.

For a long time, people couldn't understand how they did it. Because lichens grew on bare rock without evident nourishment or the production of seeds, many people – educated people – believed they were stones caught in the process of becoming plants. 'Spontaneously, inorganic stone becomes living plant!' rejoiced one observer, a Dr Hornschuch, in 1819.

Closer inspection showed that lichens were more

interesting than magical. They are in fact a partnership between fungi and algae. The fungi excrete acids which dissolve the surface of the rock, freeing minerals that the algae convert into food sufficient to sustain both. It is not a very exciting arrangement, but it is a conspicuously successful one. The world has more than twenty thousand species of lichens.

Like most things that thrive in harsh environments, lichens are slow-growing. It may take a lichen more than half a century to attain the dimensions of a shirt button. Those the size of dinner plates, writes David Attenborough, are therefore 'likely to be hundreds if not thousands of years old'. It would be hard to imagine a less fulfilling existence. 'They simply exist,' Attenborough adds, 'testifying to the moving fact that life even at its simplest level occurs, apparently, just for its own sake.'

It is easy to overlook this thought that life just is. As humans we are inclined to feel that life must have a point. We have plans and aspirations and desires. We want to take constant advantage of all the intoxicating existence we've been endowed with. But what's life to a lichen? Yet its impulse to exist, to be, is every bit as strong as ours – arguably even stronger. If I were told that I had to spend decades being a furry growth on a rock in the woods, I believe I would lose the will to go on. Lichens don't. Like virtually all living things, they will suffer any hardship, endure any insult, for a moment's additional existence. Life, in short, just wants to be. But – and here's an interesting point – for the most part it doesn't want to be much.

This is perhaps a little odd, because life has had plenty of time to develop ambitions. If you imagine the 4,500 million years of Earth's history compressed into a normal earthly day, then life begins very early, about 4 a.m., with the rise of the first simple, single-celled organisms, but then advances no further for the next sixteen hours. Not until almost eight-thirty in the evening, with the day five-sixths over, has the Earth anything to show the universe but a restless skin of microbes. Then, finally, the first sea plants appear, followed twenty minutes later by the first jellyfish and the enigmatic Ediacaran fauna first seen by Reginald Sprigg in Australia. At 9.04 p.m. trilobites swim onto the scene, followed more or less immediately by the shapely creatures of the Burgess Shale. Just before 10 p.m. plants begin to pop up on the land. Soon after, with less than two hours left in the day, the first land creatures follow.

Thanks to ten minutes or so of balmy weather, by 10.24 the Earth is covered in the great carboniferous forests whose residues give us all our coal, and the first winged insects are evident. Dinosaurs plod onto the scene just before 11 p.m. and hold sway for about three-quarters of an hour. At twenty-one minutes to midnight they vanish and the age of mammals begins. Humans emerge one minute and seventeen seconds before midnight. The whole of our recorded history, on this scale, would be no more than a few seconds, a single human lifetime barely an instant. Throughout this greatly speeded-up day, continents slide about and bang together at a clip that seems positively reckless. Mountains rise and melt away, ocean basins come and go, ice sheets advance and withdraw. And throughout the whole, about three times every minute, somewhere on the planet there is a flashbulb pop of light marking the impact of a Manson-sized meteor or larger. It's a wonder that anything at all can survive in such a pummelled and unsettled environment. In fact, not many things do for long.

Perhaps an even more effective way of grasping our extreme recentness as a part of this 4.5-billion-year-old picture is to stretch your arms to their fullest extent and imagine that width as the entire history of the Earth. On this scale, according to John McPhee in *Basin and Range*, the distance from the fingertips of one hand to the wrist of the other is Precambrian. All of complex life is in one hand, 'and in a single stroke with a medium-grained nail file you could eradicate human history.'

Fortunately, that moment hasn't happened, but the chances are good that it will. I don't wish to interject a note of gloom just at this point, but the fact is that there is one other extremely pertinent quality about life on Earth: it goes extinct. Quite regularly. For all the trouble they take to assemble and preserve themselves, species crumple and die remarkably routinely. And the more complex they get, the more quickly they appear to go extinct. Which is perhaps one reason why so much of life isn't terribly ambitious.

So any time life does something bold it is quite an event, and few occasions were more eventful than when life moved on to the next stage in our narrative and came out of the sea.

Land was a formidable environment: hot, dry, bathed in intense ultraviolet radiation, lacking the buoyancy that makes movement in water comparatively effortless. To live on land, creatures had to undergo wholesale revisions of their anatomies. Hold a fish at each end and it sags in the middle, its backbone too weak to support it. To survive out of water, marine creatures needed to come up with new load-bearing internal architecture – not the sort of adjustment that happens overnight. Above all and most obviously, any land creature would have to develop a way to take its oxygen directly from the air rather than filter it from water. These were not trivial challenges to overcome. On the other hand, there was a powerful incentive to leave the water: it was getting dangerous down there. The slow fusion of the continents into a single land mass, Pangaea, meant there was much less coastline than formerly and thus less coastal habitat. So competition was fierce. There was also an omnivorous and unsettling new type of predator on the scene, one so perfectly designed for attack that it has scarcely changed in all the long aeons since its emergence: the shark. Never would there be a more propitious time to find an alternative environment to water.

Plants began the process of land colonization about 450 million years ago, accompanied of necessity by tiny mites and other organisms which they needed to break down and recycle dead organic matter on their behalf. Larger animals took a little longer to emerge, but by about 400 million years ago they were venturing out of the water, too. Popular illustrations have encouraged us to envision the first venturesome land dwellers as a kind of ambitious fish – something like the modern mudskipper, which can hop from puddle to puddle during droughts – or even as a fully formed amphibian. In fact, the first visible mobile residents on dry land were probably much more like modern woodlice, sometimes also known as pillbugs or sow bugs. These are the little bugs (crustaceans, in fact) that are commonly thrown into confusion when you upturn a rock or log.

For those that learned to breathe oxygen from the air, times were good. Oxygen levels in the Devonian and Carboniferous periods, when terrestrial life first bloomed, were as high as 35 per cent (as opposed to nearer 20 per cent now). This allowed animals to grow remarkably large remarkably quickly.

And how, you may reasonably wonder, can scientists know what oxygen levels were like hundreds of millions of years ago? The answer lies in a slightly obscure but ingenious field known as isotope geochemistry. The long-ago seas of the Carboniferous and Devonian swarmed with tiny plankton which wrapped themselves inside tiny protective shells. Then, as now, the plankton created their shells by drawing oxygen from the atmosphere and combining it with other elements (carbon especially) to form durable compounds such as calcium carbonate. It's the same chemical trick that goes on in (and is discussed elsewhere in relation to) the long-term carbon cycle – a process that doesn't make for terribly exciting narrative but is vital for creating a habitable planet.

Eventually in this process all the tiny organisms die and drift to the bottom of the sea, where they are slowly compressed into limestone. Among the tiny atomic structures the plankton take to the grave with them are two very stable isotopes – oxygen-16 and oxygen-18. (If you have forgotten what an isotope is, it doesn't matter, though for the record it's an atom with an abnormal number of neutrons.) This is where the geochemists come in, for the isotopes accumulate at different rates depending on how much oxygen or carbon dioxide is in the atmosphere at the time of their creation. By comparing the ancient rates of deposition of the two isotopes, geochemists can read conditions in the ancient world – oxygen levels, air and ocean temperatures, the extent and timing of ice ages and much else. By combining their isotope findings with other fossil residues that indicate other conditions such as pollen levels and so on – scientists can, with considerable confidence, recreate entire landscapes that no human eye ever saw.

The principal reason oxygen levels were able to build so robustly throughout the period of early terrestrial life was that much of the world's landscape was dominated by giant tree ferns and vast swamps, which by their boggy nature disrupted the normal carbon recycling process. Instead of completely rotting down, falling fronds and other dead vegetative matter accumulated in rich, wet sediments, which were eventually squeezed into the vast coal beds that sustain much economic activity even now.

The heady levels of oxygen clearly encouraged outsized growth. The oldest indication of a surface animal yet found is a track left 350 million years ago by a millipede-like creature on a rock in Scotland. It was over a metre long.

Before the era was out some millipedes would reach lengths more than double that.

With such creatures on the prowl, it is perhaps not surprising that insects in the period evolved a trick that could keep them safely out of tongueshot: they learned to fly. Some took to this new means of locomotion with such uncanny facility that they haven't changed their techniques in all the time since. Then, as now, dragonflies could cruise at over 50 kilometres an hour, instantly stop, hover, fly backwards, and lift far more, proportionately, than any flying machine humans have come up with. 'The U.S. Air Force,' one commentator has written, 'has put them in wind tunnels to see how they do it, and despaired.' They, too, gorged on the rich air. In Carboniferous forests dragonflies grew as big as ravens. Trees and other vegetation likewise attained outsized proportions. Horsetails and tree ferns grew to heights of 15 metres, club mosses to 40 metres.

The first terrestrial vertebrates – which is to say, the first land animals from which we would derive – are something of a mystery. This is partly because of a shortage of relevant fossils, but partly also because of an idiosyncratic Swede named Erik Jarvik, whose odd interpretations and secretive manner held back progress on this question for almost half a century. Jarvik was part of a team of Scandinavian scholars who went to Greenland in the 1930s and 1940s looking for fossil fish. In particular they sought lobe-finned fish of the type that presumably were ancestral to us and all other walking creatures, known as tetrapods.

Most animals are tetrapods, and all living tetrapods have one thing in common: four limbs, each of which ends in a maximum of five fingers or toes. Dinosaurs, whales, birds, humans, even fish – all are tetrapods, which clearly suggests they come from a single common ancestor. The clue to this ancestor, it was assumed, would be found in the Devonian era, from about 400 million years ago. Before that time nothing walked on land. After that time lots of things did. Luckily the team found just such a creature, a metre-long animal called an *Ichthyostega.* The analysis of the fossil fell to Jarvik, who began his study in 1948 and kept at it for the next forty-eight years. Unfortunately, Jarvik refused to let anyone else study his tetrapod. The world's palaeontologists had to be content with two sketchy interim papers in which Jarvik noted that the creature had five fingers on each of four limbs, confirming its ancestral importance.

Jarvik died in 1998. After his death, other palaeontologists eagerly examined the specimen and found that Jarvik had severely miscounted the fingers and toes – there were actually eight on each limb – and failed to observe that the fish could not possibly have walked. The structure of the fin was such that it would have collapsed under its own weight. Needless to say, this did not do a great deal to advance our understanding of the first land animals. Today three early tetrapods are known and none has five digits. In short, we don't know quite where we came from.

But come we did, though reaching our present state of eminence has not, of course, always been straightforward. Since life on land began, it has consisted of four mega-dynasties, as they are sometimes called. The first consisted of primitive, plodding but sometimes fairly hefty amphibians and reptiles. The best-known animal of this age was the Dimetrodon, a sail-backed creature that is commonly confused with dinosaurs (including, I note, in a picture caption in the Carl Sagan book *Comet*). The Dimetrodon was in fact a synapsid. So, once upon a time, were we. Synapsids were one of the four main divisions of early reptilian life, the others being anapsids, euryapsids and diapsids. The names simply refer to the number and location of small holes found in the sides of their owners' skulls. Synapsids had one hole in their lower temples; diapsids had two; euryapsids had a single hole higher up.

Over time, each of these principal groupings split into further subdivisions, of which some prospered and some faltered. Anapsids gave rise to the turtles, which for a time, perhaps a touch improbably, appeared poised to predominate as the planet's most advanced and deadly species, before an evolutionary lurch let them settle for durability rather than dominance. The synapsids divided into four streams, only one of which survived beyond the Permian. Happily, that was the stream we belonged to, and it evolved into a family of protomammals known as therapsids. These formed Megadynasty 2.

Unfortunately for the therapsids, their cousins the diapsids were also productively evolving, in their case into dinosaurs (among other things), which gradually proved too much for the therapsids. Unable to compete head-to-head with these aggressive new creatures, the therapsids by and large vanished from the record. A very few, however, evolved into small, furry, burrowing beings that bided their time for a very long while as little mammals. The biggest of them grew no larger than a housecat and most were no bigger than mice. Eventually, this would prove their salvation, but they would have to wait nearly 150 million years for Megadynasty 3, the Age of Dinosaurs, to come to an abrupt end and make way for Megadynasty 4 and our own Age of Mammals.

Each of these massive transformations, as well as many smaller ones between and since, was dependent on that paradoxically important motor of progress: extinction. It is a curious fact that on Earth species death is, in the most literal sense, a way of life. No-one knows how many species of organisms have existed since life began. Thirty billion is a commonly cited figure, but the number has been put as high as 4,000 billion. Whatever the actual total, 99.99 per cent of all species that have ever lived are no longer with us. 'To a first approximation,' as David Raup of the University of Chicago likes to say, 'all species are extinct.' For complex organisms, the average lifespan of a species is only about four million years – roughly about where we are now.

Extinction is always bad news for the victims, of course, but it appears to be a good thing for a dynamic planet. 'The alternative to extinction is stagnation,' says Ian Tattersall of the American Museum of Natural History, 'and stagnation is seldom a good thing in any realm.' (I should perhaps note that we are speaking here of extinction as a natural, long-term process. Extinction brought about by human carelessness is another matter altogether.)

Crises in the Earth's history are invariably associated with dramatic leaps afterwards. The fall of the Ediacaran fauna was followed by the creative outburst of the Cambrian period. The Ordovician extinction of 440 million years ago cleared the oceans of a lot of immobile filter feeders and, somehow, created conditions that favoured darting fish and giant aquatic reptiles. These in turn were in an ideal position to send colonists onto dry land when another blowout in the late Devonian period gave life another sound shaking. And so it has gone at scattered intervals through history. If most of these events hadn't happened just as they did, just when they did, we almost certainly wouldn't be here now.

The Earth has seen five major extinction episodes in its time – the Ordovician, Devonian, Permian, Triassic and Cretaceous, in that order – and many smaller ones. The Ordovician (440 million years ago) and Devonian (365 million) each wiped out about 80 to 85 per cent of species. The Triassic (210 million years ago) and Cretaceous (65 million years) each wiped out 70–75 per cent of species. But the real whopper was the Permian extinction of about 245 million years ago, which raised the curtain on the long age of the dinosaurs. In the Permian, at least 95 per cent of animals known from the fossil record checked out, never to return. Even about a third of insect species went – the only occasion on which they were lost en masse. It is as close as we have ever come to total obliteration.

'It was, truly, a mass extinction, a carnage of a magnitude that had never troubled the Earth before,' says Richard Fortey. The Permian event was particularly devastating to sea creatures. Trilobites vanished altogether. Clams and sea urchins nearly went. Virtually all other marine organisms were staggered. Altogether, on land and in the water, it is thought that the Earth lost 52 per cent of its families – that's the level above genus and below order on the grand scale of life (the subject of the next chapter) – and perhaps as many as 96 per cent of all its species. It would be a long time – as much as 80 million years by one reckoning – before species totals recovered.

Two points need to be kept in mind. First, these are all just informed guesses. Estimates for the number of animal species alive at the end of the Permian range from as low as 45,000 to as high as 240,000. If you don't know how many species were alive, you can hardly specify with conviction the proportion that perished. Moreover, we are talking about the death of species, not individuals. For individuals the death toll could be much higher – in many cases, practically total. The species that survived to the next phase of life's lottery almost certainly owe their existence to a few scarred and limping survivors.

In between the big kill-offs, there have also been many smaller, less well-known extinction episodes – the Hemphillian, Frasnian, Famennian, Rancholabrean and a dozen or so others – which were not so devastating to total species numbers, but often critically hit certain populations. Grazing animals, including horses, were nearly wiped out in the Hemphillian event about five million years ago. Horses declined to a single species, which appears so sporadically in the fossil record as to suggest that for a time it teetered on the brink of oblivion. Imagine a human history without horses, without grazing animals.

In nearly every case, for both big extinctions and more modest ones, we have bewilderingly little idea of what the cause was. Even after stripping out the more crackpot notions, there are still more theories for what caused the extinction events than there have been events. At least two dozen potential culprits have been identified as causes or prime contributors, including global warming, global cooling, changing sea levels, oxygen depletion of the seas (a condition known as anoxia), epidemics, giant leaks of methane gas from the sea floor, meteor and comet impacts, runaway hurricanes of a type known as hypercanes, huge volcanic upwellings and catastrophic solar flares.

This last is a particularly intriguing possibility. Nobody knows how big solar flares can get because we have only been watching them since the beginning of the space age, but the Sun is a mighty engine and its storms are commensurately enormous. A typical solar flare – something we wouldn't even notice on Earth – will release the energy equivalent of a billion hydrogen bombs and fling into space 100 billion tonnes or so of murderous high-energy particles. The magnetosphere and atmosphere between them normally swat these back into space or steer them safely towards the poles (where they produce the Earth's comely auroras), but it is thought that an unusually big blast, say a hundred times the typical flare, could overwhelm our ethereal defences. The light show would be a glorious one, but it would almost certainly kill a very high proportion of all that basked in its glow. Moreover, and rather chillingly, according to Bruce Tsurutani of the NASA Jet Propulsion Laboratory, 'it would leave no trace in history.'

What all this leaves us with, as one researcher has put it, is 'tons of conjecture and very little evidence'. Cooling seems to be associated with at least three of the big extinction events – the Ordovician, Devonian and Permian – but beyond that little is generally accepted, including whether a particular episode happened swiftly or slowly. Scientists can't agree, for instance, whether the late Devonian extinction – the event that was followed by vertebrates moving on to the land – happened over millions of years or thousands of years or in one lively day.

One of the reasons it is so hard to produce convincing explanations for extinctions is that it is so very hard to exterminate life on a grand scale. As we have seen from the Manson impact, you can receive a ferocious blow and still stage a full, if presumably somewhat wobbly, recovery. So why, out of all the thousands of impacts Earth has endured, was the KT event of 65 million years ago, which put paid to the dinosaurs, so singularly devastating? Well, first, it was positively enormous. It struck with the force of 100 million megatonnes. Such an outburst is not easily imagined, but, as James Lawrence Powell has pointed out, if you exploded one Hiroshima-sized bomb for every person alive on Earth today you would still be about a billion bombs short of the size of the KT impact. Yet even that alone may not have been enough to wipe out 70 per cent of Earth's life, dinosaurs included.

The KT meteor had the additional advantage – advantage if you are a mammal, that is – that it landed in a shallow sea just 10 metres deep, probably at just the right angle, at a time when oxygen levels were 10 per cent higher than at present and so the world was more combustible. Above all, the floor of the sea where it landed was made of rock rich in sulphur. The result was an impact that turned an area of sea floor the size of Belgium into aerosols of sulphuric acid. For months afterwards, the Earth was subjected to rains acid enough to burn skin.

In a sense, an even greater question than 'What wiped out 70 per cent of the species that were existing at the time?' is

'How did the remaining 30 per cent survive?' Why was the event so irremediably devastating to every single dinosaur that existed, while other reptiles, like snakes and crocodiles, passed through unimpeded? So far as we can tell, no species of toad, newt, salamander or other amphibian went extinct in North America. 'Why should these delicate creatures have emerged unscathed from such an unparalleled disaster?' asks Tim Flannery in his fascinating prehistory of America, *Eternal Frontier.*

In the seas it was much the same story. All the ammonites vanished, but their cousins the nautiloids, who lived similar lifestyles, swam on. Among plankton, some species were practically wiped out – 92 per cent of foraminiferans, for instance – while other organisms like diatoms, designed to a similar plan and living alongside them, were comparatively unscathed.

These are difficult inconsistencies. As Richard Fortey observes: 'Somehow it does not seem satisfying just to call them "lucky ones" and leave it at that.' If, as seems entirely likely, the event was followed by months of dark and choking smoke, then many of the insect survivors become difficult to account for. 'Some insects, like beetles,' Fortey notes, 'could live on wood or other things lying around. But what about those like bees that navigate by sunlight and need pollen? Explaining their survival isn't so easy.'

Above all, there are the corals. Corals require algae to survive and algae require sunlight, and both together require steady minimum temperatures. Much publicity has been given in the last few years to corals dying from changes in sea temperature of only a degree or so. If they are that vulnerable to small changes, how did they survive the long impact winter?

There are also many regional variations that are hard to explain. Extinctions seem to have been far less severe in the southern hemisphere than the northern. New Zealand in particular appears to have come through largely unscathed, and yet it had almost no burrowing creatures. Even its vegetation was overwhelmingly spared, and yet the scale of conflagration elsewhere suggests that devastation was global. In short, there is just a great deal we don't know.

Some animals absolutely prospered – including, a little surprisingly, the turtles once again. As Flannery notes, the period immediately after the dinosaur extinction could well be known as the Age of Turtles. Sixteen species survived in North America and three more came into existence soon after.

Clearly it helped to be at home in water. The KT impact wiped out almost 90 per cent of land-based species but only 10 per cent of those living in fresh water. Water obviously offered protection against heat and flame, but also presumably provided more sustenance in the lean period that followed. All the land-based animals that survived had a habit of retreating to a safer environment during times of danger – into water or underground – either of which would have provided considerable shelter against the ravages without. Animals that scavenged for a living would also have enjoyed an advantage. Lizards were, and are, largely impervious to the bacteria in rotting carcasses. Indeed, often they are positively drawn to them, and for a long while there were clearly a lot of putrid carcasses about.

It is often wrongly stated that only small animals survived the KT event. In fact, among the survivors were crocodiles, which were not just large but three times larger than they are today. But on the whole, it is true, most of the survivors were small and furtive. Indeed, with the world dark and hostile, it was a perfect time to be small, warm-blooded, nocturnal, flexible in diet and cautious by nature – the very qualities that distinguished our mammalian forebears. Had our evolution been more advanced, we would probably have been wiped out. Instead, mammals found themselves in a world to which they were as well suited as anything alive.

However, it wasn't as if mammals swarmed forward to fill every niche. 'Evolution may abhor a vacuum,' wrote the palaeobiologist Steven M. Stanley, 'but it often takes a long time to fill it.' For perhaps as many as ten million years mammals remained cautiously small. In the early Tertiary, if you were the size of a bobcat you could be king.

But once they got going, mammals expanded prodigiously – sometimes to an almost preposterous degree. For a time, there were guinea pigs the size of rhinos and rhinos the size of a two-storey house. Wherever there was a vacancy in the predatory chain, mammals rose (often literally) to fill it. Early members of the raccoon family migrated to South America, discovered a vacancy, and evolved into creatures the size and ferocity of bears. Birds, too, prospered disproportionately. For millions of years, a gigantic, flightless, carnivorous bird called *Titanis* was possibly the most ferocious creature in North America. Certainly it was the most daunting bird that ever lived. It stood 3 metres high, weighed over 350 kilograms and had a beak that could tear the head off pretty much anything that irked it. Its family survived in formidable fashion for fifty million years, yet until a skeleton was discovered in Florida in 1963, we had no idea that it had ever existed.

Which brings us to another reason for our uncertainty about extinctions: the paltriness of the fossil record. We have touched already on the unlikelihood of any set of bones becoming fossilized, but the record is actually worse than you might think. Consider dinosaurs. Museums give the impression that we have a global abundance of dinosaur fossils. In fact, overwhelmingly museum displays are artificial. The giant diplodocus that dominates the entrance hall of the Natural History Museum in London and has delighted and informed generations of visitors is made entirely of plaster – built in 1903 in Pittsburgh and presented to the museum by Andrew Carnegie. The entrance hall of the American Museum of Natural History in New York is dominated by an even grander tableau: a skeleton of a large barosaurus defending her baby from attack by a darting and toothy allosaurus. It is a wonderfully impressive display – the barosaurus rises perhaps 9 metres towards the high ceiling – but also entirely fake. Every one of the several hundred bones in the display is a cast. Visit almost any large natural history museum in the world – in Paris, Vienna, Frankfurt, Buenos Aires, Mexico City – and what will greet you are antique models, not ancient bones.

The fact is, we don't really know a great deal about the dinosaurs. For the whole of the age of dinosaurs, fewer than 1,000 species have been identified (almost half of them known from a single specimen), which is about a quarter of the number of mammal species alive now. Dinosaurs, bear in mind, ruled the Earth for roughly three times as long as mammals have, so either dinosaurs were remarkably unproductive of species or we have barely scratched the surface (to use an irresistibly apt cliché).

For millions of years through the age of dinosaurs not a single fossil has yet been found. Even for the period of the late Cretaceous – the most studied prehistoric period there is, thanks to our long interest in dinosaurs and their extinction – some three-quarters of all species that lived may yet be undiscovered. Animals bulkier than the diplodocus or more forbidding than tyrannosaurus may have roamed the Earth in their thousands and we may never know it. Until very recently, everything known about the dinosaurs of this period came from only about three hundred specimens representing just sixteen species. The scantiness of the record led to the widespread belief that dinosaurs were already on their way out when the KT impact occurred.

In the late 1980s a palaeontologist from the Milwaukee Public Museum, Peter Sheehan, decided to conduct an experiment. Using 200 volunteers, he made a painstaking census of a well-defined, but also well-picked-over, area of the famous Hell Creek Formation in Montana. Sifting meticulously, the volunteers collected every last tooth and vertebra and chip of bone – everything that had been overlooked by previous diggers. The work took three years. When they had finished, they found that they had more than tripled – for the planet – the number of dinosaur fossils from the late Cretaceous. The survey established that dinosaurs remained numerous right up to the time of the KT impact. 'There is no reason to believe that the dinosaurs were dying out gradually during the last three million years of the Cretaceous,' Sheehan reported.

We are so used to the notion of our own inevitability as life's dominant species that it is hard to grasp that we are here only because of timely extraterrestrial bangs and other random flukes. The one thing we have in common with all other living things is that for nearly four billion years our ancestors have managed to slip through a series of closing doors every time we needed them to. Stephen Jay Gould expressed it succinctly in a well-known line: 'Humans are here today because our particular line never fractured – never once at any of the billion points that could have erased us from history.'

We started this chapter with three points: life wants to be; life doesn't always want to be much; life from time to time goes extinct. To this we may add a fourth: life goes on. And often, as we shall see, it goes on in ways that are decidedly amazing.

23

THE RICHNESS OF BEING

Here and there in the Natural History Museum in London, built into recesses along the underlit corridors or standing between glass cases of minerals and ostrich eggs and a century or so of other productive clutter, are secret doors – at least secret in the sense that there is nothing about them to attract the visitor's notice. Occasionally you might see someone with the distracted manner and interestingly wilful hair that mark the scholar emerge from one of the doors and hasten down a corridor, probably to disappear through another door a little further on, but this is a relatively rare event. For the most part the doors stay shut, giving no hint that beyond them exists another – a parallel – Natural History Museum as vast as, and in many ways more wonderful than, the one the public knows and adores.

The Natural History Museum contains some seventy million objects from every realm of life and every corner of the planet, with another hundred thousand or so added to the collection each year, but it is really only behind the scenes that you get a sense of what a treasure house this is. In cupboards and cabinets and long rooms full of close-packed shelves are kept tens of thousands of pickled animals in bottles, millions of insects pinned to squares of card, drawers of shiny molluscs, bones of dinosaurs, skulls of early humans, endless folders of neatly pressed plants. It is a little like wandering through Darwin's brain. The spirit room alone holds 15 miles of shelving containing jar upon jar of animals preserved in methylated spirit.

Back here are specimens collected by Joseph Banks in Australia, Alexander von Humboldt in Amazonia and Darwin on the *Beagle* voyage – and much else that is either very rare or historically important or both. Many people would love to get their hands on these things. A few actually have. In 1954 the museum acquired an outstanding ornithological collection from the estate of a devoted collector named Richard Meinertzhagen, author of *Birds of Arabia*, among other scholarly works. Meinertzhagen had been a faithful attendee of the museum for years, coming almost daily to take notes for the production of his books and monographs. When the crates arrived, the curators excitedly levered them open to see what they had been left and were surprised, to put it mildly, to discover that a very large number of specimens bore the museum's own labels. Mr Meinertzhagen, it turned out, had been helping himself to their collections for years. It explained his habit of wearing a large overcoat even during warm weather.

A few years later a charming old regular in the molluscs department – 'quite a distinguished gentleman', I was told – was caught inserting valued sea shells into the hollow legs of his Zimmer frame.

'I don't suppose there's anything in here that somebody somewhere doesn't covet,' Richard Fortey said with a thoughtful air as he gave me a tour of the beguiling world that is the behind-the-scenes part of the museum. We wandered through a confusion of departments where people sat at large tables doing intent, investigative things with arthropods and palm fronds and boxes of yellowed bones. Everywhere there was an air of unhurried thoroughness, of people being engaged in a gigantic endeavour that could never be completed and mustn't be rushed. In 1967, I had read, the museum issued its report on the John Murray Expedition, an Indian Ocean survey, forty-four years after the expedition had concluded. This is a world where things move at their own pace, including a tiny lift Fortey and I shared with a scholarly-looking elderly man with whom Fortey chatted genially and familiarly as we proceeded upwards at about the rate that sediments are laid down.

When the man departed, Fortey said to me: 'That was a very nice chap named Norman who's spent forty-two years studying one species of plant, St John's wort. He retired in 1989, but he still comes in every week.'

'How do you spend forty-two years on one species of plant?' I asked.

'It's remarkable, isn't it?' Fortey agreed. He thought for a moment. 'He's very thorough, apparently.' The lift door opened to reveal a bricked-over opening. Fortey looked confounded. 'That's very strange,' he said. 'That used to be Botany back there.' He punched a button for another floor and we found our way at length to Botany by means of back staircases and discreet trespass through yet more departments where investigators toiled lovingly over once-living objects. And so it was that I was introduced to Len Ellis and the quiet world of bryophytes – mosses to the rest of us.

When Emerson poetically noted that mosses favour the north sides of trees ('The moss upon the forest bark, was pole-star when the night was dark') he really meant lichens, for in the nineteenth century mosses and lichens weren't distinguished. True mosses aren't actually fussy about where they grow, so they are no good as natural compasses. In fact, mosses aren't actually much good for anything. 'Perhaps no great group of plants has so few uses, commercial or economic, as the mosses,' wrote Henry S. Conard, perhaps just a touch sadly, in *How to Know the Mosses and Liverworts*, published in 1956 and still to be found on many library shelves as almost the only attempt to popularize the subject.

They are, however, prolific. Even with lichens removed, bryophytes is a busy realm, with over ten thousand species contained within some seven hundred genera. The plump and stately *Moss Flora of Britain and Ireland* by A. J. E. Smith runs to seven hundred pages, and Britain and Ireland are by no means outstandingly mossy places. 'The tropics are where you find the variety,' Len Ellis told me. A quiet, spare man, he has been at the Natural History Museum for twenty-seven years and curator of the department since 1990. 'You can go out into a place like the rainforests of Malaysia and find new varieties with relative ease. I did that myself not long ago. I looked down and there was a species that had never been recorded.'

'So we don't know how many species are still to be discovered?'

'Oh, no. No idea.'

You might not think there would be that many people in the world prepared to devote lifetimes to the study of something so inescapably low-key, but in fact moss people number in the hundreds and they feel very strongly about their subject. 'Oh, yes,' Ellis told me, 'the meetings can get very lively at times.'

I asked him for an example of controversy.

'Well, here's one inflicted on us by one of your countrymen,' he said, smiling lightly, and opened a hefty reference work containing illustrations of mosses whose most notable characteristic to the uninstructed eye was their uncanny similarity one to another. 'That,' he said, tapping a moss, 'used to be one genus, *Drepanocladus*. Now it's been reorganized into three: *Drepanocladus*, *Warnstorfia* and *Hamatacoulis*.'

'And did that lead to blows?' I asked, perhaps a touch hopefully.

'Well, it made sense. It made perfect sense. But it meant a lot of reordering of collections and it put all the books out of date for a time, so there was a bit of, you know, grumbling.'

Mosses offer mysteries as well, he told me. One famous case – famous to moss people, anyway – involved a retiring type called *Hyophila stanfordensis*, which was discovered on the campus of Stanford University in California and later also found growing beside a path in Cornwall, but has never been encountered anywhere in between. How it came to exist in two such unconnected locations is anybody's guess. 'It's now known as *Hennediella stanfordensis*,' Ellis said. 'Another revision.'

We nodded thoughtfully.

When a new moss is found it must be compared with all other mosses to make sure that it hasn't been recorded already. Then a formal description must be written and illustrations prepared, and the result published in a respectable journal. The whole process seldom takes less than six months. The twentieth century was not a great age for moss taxonomy. Much of the century's work was devoted to untangling the confusions and duplications left behind by the nineteenth century.

That was the golden age of moss collecting. (You may recall that Charles Lyell's father was a great moss man.) One aptly named Englishman, George Hunt, hunted British mosses so assiduously that he probably contributed to the extinction of several species. But it is thanks to such efforts that Len Ellis's collection is one of the world's most comprehensive. All 780,000 of his specimens are pressed into large folded sheets of heavy paper, some very old and covered with spidery Victorian script. Some, for all we knew, might have been in the hand of Robert Brown, the great Victorian botanist, unveiler of Brownian motion and the nucleus of cells, who founded and ran the museum's botany department for its first thirty-one years until his death in 1858. All the specimens are kept in lustrous old mahogany cabinets so strikingly fine that I remarked upon them.

'Oh, those were Sir Joseph Banks's, from his house in Soho Square,' Ellis said casually, as if identifying a recent purchase from Ikea. 'He had them built to hold his specimens from the *Endeavour* voyage.' He regarded the cabinets thoughtfully, as if for the first time in a long while. 'I don't know how *we* ended up with them in bryology,' he added.

This was an amazing disclosure. Joseph Banks was England's greatest botanist and the *Endeavour* voyage – that is, the one on which Captain Cook charted the 1769 transit of Venus and claimed Australia for the crown, among rather a lot else – was the greatest botanical expedition in history. Banks paid £10,000, about £600,000 in today's money, to take himself and a party of nine others – a naturalist, a secretary, three artists and four servants – on the three-year adventure around the world. Goodness knows what the bluff Captain Cook made of such a velvety and pampered assemblage, but he seems to have liked Banks well enough and could not but admire his talents in botany – a feeling shared by posterity.

Never before or since has a botanical party enjoyed greater triumphs. Partly it was because the voyage took in so many new or little-known places – Tierra del Fuego, Tahiti, New Zealand, Australia, New Guinea – but mostly it was because Banks was such an astute and inventive collector. Even when unable to go ashore at Rio de Janeiro because of a quarantine, he sifted through a bale of fodder sent for the ship's livestock and made new discoveries. Nothing, it seems, escaped his notice. Altogether he brought back thirty thousand plant specimens, including fourteen hundred not seen before – enough to increase by about a quarter the number of known plants in the world.

But Banks's grand cache was only part of the total haul in what was an almost absurdly acquisitive age. Plant collecting in the eighteenth century became a kind of international mania. Glory and wealth alike awaited those who could find new species, and botanists and adventurers went to the most incredible lengths to satisfy the world's craving for horticultural novelty. Thomas Nuttall, the man who named the wisteria after Caspar Wistar, came to America as an uneducated printer but discovered a passion for plants and walked halfway across the country and back again, collecting hundreds of growing things never seen before. John Fraser, for whom is named the Fraser fir, spent years in the wilderness collecting on behalf of Catherine the Great and emerged at length to find that Russia had a new tsar who thought he was mad and refused to honour his contract. Fraser took everything to Chelsea, where he opened a nursery and made a handsome living selling rhododendrons, azaleas, magnolias, Virginia creepers, asters and other colonial exotica to a delighted English gentry.

Huge sums could be made with the right finds. John Lyon, an amateur botanist, spent two hard and dangerous years collecting specimens, but cleared almost £125,000 in today's money for his efforts. Many, however, just did it for the love of botany. Nuttall gave most of what he found to the Liverpool Botanic Gardens. Eventually he became director of Harvard's Botanic Garden and author of the encyclopedic *Genera of North American Plants* (which he not only wrote but also largely typeset).

And that was just plants. There was also all the fauna of the new worlds – kangaroos, kiwis, raccoons, bobcats, mosquitoes and other curious forms beyond imagining. The volume of life on Earth was seemingly infinite, as Jonathan Swift noted in some famous lines:

So, naturalists observe, a flea
Hath smaller fleas that on him prey;
And these have smaller still to bite 'em;
And so proceed ad infinitum.

All this new information needed to be filed, ordered and compared with what was known. The world was desperate for a workable system of classification. Fortunately there was a man in Sweden who stood ready to provide it.

His name was Carl Linné (later changed, with permission, to the more aristocratic *von* Linné), but he is remembered now by the Latinized form Carolus Linnaeus. He was born in 1707 in the village of Råshult in southern Sweden, the son of a poor but ambitious Lutheran curate, and was such a sluggish student that his exasperated father apprenticed him (or, by some accounts, nearly apprenticed him) to a cobbler. Appalled at the prospect of spending a lifetime banging tacks into leather, young Linné begged for another chance, which was granted, and he never thereafter wavered from academic distinction. He studied medicine in Sweden and Holland, though his passion became the natural world. In the early 1730s, still in his twenties, he began to produce catalogues of the world's plant and animal species, using a system of his own devising, and gradually his fame grew.

Rarely has a man been more comfortable with his own greatness. He spent much of his leisure time penning long and flattering portraits of himself, declaring that there had never 'been a greater botanist or zoologist', and that his system of classification was 'the greatest achievement in the realm of science'. Modestly, he suggested that his gravestone should bear the inscription *Princeps Botanicorum*, 'Prince of Botanists'. It was never wise to question his generous self-assessments. Those who did so were apt to find they had weeds named after them.

Linnaeus's other striking quality was an abiding – at times, one might say, a feverish – preoccupation with sex. He was particularly struck by the similarity between certain bivalves and the female pudenda. To the parts of one species of clam he gave the names 'vulva', 'labia', 'pubes', 'anus' and 'hymen'. He grouped plants by the nature of their reproductive organs and endowed them with an arrestingly anthropomorphic amorousness. His descriptions of flowers and their behaviour are full of references to 'promiscuous intercourse', 'barren concubines' and 'the bridal bed'. In spring, he wrote in one oft-quoted passage,

> Love comes even to the plants. Males and females . . . hold their nuptials . . . showing by their sexual organs which are males, which females. The flowers' leaves serve as a bridal bed, which the Creator has so gloriously arranged, adorned with such noble bed curtains, and perfumed with so many soft scents that the bridegroom with his bride might there celebrate their nuptials with so much the greater solemnity. When the bed has thus been made ready, then is the time for the bridegroom to embrace his beloved bride and surrender himself to her.

He named one genus of plants *Clitoria*. Not surprisingly, many people thought him strange. But his system of classification was irresistible. Before Linnaeus, plants were given names that were expansively descriptive. The common ground cherry was called *Physalis amno ramosissime ramis angulosis glabris foliis dentoserratis*. Linnaeus lopped it back to *Physalis angulata*, which name it still uses. The plant world was equally disordered by inconsistencies of naming. A botanist could not be sure if *Rosa sylvestris alba cum rubore, folio glabro* was the same plant that others called *Rosa sylvestris inodora seu canina*. Linnaeus solved the puzzlement by calling it simply *Rosa canina*. To make these excisions useful and agreeable to all required much more than simply being decisive. It required an instinct – a genius, in fact – for spotting the salient qualities of a species.

The Linnaean system is so well established that we can hardly imagine an alternative, but before Linnaeus, systems of classification were often highly whimsical. Animals might be categorized by whether they were wild or domesticated, terrestrial or aquatic, large or small, even whether they were thought handsome and noble or of no consequence. Buffon arranged his animals by their utility to man. Anatomical considerations barely came into it. Linnaeus made it his life's work to rectify this deficiency by classifying all that was alive according to its physical attributes. Taxonomy – which is to say the science of classification – has never looked back.

It all took time, of course. The first edition of his great *Systema Naturae* in 1735 was just fourteen pages long. But it grew and grew, until by the twelfth edition – the last that Linnaeus would live to see – it extended to three volumes and 2,300 pages. In the end he named or recorded some thirteen thousand species of plant and animal. Other works were more comprehensive – John Ray's three-volume *Historia Generalis Plantarum* in England, completed a generation earlier, covered no fewer than 18,625 species of plants alone – but what Linnaeus had that no-one else could touch was consistency, order, simplicity and timeliness. Though his work dates from the 1730s, it didn't become widely known in England until the 1760s, just in time to make Linnaeus a kind of father figure to British naturalists. Nowhere was his system adopted with greater enthusiasm (which is why, for one thing, the Linnaean Society has its home in London and not Stockholm).

Linnaeus was not flawless. He made room for mythical beasts and 'monstrous humans' whose descriptions he gullibly accepted from seamen and other imaginative travellers. Among these were a wild man, *Homo ferus*, who walked on all fours and had not yet mastered the art of speech, and *Homo caudatus*, 'man with a tail'. But then it was, as we should not forget, an altogether more credulous age. Even the great Joseph Banks took a keen and believing interest in a series of reported sightings of mermaids off the Scottish coast at the end of the eighteenth century. For the most part, however, Linnaeus's lapses were offset by sound and often brilliant taxonomy. Among other accomplishments, he saw that whales belonged with cows, mice and other common terrestrial animals in the order quadrupedia (later changed to mammalia), which no-one had done before.

In the beginning, Linnaeus intended to give each plant only a genus name and a number – *Convolvulus 1*, *Convolvulus 2* and so on – but he soon realized that that was unsatisfactory and hit on the binomial arrangement that remains at the heart of the system to this day. The intention originally was to use the binomial system for everything – rocks, minerals, diseases, winds, whatever existed in nature. Not everyone embraced the system warmly. Many were disturbed by its tendency towards indelicacy, which was slightly ironic as before Linnaeus the common names of many plants and animals had been heartily vulgar. The dandelion was long popularly known as the 'pissabed' because of its supposed diuretic properties, and other names in everyday use included *mare's fart*, *naked ladies*, *twitch-ballock*, *hound's piss*, *open arse* and *bum-towel*. One or two of these earthy appellations may unwittingly survive in English yet. The 'maidenhair' in maidenhair moss, for instance, does *not* refer to the hair on the maiden's head. At all events, it had long been felt that the natural sciences would be appreciably dignified by a dose of classical renaming, so there was a certain dismay in discovering that the self-appointed Prince of Botany had sprinkled his texts with such designations as *Clitoria*, *Fornicata* and *Vulva*.

Over the years many of these were quietly dropped (though not all: the common slipper limpet still answers on formal occasions to *Crepidula fornicata*) and many other refinements introduced as the needs of the natural sciences grew more specialized. In particular, the system was bolstered by the gradual introduction of additional hierarchies. *Genus* (plural *genera*) and *species* had been employed by naturalists for over a hundred years before Linnaeus, and *order*, *class* and *family* in their biological senses all came into use in the 1750s and 60s. But *phylum* wasn't coined until 1876 (by the German Ernst Haeckel), and *family* and *order* were treated as interchangeable until early in the twentieth century. For a time zoologists used *family* where botanists placed *order*, to the occasional confusion of nearly everyone.*

Linnaeus had divided the animal world into six categories: mammals, reptiles, birds, fishes, insects and 'vermes', or worms, for everything that didn't fit into the first five. From the outset it was evident that putting lobsters and shrimp into the same category as worms was unsatisfactory, and various new categories such as mollusca and crustacea were created. Unfortunately these new classifications were not uniformly applied from nation to nation. In an attempt to re-establish order, the British in 1842 proclaimed a new set of rules called the Stricklandian Code, but the French saw this as high-handed, and the Société Zoologique countered with its own conflicting code. Meanwhile, the American Ornithological Society, for obscure reasons, decided to use the 1758 edition of *Systema Naturae* as the basis for all its naming, rather than the 1766 edition used elsewhere, which meant that many American birds spent the nineteenth century logged in different genera from their avian cousins

---

*To illustrate, humans are in the domain eucarya, in the kingdom animalia, in the phylum chordata, in the subphylum vertebrata, in the class mammalia, in the order primates, in the family hominidae, in the genus *Homo*, in the species *sapiens*. (The convention, I'm informed, is to italicize genus and species names, but not those of higher divisions.) Some taxonomists employ further subdivisions: tribe, suborder, infraorder, parvorder and more.

in Europe. Not until 1902, at an early meeting of the International Congress of Zoology, did naturalists begin at last to show a spirit of compromise and adopt a universal code.

Taxonomy is described sometimes as a science and sometimes as an art, but really it's a battleground. Even today there is more disorder in the system than most people realize. Take the category of the phylum, the division that describes the basic body plans of organisms. A few phyla are generally well known, such as molluscs (the home of clams and snails), arthropods (insects and crustaceans) and chordates (us and all other animals with a backbone or proto-backbone); thereafter, things move swiftly in the direction of obscurity. Among the obscure we might list gnathostomulida (marine worms), cnidaria (jellyfish, medusae, anemones and corals) and the delicate priapulida (or little 'penis worms'). Familiar or not, these are elemental divisions. Yet there is surprisingly little agreement on how many phyla there are or ought to be. Most biologists fix the total at about thirty, but some opt for a number in the low twenties while Edward O. Wilson in *The Diversity of Life* puts the number at a surprisingly robust eighty-nine. It depends on where you decide to make your divisions – whether you are a 'lumper' or a 'splitter', as they say in the biological world.

At the more workaday level of species, the possibilities for disagreements are even greater. Whether a species of grass should be called *Aegilops incurva*, *Aegilops incurvata* or *Aegilops ovata* may not be a matter that would stir many non-botanists to passion, but it can be a source of very lively heat in the right quarters. The problem is that there are five thousand species of grass and many of them look awfully alike even to people who know grass. In consequence, some species have been found and named at least twenty times, and there are hardly any, it appears, that haven't been independently identified at least twice. The two-volume *Manual of the Grasses of the United States* devotes two hundred closely typeset pages to sorting out all the synonymies, as the biological world refers to its inadvertent but quite common duplications. And that is just for the grasses of a single country.

To deal with disagreements on the global stage, a body known as the International Association for Plant Taxonomy arbitrates on questions of priority and duplication. At intervals it hands down decrees, declaring that *Zauschneria californica* (a common plant in rock gardens) is to be known henceforth as *Epilobium canum*; or that *Aglaothamnion tenuissimum* may now be regarded as conspecific with *Aglaothamnion byssoides*, but not with *Aglaothamnion pseudobyssoides*. Normally these are small matters of tidying up that attract little notice, but when they touch on beloved garden plants, as they sometimes do, shrieks of outrage inevitably follow. In the late 1980s the common chrysanthemum was banished (on apparently sound scientific principles) from the genus of the same name and relegated to the comparatively drab and undesirable world of the genus *Dendranthema*.

Chrysanthemum breeders are a proud and numerous lot, and they protested to the real-if-improbable-sounding Committee on Spermatophyta. (There are also committees for Pteridophyta, Bryophyta and Fungi, among others, all reporting to an executive called the Rapporteur-Général; this is truly an institution to cherish.) Although the rules of nomenclature are supposed to be rigidly applied, botanists are not indifferent to sentiment, and in 1995 the decision was reversed. Similar adjudications have saved petunias, euonymus, and a popular species of amaryllis from demotion, but not many species of geraniums, which some years ago were transferred, amid howls, to the genus

Pelargonium. The disputes are entertainingly surveyed in Charles Elliott's *The Potting-Shed Papers*.

Disputes and reorderings of much the same type can be found in all the other realms of the living, so keeping an overall tally is not nearly as straightforward a matter as you might suppose. In consequence, the rather amazing fact is that we don't have the faintest idea – 'not even to the nearest order of magnitude', in the words of Edward O. Wilson – of the number of things that live on our planet. Estimates range from three million to two hundred million. More extraordinary still, according to a report in *The Economist* as much as 97 per cent of the world's plant and animal species may still await discovery.

Of the organisms that we *do* know about, more than 99 in 100 are only sketchily described – 'a scientific name, a handful of specimens in a museum, and a few scraps of description in scientific journals' is how Wilson describes the state of our knowledge. In *The Diversity of Life*, he estimated the number of known species of all types – plants, insects, microbes, algae, everything – at 1.4 million, but added that that was just a guess. Other authorities have put the number of known species slightly higher, at around 1.5 million to 1.8 million, but there is no central registry of these things, so nowhere to check numbers. In short, the remarkable position in which we find ourselves is that we don't actually know what we actually know.

In principle you ought to be able to go to experts in each area of specialization, ask how many species there are in their fields, then add the totals. Many people have in fact done so. The problem is that seldom do any two come up with matching figures. Some sources put the number of known types of fungi at seventy thousand, others at a hundred thousand – nearly half as many again. You can find confident assertions that the number of described earthworm species is four thousand and equally confident assertions that the figure is twelve thousand. For insects, the numbers run from 750,000 to 950,000 species. These are, you understand, supposedly the *known* number of species. For plants, the commonly accepted numbers range from 248,000 to 265,000. That may not seem too vast a discrepancy, but it's more than twenty times the number of flowering plants in the whole of North America.

Putting things in order is not the easiest of tasks. In the early 1960s, Colin Groves of the Australian National University began a systematic survey of the 250-plus known species of primate. Oftentimes it turned out that the same species had been described more than once – sometimes several times – without any of the discoverers realizing that they were dealing with an animal that was already known to science. It took Groves four decades to untangle everything, and that was with a comparatively small group of easily distinguished, generally non-controversial creatures. Goodness knows what the results would be if anyone attempted a similar exercise with the planet's estimated twenty thousand types of lichens, fifty thousand species of mollusc or four-hundred-thousand-plus beetles.

What is certain is that there is a great deal of life out there, though the actual quantities are necessarily estimates based on extrapolations – sometimes exceedingly expansive extrapolations. In a well-known exercise in the 1980s, Terry Erwin of the Smithsonian Institution saturated a stand of nineteen rainforest trees in Panama with an insecticide fog, then collected everything that fell into his nets from the canopy. Among his haul (actually hauls, since he repeated the experiment seasonally to make sure he caught migrant species) were twelve hundred types of beetle. On the basis of the distribution of beetles elsewhere, the number of other tree species in the forest, the number of forests in the world, the number of other insect types, and so on up a long chain of variables, he estimated a figure of 30 million species of insects for the entire planet – a figure he later said was too conservative. Others using the same or similar data have come up with figures of 13 million, 80 million or 100 million insect types, underlining the conclusion that, however carefully arrived at, such figures inevitably owe at least as much to supposition as to science.

According to the *Wall Street Journal*, the world has 'about 10,000 active taxonomists' – not a great number when you consider how much there is to be recorded. But, the *Journal* adds, because of the cost (about £1,250 per species) and paperwork, only about fifteen thousand new species of all types are logged per year.

'It's not a biodiversity crisis, it's a taxonomist crisis!' barks Koen Maes, Belgian-born head of invertebrates at the Kenyan National Museum in Nairobi, whom I met briefly on a visit to the country in the autumn of 2002. There were no specialized taxonomists in the whole of Africa, he told me. 'There was one in the Ivory Coast, but I think he has retired,' he said. It takes eight to ten years to train a taxonomist, but none are coming along in Africa. 'They are the real fossils,' Maes added. He himself was to be let go at the end of the year, he said. After seven years in Kenya, his contract was not being renewed. 'No funds,' Maes explained.

Writing in the journal *Nature* a few months earlier, the British biologist G. H. Godfray noted that there is a chronic 'lack of prestige and resources' for taxonomists everywhere. In consequence, 'many species are being described poorly in isolated publications, with no attempt to relate a new taxon* to existing species and classifications.' Moreover, much of taxonomists' time is taken up not with describing new

*The formal word for a zoological category, such as *phylum* or *genus*. The plural is *taxa*.

species but simply with sorting out old ones. Many, according to Godfray, 'spend most of their career trying to interpret the work of nineteenth-century systematicists: deconstructing their often inadequate published descriptions or scouring the world's museums for type material that is often in very poor condition'. Godfray particularly stresses the absence of attention being paid to the systematizing possibilities of the internet. The fact is that taxonomy, by and large, is still quaintly wedded to paper.

In an attempt to haul things into the modern age, in 2001 Kevin Kelly, co-founder of *Wired* magazine, launched an enterprise called the All Species Foundation with the aim of finding and recording on a database every living organism. The cost of such an exercise has been estimated at anywhere from £1.3 billion to as much as £30 billion. As of the spring of 2002, the foundation had just £750,000 in funds and four full-time employees.

If, as the numbers suggest, we have perhaps a hundred million species of insects yet to find, and if our rates of discovery continue at the present pace, we should have a definitive total for insects in a little over fifteen thousand years. The rest of the animal kingdom may take a little longer.

So why do we know as little as we do? There are nearly as many reasons as there are animals left to count, but here are a few of the principal causes.

Most living things are small and easily overlooked. In practical terms, this is not always a bad thing. You might not slumber quite so contentedly if you were aware that your mattress is home to perhaps two million microscopic mites, which come out in the wee hours to sup on your sebaceous oils and feast on all those lovely, crunchy flakes of skin that you shed as you doze and toss. Your pillow alone may be home to forty thousand of them. (To them your head is just one large oily bon-bon.) And don't think a clean pillowcase will make a difference. To something on the scale of bed mites, the weave of the tightest human fabric looks like ship's rigging. Indeed, if your pillow is six years old – which is apparently about the average age for a pillow – it has been estimated that one tenth of its weight will be made up of 'sloughed skin, living mites, dead mites and mite dung', to quote the man who did the measuring, Dr John Maunder of the British Medical Entomology Centre. (But at least they are *your* mites. Think of what you snuggle up with each time you climb into a hotel bed.)* These mites have been with us since time immemorial, but they weren't discovered until 1965.

If creatures as intimately associated with us as bed mites escaped our notice until the age of colour television, it's hardly surprising that most of the rest of the small-scale world is barely known to us. Go out into the woods – any woods at all – bend down and scoop up a handful of soil, and you will be holding up to ten billion bacteria, most of them unknown to science. Your sample will also contain perhaps a million plump yeasts, some two hundred thousand hairy little fungi known as moulds, perhaps ten thousand protozoans (of which the most familiar is the amoeba) and assorted rotifers, flatworms, roundworms and other microscopic creatures known collectively as cryptozoa. A large portion of these will also be unknown.

The most comprehensive handbook of micro-organisms, *Bergey's Manual of Systematic Bacteriology*, lists about four thousand types of bacteria. In the 1980s, a pair of Norwegian scientists, Jostein Goksøyr and Vigdis Torsvik,

*We are actually getting worse at some matters of hygiene. Dr Maunder believes that the move towards low-temperature washing-machine detergents has encouraged bugs to proliferate. As he puts it: 'If you wash lousy clothing at low temperatures, all you get is cleaner lice.'

collected a gram of random soil from a beech forest near their lab in Bergen and carefully analysed its bacterial content. They found that this single small sample contained between four thousand and five thousand separate bacterial species, more than in the whole of *Bergey's Manual*. They then travelled to a coastal location a few miles away, scooped up another gram of earth and found that it contained four to five thousand *other* species. As Edward O. Wilson observes: 'If over 9,000 microbial types exist in two pinches of substrate from two localities in Norway, how many more await discovery in other, radically different habitats?' Well, according to one estimate, it could be as many as 400 million.

We don't look in the right places. In *The Diversity of Life*, Wilson describes how one botanist spent a few days tramping around 10 hectares of jungle in Borneo and discovered a thousand new species of flowering plant – more than are found in the whole of North America. The plants weren't hard to find. It's just that no-one had looked there before. Koen Maes of the Kenyan National Museum told me that he went to one cloud forest, as mountaintop forests are known in Kenya, and in half an hour 'of not particularly dedicated looking' found four new species of millipedes, three representing new genera, and one new species of tree. 'Big tree,' he added, and shaped his arms as if about to dance with a very large partner. Cloud forests are found on the tops of plateaux and have sometimes been isolated for millions of years. 'They provide the ideal climate for biology and they have hardly been studied,' he said.

Overall, tropical rainforests cover only about 6 per cent of Earth's surface, but they harbour more than half of its animal life and about two-thirds of its flowering plants – and most of this life remains unknown to us because too few researchers spend time in them. Not incidentally, much of this could be quite valuable. At least 99 per cent of flowering plants have never been tested for their medicinal properties. Because they can't flee from predators, plants have had to contrive elaborate chemical defences, and so are particularly rich in intriguing compounds. Even now, nearly a quarter of all prescribed medicines are derived from just forty plants, with another 16 per cent coming from animals or microbes, so there is a serious risk with every hectare of forest felled of losing medically vital possibilities. Using a method called combinatorial chemistry, chemists can generate 40,000 compounds at a time in labs, but these products are random and not uncommonly useless, whereas any natural molecule will have already passed what *The Economist* calls 'the ultimate screening programme: over three and a half billion years of evolution'.

Looking for the unknown isn't simply a matter of travelling to remote or distant places, however. In his book *Life: An Unauthorised Biography*, Richard Fortey notes how one ancient bacterium was found on the wall of a country pub 'where men had urinated for generations' – a discovery that would seem to involve rare amounts of luck *and* devotion and possibly some other quality not specified.

There aren't enough specialists. The stock of things to be found, examined and recorded very much outruns the supply of scientists available to do it. Take the hardy and little-known organisms known as bdelloid rotifers. These are microscopic animals that can survive almost anything. When conditions are tough, they curl up into a compact shape, switch off their metabolism and wait for better times. In this state, you can drop them into boiling water or freeze them almost to absolute zero – that is, the level where even atoms give up – and, when this torment has finished and they are returned to a more pleasing environment, they will uncurl and move on as if nothing has happened. So far, about 500 species have been identified (though other sources say 360), but nobody has any idea, even remotely, how many there may be altogether. For years almost all that was known about them was thanks to the work of a devoted amateur, a London clerical worker named David Bryce who studied them in his spare time. They can be found all over the world, but you could have all the bdelloid rotifer experts in the world to dinner and not have to borrow plates from the neighbours.

Even creatures as important and ubiquitous as fungi (and fungi are both) attract comparatively little notice. Fungi are everywhere and come in many forms – as mushrooms, moulds, mildews, yeasts and puffballs, to name but a sampling – and they exist in volumes that most of us little suspect. Gather together all the fungi found in a typical hectare of meadowland and you would have 2,800 kilograms of the stuff. These are not marginal organisms. Without fungi there would be no potato blights, Dutch elm disease, jock itch or athlete's foot, but also no yogurts or beers or cheeses. Altogether about seventy thousand species of fungi have been identified, but it is thought the total number could be as high as 1.8 million. A lot of mycologists work in industry, making cheeses and yogurts and the like, so it is hard to say how many are actively involved in research, but we can safely take it that there are more species of fungi to be found than there are people to find them.

The world is a really big place. We have been gulled by the ease of air travel and other forms of communication into thinking that the world is not all that big, but at ground level, where researchers must work, it is actually enormous – enormous enough to be full of surprises. The okapi, the nearest living relative of the giraffe, is now known to exist in substantial numbers in the rainforests of Zaire – the total population is estimated at perhaps thirty thousand – yet its existence wasn't even suspected until the twentieth century. The large, flightless New Zealand bird called the takahe had been presumed extinct for two hundred years before being found living in a rugged area of the country's South Island. In 1995 a team of French and British scientists in Tibet, who were lost in a snowstorm in a remote valley, came across a breed of horse, called the Riwoche, that had previously been known only from prehistoric cave drawings. The valley's inhabitants were astonished to learn that the horse was considered a rarity in the wider world.

Some people think even greater surprises may await us. 'A leading British ethno-biologist', wrote *The Economist* in 1995, 'thinks a megatherium, a sort of giant ground sloth which may stand as high as a giraffe . . . may lurk in the fastnesses of the Amazon basin.' Perhaps significantly, the ethno-biologist wasn't named; perhaps even more significantly, nothing more has been heard of him or his giant sloth. No-one, however, can categorically say that no such thing is there until every jungly glade has been investigated, and we are a long way from achieving that.

But even if we groomed thousands of fieldworkers and dispatched them to the furthest corners of the world, it would not be effort enough, for wherever life can be, it is. Life's extraordinary fecundity is amazing, even gratifying, but also problematic. To survey it all, you would have to turn over every rock, sift through the litter on every forest floor, sieve unimaginable quantities of sand and dirt, climb into every forest canopy and devise much more efficient ways to examine the seas. Even then you would overlook whole ecosystems. In the 1980s, amateur cave explorers entered a deep cave in Romania that had been sealed off from the outside world for a long but unknown period and found thirty-three species of insects and other small creatures – spiders, centipedes, lice – all blind, colourless and new to science. They were living off the microbes in the surface scum of pools, which in turn were feeding on hydrogen sulphide from hot springs.

Our instinct may be to see the impossibility of tracking everything down as frustrating, dispiriting, perhaps even appalling, but it can just as well be viewed as almost unbearably exciting. We live on a planet that has a more or less infinite capacity to surprise. What reasoning person could possibly want it any other way?

What is nearly always most arresting in any ramble through the scattered disciplines of modern science is realizing how many people have been willing to devote lifetimes to the most sumptuously esoteric lines of enquiry. In one of his essays, Stephen Jay Gould notes how a hero of his named Henry Edward Crampton spent fifty years, from 1906 to his death in 1956, quietly studying a genus of land snails called *Partula* in Polynesia. Over and over, year after year, Crampton measured to the tiniest degree – to eight decimal places – the whorls and arcs and gentle curves of numberless *Partula*, compiling the results into fastidiously detailed tables. A single line of text in a Crampton table could represent weeks of measurement and calculation.

Only slightly less devoted, and certainly more unexpected, was Alfred C. Kinsey, who became famous for his studies of human sexuality in the 1940s and 1950s. Before his mind became filled with sex, so to speak, Kinsey was an entomologist, and a dogged one at that. In one expedition lasting two years, he hiked 4,000 kilometres to assemble a collection of three hundred thousand wasps. How many stings he collected along the way is not, alas, recorded.

Something that had been puzzling me was the question of how you assured a chain of succession in these arcane fields. Clearly there cannot be many institutions in the world that require or are prepared to support specialists in barnacles or Pacific snails. As we parted at the Natural History Museum in London, I asked Richard Fortey how science ensures that when one person goes there's someone ready to take his place.

He chuckled rather heartily at my naivety. 'I'm afraid it's not as if we have substitutes sitting on the bench somewhere waiting to be called in to play. When a specialist retires or, even more unfortunately, dies, that can bring a stop to things in that field, sometimes for a very long while.'

'And I suppose that's why you value someone who spends forty-two years studying a single species of plant, even if it doesn't produce anything terribly new?'

'Precisely,' he said, 'precisely.' And he really seemed to mean it.

24

CELLS

It starts with a single cell. The first cell splits to become two and the two become four and so on. After just forty-seven doublings, you have 10,000 trillion (10,000,000,000,000,000) cells in your body and are ready to spring forth as a human being.* And every one of those cells knows exactly what to do to preserve and nurture you from the moment of conception to your last breath.

You have no secrets from your cells. They know far more about you than you do. Each one carries a copy of the complete genetic code – the instruction manual for your body – so it knows how to do not only its own job but every other job in the body too. Never in your life will you have to remind a cell to keep an eye on its adenosine triphosphate levels or to find a place for the extra squirt of folic acid that's just unexpectedly turned up. It will do that for you, and millions more things besides.

Every cell in nature is a thing of wonder. Even the

*Actually, quite a lot of cells are lost in the process of development, so the number you emerge with is really just a guess. Depending on which source you consult, it can vary by several orders of magnitude. The figure of 10,000 trillion (or ten quadrillion) is from Margulis and Sagan, *Microcosmos*.

simplest are far beyond the limits of human ingenuity. To build the most basic yeast cell, for example, you would have to miniaturize about the same number of components as are found in a Boeing 777 jetliner and fit them into a sphere just 5 microns across; then somehow you would have to persuade that sphere to reproduce.

But yeast cells are as nothing compared with human cells, which are not just more varied and complicated, but vastly more fascinating because of their complex interactions.

Your cells are a country of 10,000 trillion citizens, each devoted in some intensively specific way to your overall well-being. There isn't a thing they don't do for you. They let you feel pleasure and form thoughts. They enable you to stand and stretch and caper. When you eat, they extract the nutrients, distribute the energy, and carry off the wastes – all those things you learned about in school biology – but they also remember to make you hungry in the first place and reward you with a feeling of well-being afterwards so that you won't forget to eat again. They keep your hair growing, your ears waxed, your brain quietly purring. They manage every corner of your being. They will jump to your defence the instant you are threatened. They will unhesitatingly die for you – billions of them do so daily. And not once in all your years have you thanked even one of them. So let us take a moment now to regard them with the wonder and appreciation they deserve.

We understand a little of how cells do the things they do – how they lay down fat or manufacture insulin or engage in many of the other acts necessary to maintain a complicated entity like yourself – but only a little. You have at least 200,000 different types of protein labouring away inside you and so far we understand what no more than about 2 per cent of them do. (Others put the figure at more like 50 per cent; it depends, apparently, on what you mean by 'understand'.)

Surprises at the cellular level turn up all the time. In nature, nitric oxide is a formidable toxin and a common component of air pollution. So scientists were naturally a little surprised when, in the mid-1980s, they found it being produced in a curiously devoted manner in human cells. Its purpose was at first a mystery, but then scientists began to find it all over the place – controlling the flow of blood and the energy levels of cells, attacking cancers and other pathogens, regulating the sense of smell, even assisting in penile erections. It also explained why nitroglycerine, the well-known explosive, soothes the heart pain known as angina. (It is converted into nitric oxide in the bloodstream, relaxing the muscle linings of vessels, allowing blood to flow more freely.) In barely the space of a decade this one gassy substance went from extraneous toxin to ubiquitous elixir.

You possess 'some few hundred' different types of cell, according to the Belgian biochemist Christian de Duve, and they vary enormously in size and shape, from nerve cells whose filaments can stretch to over a metre to tiny, disc-shaped red blood cells to the rod-shaped photocells that help to give us vision. They also come in a sumptuously wide range of sizes – nowhere more strikingly than at the moment of conception, when a single beating sperm confronts an egg 85,000 times bigger than it (which rather puts the notion of male conquest into perspective). On average, however, a human cell is about 20 microns wide – that is, about two-hundredths of a millimetre – which is too small to be seen but roomy enough to hold thousands of complicated structures like mitochondria, and millions upon millions of molecules. In the most literal way, cells also vary in liveliness. Your skin cells are all dead. It's a somewhat galling notion to reflect that every inch of your surface is deceased. If you are an average-sized adult you are lugging around over 2 kilograms of dead skin, of which several billion tiny fragments are sloughed off each day. Run a finger along a dusty shelf and you are drawing a pattern very largely in old skin.

Most living cells seldom last more than a month or so, but there are some notable exceptions. Liver cells can survive for years, though the components within them may be renewed every few days. Brain cells last as long as you do. You are issued with a hundred billion or so at birth and that is all you are ever going to get. It has been estimated that you lose five hundred of them an hour, so if you have any serious thinking to do there really isn't a moment to waste. The good news is that the individual components of your brain cells are constantly renewed so that, as with the liver cells, no part of them is actually likely to be more than about a month old. Indeed, it has been suggested that there isn't a single bit of any of us – not so much as a stray molecule – that was part of us nine years ago. It may not feel like it, but at the cellular level we are all youngsters.

The first person to describe a cell was Robert Hooke, whom we last encountered squabbling with Isaac Newton over credit for the invention of the inverse square law. Hooke achieved many things in his sixty-eight years – he was both an accomplished theoretician and a dab hand at making ingenious and useful instruments – but nothing he did brought him greater admiration than his popular book *Micrographia: or Some Physiological Descriptions of Miniature Bodies Made by Magnifying Glasses*, published in 1665. It revealed to an enchanted public a universe of the very small that was far more diverse, crowded and finely structured than anyone had ever come close to imagining.

Among the microscopic features first identified by Hooke were little chambers in plants that he called 'cells' because they reminded him of monks' cells. Hooke calculated that a one-inch square of cork would contain 1,259,712,000 of these tiny chambers – the first appearance of such a very large number anywhere in science. Microscopes by this time had been around for a generation or so, but what set Hooke's apart were their technical supremacy. They achieved magnifications of thirty times, making them the last word in seventeenth-century optical technology.

So it came as something of a shock when just a decade later Hooke and the other members of London's Royal Society began to receive drawings and reports from an unlettered linen draper in the Dutch city of Delft employing magnifications of up to 275 times. The draper's name was Antoni van Leeuwenhoek. Though he had little formal education and no background in science, he was a perceptive and dedicated observer and a technical genius.

To this day it is not known how he got such magnificent magnifications from such simple handheld devices, which were little more than modest wooden dowels with a tiny bubble of glass embedded in them, far more like magnifying glasses than what most of us think of as microscopes, but really not much like either. Leeuwenhoek made a new instrument for every experiment he performed and was extremely secretive about his techniques, though he did sometimes offer tips to the British on how they might improve their resolutions.*

Over a period of fifty years – beginning, remarkably enough, when he was already past forty – Leeuwenhoek made almost two hundred reports to the Royal Society, all

*Leeuwenhoek was close friends with another Delft notable, the artist Jan Vermeer. In the mid-1600s Vermeer, who previously had been a competent but not outstanding artist, suddenly developed the mastery of light and perspective for which he has been celebrated ever since. Though it has never been proved, it has long been suspected that he used a camera obscura, a device for projecting images onto a flat surface through a lens. No such device was listed among Vermeer's personal effects after his death, but it happens that the executor of Vermeer's estate was none other than Antoni van Leeuwenhoek, the most secretive lens-maker of his day.

written in Low Dutch, the only tongue of which he was master. He offered no interpretations, but simply the facts of what he had found, accompanied by exquisite drawings. He sent reports on almost everything that could be usefully examined – bread mould, a bee's stinger, blood cells, teeth, hair, his own saliva, excrement and semen (these last with fretful apologies for their inescapably unsavoury nature) – nearly all of which had never been seen microscopically before.

After he reported finding 'animalcules' in a sample of pepper-water in 1676, the members of the Royal Society spent a year with the best devices English technology could produce searching for the 'little animals' before finally getting the magnification right. What Leeuwenhoek had found were protozoa. He calculated that there were 8,280,000 of these tiny beings in a single drop of water – more than the number of people in Holland. The world teemed with life in ways and numbers that no-one had previously suspected.

Inspired by Leeuwenhoek's fantastic findings, others began to peer into microscopes with such keenness that they sometimes found things that weren't in fact there. One respected Dutch observer, Nicolaus Hartsoecker, was convinced he saw 'tiny preformed men' in sperm cells. He called the little beings 'homunculi', and for some time many people believed that all humans – indeed, all creatures – were simply vastly inflated versions of tiny but complete precursor beings. Leeuwenhoek himself occasionally got carried away with his enthusiasms. In one of his least successful experiments he tried to study the explosive properties of gunpowder by observing a small blast at close range; he nearly blinded himself in the process.

In 1683 Leeuwenhoek discovered bacteria – but that was about as far as progress could get for the next century and a half, because of the limitations of microscope technology. Not until 1831 would anyone first see the nucleus of a cell

– it was found by the Scottish botanist Robert Brown, that frequent but always shadowy visitor to the history of science. Brown, who lived from 1773 to 1858, called it *nucleus* from the Latin *nucula*, meaning little nut or kernel. Only in 1839, however, did anyone realize that *all* living matter is cellular. It was Theodor Schwann, a German, who had this insight, and it was not only comparatively late, as scientific insights go, but not widely embraced at first. It wasn't until the 1860s, and some landmark work by Louis Pasteur in France, that it was shown conclusively that life cannot arise spontaneously but must come from pre-existing cells. The belief became known as the 'cell theory', and it is the basis of all modern biology.

The cell has been compared to many things, from 'a complex chemical refinery' (by the physicist James Trefil) to 'a vast, teeming metropolis' (the biochemist Guy Brown). A cell is both of those things and neither. It is like a refinery in that it is devoted to chemical activity on a grand scale and like a metropolis in that it is crowded and busy and filled with interactions that seem confused and random but clearly have some system to them. But it is a much more nightmarish place than any city or factory that you have ever seen. To begin with there is no up or down inside the cell (gravity doesn't meaningfully apply at the cellular scale), and not an atom's width of space is unused. There is activity *every*where and a ceaseless thrum of electrical energy. You may not feel terribly electrical, but you are. The food we eat and the oxygen we breathe are combined in the cells into electricity. The reason we don't give each other massive shocks or scorch the sofa when we sit down is that it is all happening on a tiny scale: a mere 0.1 volts travelling distances measured in nanometres. However, scale that up and it would translate as a jolt of 20 million volts per metre, about the same as the charge carried by the main body of a thunderstorm.

Whatever their size or shape, nearly all your cells are built to fundamentally the same plan: they have an outer casing or membrane, a nucleus wherein resides the necessary genetic information to keep you going, and a busy space between the two called the cytoplasm. The membrane is not, as most of us imagine it, a durable, rubbery casing, something that you would need a sharp pin to prick. Rather, it is made up of a type of fatty material known as a lipid, which has the approximate consistency 'of a light grade of machine oil', to quote Sherwin B. Nuland. If that seems surprisingly insubstantial, bear in mind that at the microscopic level things behave differently. To anything on a molecular scale water becomes a kind of heavy-duty gel and a lipid is like iron.

If you could visit a cell, you wouldn't like it. Blown up to a scale at which atoms were about the size of peas, a cell itself would be a sphere roughly half a mile across, and supported by a complex framework of girders called the cytoskeleton. Within it, millions upon millions of objects – some the size of basketballs, others the size of cars – would whiz about like bullets. There wouldn't be a place you could stand without being pummelled and ripped thousands of times every second from every direction. Even for its full-time occupants the inside of a cell is a hazardous place. Each strand of DNA is on average attacked or damaged once every 8.4 seconds – ten thousand times in a day – by chemicals and other agents that whack into or carelessly slice through it, and each of these wounds must be swiftly stitched up if the cell is not to perish.

The proteins are especially lively, spinning, pulsating and flying into each other up to a billion times a second. Enzymes, themselves a type of protein, dash everywhere, performing up to a thousand tasks a second. Like greatly speeded-up worker ants, they busily build and rebuild molecules, hauling a piece off this one, adding a piece to that one. Some monitor passing proteins and mark with a chemical those that are irreparably damaged or flawed. Once so selected, the doomed proteins proceed to a structure called a proteasome, where they are stripped down and their components used to build new proteins. Some types of protein exist for less than half an hour; others survive for weeks. But all lead existences that are inconceivably frenzied. As de Duve notes, 'the molecular world must necessarily remain entirely beyond the powers of our imagination owing to the incredible speed with which things happen in it.'

But slow things down, to a speed at which the interactions can be observed, and things don't seem quite so unnerving. You can see that a cell is just millions of objects – lysosomes, endosomes, ribosomes, ligands, peroxisomes, proteins of every size and shape – bumping into millions of other objects and performing mundane tasks: extracting energy from nutrients, assembling structures, getting rid of waste, warding off intruders, sending and receiving messages, making repairs. Typically a cell will contain some twenty thousand different types of protein, and of these about two thousand types will each be represented by at least fifty thousand molecules. 'This means,' says Nuland, 'that even if we count only those molecules present in amounts of more than 50,000 each, the total is still a very minimum of 100 million protein molecules in each cell. Such a staggering figure gives some idea of the swarming immensity of biochemical activity within us.'

It is all an immensely demanding process. Your heart must pump 343 litres of blood an hour, over 8,000 litres every day, 3 million litres in a year – that's enough to fill four Olympic-sized swimming pools – to keep all those cells freshly oxygenated. (And that's at rest. During exercise the rate can increase as much as sixfold.) The oxygen is taken up by the mitochondria. These are the cells' power stations and there are about a thousand of them in a typical cell, though the number varies considerably depending on what a cell does and how much energy it requires.

You may recall from an earlier chapter that the mitochondria are thought to have originated as captive bacteria and that they now live essentially as lodgers in our cells, preserving their own genetic instructions, dividing to their own timetable, speaking their own language. You may also recall that we are at the mercy of their goodwill. Here's why. Virtually all the food and oxygen you take into your body are delivered, after processing, to the mitochondria, where they are converted into a molecule called adenosine triphosphate, or ATP.

You may not have heard of ATP, but it is what keeps you going. ATP molecules are essentially little battery packs that move through the cell providing energy for all the cell's processes, and you get through a *lot* of it. At any given moment, a typical cell in your body will have about one billion ATP molecules in it, and in two minutes every one of them will have been drained dry and another billion will have taken their place. Every day you produce and use up a volume of ATP equivalent to about half your body weight. Feel the warmth of your skin. That's your ATP at work.

When cells are no longer needed, they die with what can only be called great dignity. They take down all the struts and buttresses that hold them together and quietly devour their component parts. The process is known as apoptosis or programmed cell death. Every day billions of your cells die for your benefit and billions of others clean up the mess. Cells can also die violently – for instance, when infected – but mostly they die because they are told to. Indeed, if not told to live – if not given some kind of active instruction from another cell – cells automatically kill themselves. Cells need a lot of reassurance.

When, as occasionally happens, a cell fails to expire in the prescribed manner, but rather begins to divide and proliferate wildly, we call the result cancer. Cancer cells are really just confused cells. Cells make this mistake fairly regularly, but the body has elaborate mechanisms for dealing with it. It is only very rarely that the process spirals out of control. On average, humans suffer one fatal malignancy for each 100 million billion cell divisions. Cancer is bad luck in every possible sense of the term.

The wonder of cells is not that things occasionally go wrong, but that they manage everything so smoothly for decades at a stretch. They do so by constantly sending and monitoring streams of messages – a cacophony of messages – from all around the body: instructions, queries, corrections, requests for assistance, updates, notices to divide or expire. Most of these signals arrive by means of couriers called hormones, chemical entities such as insulin, adrenaline, oestrogen and testosterone that convey information from remote outposts like the thyroid and endocrine glands. Still other messages arrive by telegraph from the brain or from regional centres in a process called paracrine signalling. Finally, cells communicate directly with their neighbours to make sure their actions are co-ordinated.

What is perhaps most remarkable is that it is all just random frantic action, a sequence of endless encounters directed by nothing more than elemental rules of attraction and repulsion. There is clearly no thinking presence behind any of the actions of the cells. It all just happens, smoothly and repeatedly and so reliably that seldom are we even conscious of it; yet somehow all this produces not just order within the cell but a perfect harmony right across the organism. In ways that we have barely begun to understand, trillions upon trillions of reflexive chemical reactions add up to a mobile, thinking, decision-making you – or, come to that, a rather less reflective but still incredibly organized dung beetle. Every living thing, never forget, is a wonder of atomic engineering.

Indeed, some organisms that we think of as primitive enjoy a level of cellular organization that makes our own look carelessly pedestrian. Disassemble the cells of a sponge (by passing them through a sieve, for instance), then dump them into a solution and they will find their way back together and build themselves into a sponge again. You can do this to them over and over and they will doggedly reassemble because, like you and me and every other living thing, they have one overwhelming impulse: to continue to be.

And that's because of a curious, determined, barely understood molecule that is itself not alive and for the most part doesn't do anything at all. We call it DNA, and to begin to understand its supreme importance to science and to us we need to go back 160 years or so to Victorian England to the moment when the naturalist Charles Darwin had what has been called 'the single best idea that anyone has ever had' – and then, for reasons that take a little explaining, locked it away in a drawer for the next fifteen years.

25

DARWIN'S SINGULAR NOTION

In the late summer or early autumn of 1859, Whitwell Elwin, editor of the respected British journal the *Quarterly Review*, was sent an advance copy of a new book by the naturalist Charles Darwin. Elwin read the book with interest, and agreed that it had merit, but feared that the subject matter was too narrow to attract a wide audience. He urged Darwin to write a book about pigeons instead. 'Everyone is interested in pigeons,' he observed helpfully.

Elwin's sage advice was ignored and *On the Origin of Species by Means of Natural Selection, or the Preservation of Favoured Races in the Struggle for Life* was published in late November 1859, priced at 15 shillings. The first edition of 1,250 copies sold out on the first day. It has never been out of print, and scarcely out of controversy, in all the time since – not bad going for a man whose principal other interest was earthworms and who, but for a single impetuous decision to sail around the world, would very probably have passed his life as an anonymous country parson known for – well, for an interest in earthworms.

Charles Robert Darwin was born on 12 February 1809*

*An auspicious date in history: on the same day in Kentucky, Abraham Lincoln was born.

in Shrewsbury, a sedate market town in the west Midlands. His father was a prosperous and well-regarded physician. His mother, who died when Charles was only eight, was the daughter of Josiah Wedgwood, of pottery fame.

Darwin enjoyed every advantage of upbringing, but continually pained his widowed father with his lacklustre academic performance. 'You care for nothing but shooting, dogs, and rat-catching, and you will be a disgrace to yourself and all your family,' wrote the elder Darwin in a line that nearly always appears just about here in any review of Charles's early life. Although his inclination was to natural history, for his father's sake he tried to study medicine at Edinburgh University, but couldn't bear the blood and suffering. The experience of witnessing an operation on an understandably distressed child – this was in the days before anaesthetics, of course – left him permanently traumatized. He tried law instead, but found that insupportably dull and finally managed, more or less by default, to acquire a degree in divinity from Cambridge.

A life in a rural vicarage seemed to await him when from out of the blue there came a more tempting offer. Darwin was invited to sail on the naval survey ship HMS *Beagle*, essentially as dinner company for the captain, Robert FitzRoy, whose rank precluded his socializing with anyone other than a gentleman. FitzRoy, who was very odd, chose Darwin in part because he liked the shape of Darwin's nose. (It betokened depth of character, he believed.) Darwin was not FitzRoy's first choice, but got the nod when FitzRoy's preferred companion dropped out. From a twenty-first century perspective the two men's most striking shared feature was their extreme youthfulness. At the time of sailing, FitzRoy was only twenty-three, Darwin just twenty-two.

FitzRoy's formal assignment was to chart coastal waters, but his hobby – passion, really – was to seek out evidence for a literal, biblical interpretation of creation. That Darwin was trained for the ministry was central to FitzRoy's decision to have him aboard. That Darwin subsequently proved to be not only liberal of view but less than wholeheartedly devoted to Christian fundamentals became a source of lasting friction between them.

Darwin's time aboard the *Beagle*, from 1831 to 1836, was obviously the formative experience of his life, but also one of the most trying. He and his captain shared a small cabin, which can't have been easy as FitzRoy was subject to fits of fury followed by spells of simmering resentment. He and Darwin constantly engaged in quarrels, some 'bordering on insanity', as Darwin later recalled. Ocean voyages tended to become melancholy undertakings at the best of times – the previous captain of the *Beagle* had put a bullet through his brain during a moment of lonely gloom – and FitzRoy came from a family well known for a depressive instinct. His uncle, Viscount Castlereagh, had slit his throat the previous decade while serving as Chancellor of the Exchequer. (FitzRoy would himself commit suicide by the same method in 1865.) Even in his calmer moods, FitzRoy proved strangely unknowable. Darwin was astounded to learn upon the conclusion of their voyage that almost at once FitzRoy married a young woman to whom he had long been betrothed. In five years in Darwin's company, he had not once hinted at an attachment or even mentioned her name.

In every other respect, however, the *Beagle* voyage was a triumph. Darwin experienced adventure enough to last a lifetime and accumulated a hoard of specimens sufficient to make his reputation and keep him occupied for years. He found a magnificent trove of giant ancient fossils, including the finest megatherium known to date; survived a lethal earthquake in Chile; discovered a new species of dolphin (which he dutifully named *Delphinus fitzroyi*); conducted diligent and useful geological investigations throughout the Andes; and developed a new and much-admired theory for the formation of coral atolls, which suggested, not incidentally, that atolls could not form in less than a million years – the first hint of his longstanding attachment to the extreme antiquity of earthly processes. In 1836, aged twenty-seven, he returned home, having been away for five years and two days. He never left England again.

One thing Darwin didn't do on the voyage was propound the theory (or even a theory) of evolution. For a start, evolution as a concept was already decades old by the 1830s. Darwin's own grandfather, Erasmus, had paid tribute to evolutionary principles in a poem of inspired mediocrity called 'The Temple of Nature' years before Charles was even born. It wasn't until the younger Darwin was back in England and read Thomas Malthus's *Essay on the Principle of Population* (which proposed that increases in food supply could never keep up with population growth for mathematical reasons) that the idea began to percolate through his mind that life is a perpetual struggle and that natural selection was the means by which some species prospered while others failed. Specifically, what Darwin saw was that all organisms compete for resources, and those that had some innate advantage would prosper and pass on that advantage to their offspring. By such means would species continuously improve.

It seems an awfully simple idea – it is an awfully simple idea – but it explained a great deal, and Darwin was prepared to devote his life to it. 'How stupid of me not to have thought of it!' T. H. Huxley cried upon reading *On the Origin of Species*. It is a view that has been echoed ever since.

Interestingly, Darwin didn't use the phrase 'survival of the fittest' in any of his work (though he did express his admiration for it). The expression was coined, in 1864, five years after the publication of *On the Origin of Species*, by Herbert Spencer in *Principles of Biology*. Nor did he employ the word 'evolution' in print until the sixth edition of *Origin* (by which time its use had become too widespread to resist), preferring instead 'descent with modification'. Nor, above all, were his conclusions in any way inspired by his noticing, during his time in the Galápagos Islands, an interesting diversity in the beaks of finches. The story as conventionally told (or, at least, as frequently remembered by many of us) is that Darwin, while travelling from island to island, noticed that each of the finches' beaks were marvellously adapted for exploiting local resources – that on one island beaks were sturdy and short and good for cracking nuts, while on the next island beaks were perhaps long and thin and well suited for winkling food out of crevices – and it was this that set him to thinking that perhaps the birds had not been created this way, but had in a sense created themselves.

In fact, the birds *had* created themselves, but it wasn't Darwin who noticed it. At the time of the *Beagle* voyage, Darwin was fresh out of university and not yet an accomplished naturalist, and so failed to see that the Galápagos birds were all of a type. It was his friend the ornithologist John Gould who realized that what Darwin had found was lots of finches with different talents. Unfortunately, in his inexperience Darwin had not noted which birds came from which islands. (He had made a similar error with tortoises.) It took years to sort the muddles out.

Because of these various oversights, and the need to sort through crates and crates of other *Beagle* specimens, it wasn't until 1842, five years after his return to England, that Darwin finally began to sketch out the rudiments of his new theory. These he expanded into a 230-page 'sketch' two years later. And then he did an extraordinary thing: he put his notes away and for the next decade and a half busied himself with other matters. He fathered ten children, devoted nearly eight years to writing an exhaustive opus on barnacles ('I hate a barnacle as no man ever did before,' he sighed, understandably, upon the work's conclusion) and fell prey to strange disorders that left him chronically listless, faint, and 'flurried', as he put it. The symptoms nearly always included a terrible nausea and generally also incorporated palpitations, migraines, exhaustion, trembling, spots before the eyes, shortness of breath, 'swimming of the head' and, not surprisingly, depression.

The cause of the illness has never been established. The most romantic and perhaps likely of the many suggested possibilities is that he suffered from Chagas's disease, a lingering tropical malady that he could have acquired from the bite of a Benchuga bug in South America. A more prosaic explanation is that his condition was psychosomatic. In either case, the misery was not. Often he could work for no more than twenty minutes at a stretch, sometimes not even that.

Much of the rest of his time was devoted to a series of increasingly desperate treatments – icy plunge baths, dousings in vinegar, draping himself with 'electric chains' that subjected him to small jolts of current. He became something of a hermit, seldom leaving his home in Kent, Down House. One of his first acts upon moving to the house was to erect a mirror outside his study window so that he could identify, and if necessary avoid, callers.

Darwin kept his theory to himself because he well knew the storm it would cause. In 1844, the year he locked his notes away, a book called *Vestiges of the Natural History of Creation* roused much of the thinking world to fury by suggesting that humans might have evolved from lesser primates without the assistance of a divine creator. Anticipating the outcry, the author had taken careful steps to conceal his identity, which he kept a secret from even his closest friends for the next forty years. Some wondered if Darwin himself might be the author. Others suspected

Prince Albert. In fact, the author was a successful and generally unassuming Scottish publisher named Robert Chambers whose reluctance to reveal himself had a practical dimension as well as a personal one: his firm was a leading publisher of Bibles.* *Vestiges* was warmly blasted from pulpits throughout Britain and far beyond, but also attracted a good deal of more scholarly ire. The *Edinburgh Review* devoted nearly an entire issue – eighty-five pages – to pulling it to pieces. Even T. H. Huxley, a believer in evolution, attacked the book with some venom, unaware that the author was a friend.

Darwin's own manuscript might have remained locked away till his death but for an alarming blow that arrived from the Far East in the early summer of 1858 in the form of a packet containing a friendly letter from a young naturalist named Alfred Russel Wallace and the draft of a paper, 'On the Tendency of Varieties to Depart Indefinitely from the Original Type', outlining a theory of natural selection that was uncannily similar to Darwin's secret jottings. Even some of the phrasing echoed Darwin's own. 'I never saw a more striking coincidence,' Darwin reflected in dismay. 'If Wallace had my manuscript sketch written out in 1842, he could not have made a better short abstract.'

Wallace didn't drop into Darwin's life quite as unexpectedly as is sometimes suggested. The two were already corresponding, and Wallace had more than once generously sent Darwin specimens that he thought might be of interest. In the process of these exchanges Darwin had discreetly warned Wallace that he regarded the subject of species creation as his exclusive territory. 'This summer will make

*Darwin was one of the few to guess correctly. He happened to be visiting Chambers one day when an advance copy of the sixth edition of *Vestiges* was delivered. The keenness with which Chambers checked the revisions was something of a giveaway, though it appears the two men did not discuss it.

the 20th year (!) since I opened my first note-book, on the question of how & in what way do species & varieties differ from each other,' he had written to Wallace some time earlier. 'I am now preparing my work for publication,' he added, even though he wasn't really.

Wallace failed to grasp what Darwin was trying to tell him – and in any case, of course, he could have had no idea that his own theory was so nearly identical to one that Darwin had been evolving, as it were, for two decades.

Darwin was placed in an agonizing quandary. If he rushed into print to preserve his priority, he would be taking advantage of an innocent tip-off from a distant admirer. But if he stepped aside, as gentlemanly conduct arguably required, he would lose credit for a theory that he had independently propounded. Wallace's theory was, by Wallace's own admission, the result of a flash of insight; Darwin's was the product of years of careful, plodding, methodical thought. It was all crushingly unfair.

To compound his misery, Darwin's youngest son, also named Charles, had contracted scarlet fever and was critically ill. At the height of the crisis, on 28 June, the child died. Despite the distraction of his son's illness, Darwin found time to dash off letters to his friends Charles Lyell and Joseph Hooker, offering to step aside but noting that to do so would mean that all his work, 'whatever it may amount to, will be smashed'. Lyell and Hooker came up with the compromise solution of presenting a summary of Darwin's and Wallace's ideas together. The venue they settled on was a meeting of the Linnaean Society, which at the time was struggling to find its way back into fashion as a seat of scientific eminence. On 1 July 1858, Darwin's and Wallace's theory was unveiled to the world. Darwin himself was not present. On the day of the meeting, he and his wife were burying their son.

The Darwin–Wallace presentation was one of seven that

evening – one of the others was on the flora of Angola – and if the thirty or so people in the audience had any idea that they were witnessing the scientific highlight of the century, they showed no sign of it. No discussion followed. Nor did the event attract much notice elsewhere. Darwin cheerfully noted later that only one person, a Professor Haughton of Dublin, mentioned the two papers in print and his conclusion was that 'all that was new in them was false, and what was true was old.'

Wallace, still in the distant east, learned of these manoeuvrings long after the event, but was remarkably equable, and seemed pleased to have been included at all. He even referred to the theory for ever after as 'Darwinism'.

Much less amenable to Darwin's claim of priority was a Scottish gardener named Patrick Matthew who had, rather remarkably, also come up with the principles of natural selection more than twenty years earlier – in fact, in the very year that Darwin had set sail in the *Beagle*. Unfortunately, Matthew had published these views in a book called *Naval Timber and Arboriculture*, which had been missed not just by Darwin, but by the entire world. Matthew kicked up in a lively manner, with a letter to *Gardener's Chronicle*, when he saw Darwin gaining credit everywhere for an idea that really was his. Darwin apologized without hesitation, though he did note for the record: 'I think that no one will feel surprised that neither I, nor apparently any other naturalist, has heard of Mr Matthew's views, considering how briefly they are given, and they appeared in the Appendix to a work on Naval Timber and Arboriculture.'

Wallace continued for another fifty years as a naturalist and thinker, occasionally a very good one, but increasingly fell from scientific favour by taking up dubious interests such as spiritualism and the possibility of life existing elsewhere in the universe. So the theory became, essentially by default, Darwin's alone.

Darwin never ceased being tormented by his ideas. He referred to himself as 'the Devil's Chaplain' and said that revealing the theory felt 'like confessing a murder'. Apart from all else, he knew it deeply pained his beloved and pious wife. Even so, he set to work at once expanding his manuscript into a book-length work. Provisionally he called it *An Abstract of an Essay on the Origin of Species and Varieties through Natural Selection* – a title so tepid and tentative that his publisher, John Murray, decided to issue just 500 copies. But once presented with the manuscript, and a slightly more arresting title, Murray reconsidered and increased the initial print run to 1,250.

On the Origin of Species was an immediate commercial success, but rather less of a critical one. Darwin's theory presented two intractable difficulties. It needed far more time than Lord Kelvin was willing to concede, and it was scarcely supported by fossil evidence. Where, asked Darwin's more thoughtful critics, were the transitional forms that his theory so clearly called for? If new species were continuously evolving, then there ought to be lots of intermediate forms scattered across the fossil record, but there were not.* In fact, the record as it existed then (and for a long time afterwards) showed no life at all right up to the moment of the famous Cambrian explosion.

But now here was Darwin, without any evidence, insisting that the earlier seas must have had abundant life and that we just hadn't found it yet because, for whatever reason, it hadn't been preserved. It simply could not be otherwise, Darwin maintained. 'The case at present must remain

*By coincidence, in 1861, at the height of the controversy, just such evidence turned up when workers in Bavaria found the bones of an ancient archaeopteryx, a creature halfway between a bird and a dinosaur. (It had feathers, but it also had teeth.) It was an impressive and helpful find, and its significance much debated, but a single discovery could hardly be considered conclusive.

inexplicable; and may be truly urged as a valid argument against the views here entertained,' he allowed most candidly, but he refused to allow an alternative possibility. By way of explanation he speculated – inventively but incorrectly – that perhaps the Precambrian seas had been too clear to lay down sediments and thus had preserved no fossils.

Even Darwin's closest friends were troubled by the blitheness of some of his assertions. Adam Sedgwick, who had taught Darwin at Cambridge and taken him on a geological tour of Wales in 1831, said the book gave him 'more pain than pleasure'. Louis Agassiz, the celebrated Swiss palaeontologist, dismissed it as poor conjecture. Even Lyell concluded gloomily: 'Darwin goes too far.'

T. H. Huxley disliked Darwin's insistence on huge amounts of geological time because Huxley was a saltationist, which is to say a believer in the idea that evolutionary changes happen not gradually but suddenly. Saltationists (the word comes from the Latin for 'leap') couldn't accept that complicated organs could ever emerge in slow stages. What good, after all, is one-tenth of a wing or half an eye? Such organs, they thought, made sense only if they appeared in a finished state.

The belief was a little surprising in as radical a spirit as Huxley because it closely recalled a very conservative religious notion first put forward by the English theologian William Paley in 1802 and known as argument from design. Paley contended that if you found a pocket-watch on the ground, even if you had never seen such a thing before, you would instantly perceive that it had been made by an intelligent entity. So it was, he believed, with nature: its complexity was proof of its design. The notion was a powerful one in the nineteenth century, and it gave Darwin trouble too. 'The eye to this day gives me a cold shudder,' he acknowledged in a letter to a friend. In the *Origin* he conceded that it 'seems, I freely confess, absurd in the highest possible degree' that natural selection could produce such an instrument in gradual steps.

Even so, and to the unending exasperation of his supporters, Darwin not only insisted that all change was gradual, but in nearly every edition of *Origin* stepped up the amount of time he supposed necessary to allow evolution to progress, which pushed his ideas increasingly out of favour. 'Eventually,' according to the scientist and historian Jeffrey Schwartz, 'Darwin lost virtually all the support that still remained among the ranks of fellow natural historians and geologists.'

Ironically, considering that Darwin called his book *On the Origin of Species*, the one thing he couldn't explain was how species originated. Darwin's theory suggested a mechanism for how a species might become stronger or better or faster – in a word, fitter – but gave no indication of how it might throw up a new species. A Scottish engineer, Fleeming Jenkin, considered the problem and noted an important flaw in Darwin's argument. Darwin believed that any beneficial trait that arose in one generation would be passed on to subsequent generations, thus strengthening the species. Jenkin pointed out that a favourable trait in one parent wouldn't become dominant in succeeding generations, but in fact would be diluted through blending. If you pour whisky into a tumbler of water, you don't make the whisky stronger, you make it weaker. And if you pour that dilute solution into another glass of water, it becomes weaker still. In the same way, any favourable trait introduced by one parent would be successively watered down by subsequent matings until it ceased to be apparent at all. Thus Darwin's theory was a recipe not for change, but for constancy. Lucky flukes might arise from time to time, but they would soon vanish under the general impulse to bring everything back to a stable mediocrity. If natural selection were to work, some alternative, unconsidered mechanism was required.

Unknown to Darwin and everyone else, 1,200 kilometres away in a tranquil corner of Middle Europe a retiring monk named Gregor Mendel was coming up with the solution.

Mendel was born in 1822 to a humble farming family in a backwater of the Austrian empire in what is now the Czech Republic. Schoolbooks once portrayed him as a simple but observant provincial monk whose discoveries were largely serendipitous – the result of noticing some interesting traits of inheritance while pottering about with pea plants in the monastery's kitchen garden. In fact, Mendel was a trained scientist – he had studied physics and mathematics at the Olmütz Philosophical Institute and University of Vienna – and he brought scientific discipline to all he did. Moreover, the monastery at Brno where he lived from 1843 was known as a learned institution. It had a library of twenty thousand books and a tradition of careful scientific investigation.

Before embarking on his experiments, Mendel spent two years preparing his control specimens, seven varieties of pea, to make sure they bred true. Then, helped by two full-time assistants, he repeatedly bred and cross-bred hybrids from thirty thousand pea plants. It was delicate work, requiring the three men to take the most exacting pains to avoid accidental cross-fertilization and to note every slight variation in the growth and appearance of seeds, pods, leaves, stems and flowers. Mendel knew what he was doing.

He never used the word 'gene' – it wasn't coined until 1913, in an English medical dictionary – though he did invent the terms 'dominant' and 'recessive'. What he established was that every seed contained two 'factors' or *Elemente*, as he called them – a dominant one and a recessive one – and these factors, when combined, produced predictable patterns of inheritance.

The results he converted into precise mathematical formulae. Altogether Mendel spent eight years on the experiments, then confirmed his results with similar experiments on flowers, corn and other plants. If anything, Mendel was *too* scientific in his approach, for when he presented his findings at the February and March meetings of the Natural History Society of Brno in 1865, the audience of about forty listened politely but was conspicuously unmoved, even though the breeding of plants was a matter of great practical interest to many of the members.

When Mendel's report was published, he eagerly sent a copy to the great Swiss botanist Karl-Wilhelm von Nägeli, whose support was more or less vital for the theory's prospects. Unfortunately, Nägeli failed to perceive the importance of what Mendel had found. He suggested that Mendel try breeding hawkweed. Mendel dutifully obeyed, but quickly realized that hawkweed had none of the requisite features for studying heritability. It was evident that Nägeli had not read the paper closely, or possibly at all. Frustrated, Mendel retired from investigating heritability and spent the rest of his life growing outstanding vegetables and studying bees, mice and sunspots, among much else. Eventually he was made abbot.

Mendel's findings weren't quite as widely ignored as is sometimes suggested. His study received a glowing entry in the *Encyclopaedia Britannica* – then a more leading record of scientific thought than now – and was cited repeatedly in an important paper by the German Wilhelm Olbers Focke. Indeed, it was because Mendel's ideas never entirely sank below the waterline of scientific thought that they were so easily recovered when the world was ready for them.

Together, without realizing it, Darwin and Mendel laid the groundwork for all of life sciences in the twentieth century. Darwin saw that all living things are connected, that ultimately they 'trace their ancestry to a single, common source'; Mendel's work provided the mechanism to explain how that could happen. The two men could easily have helped each other. Mendel owned a German edition of the *Origin of Species*, which he is known to have read, so he must have realized the applicability of his work to Darwin's, yet he appears to have made no effort to get in touch. And Darwin, for his part, is known to have studied Focke's influential paper with its repeated references to Mendel's work, but didn't connect them to his own studies.

The one thing everyone thinks featured in Darwin's argument, that humans are descended from apes, didn't feature at all except as one passing allusion. Even so, it took no great leap of imagination to see the implications for human development in Darwin's theories, and it became an immediate talking point.

The showdown came on Saturday, 30 June 1860, at a meeting of the British Association for the Advancement of Science in Oxford. Huxley had been urged to attend by Robert Chambers, author of *Vestiges of the Natural History of Creation*, though he was still unaware of Chambers's connection to that contentious tome. Darwin, as ever, was absent. The meeting was held at the Oxford Zoological Museum. More than a thousand people crowded into the chamber; hundreds more were turned away. People knew that something big was going to happen, though they had first to wait while a slumber-inducing speaker named John William Draper of New York University bravely slogged his way through two hours of introductory remarks on 'The Intellectual Development of Europe Considered with Reference to the Views of Mr Darwin.'

Finally, the Bishop of Oxford, Samuel Wilberforce, rose to speak. Wilberforce had been briefed (or so it is generally assumed) by the ardent anti-Darwinian Richard Owen, who had been a guest in his home the night before. As nearly always with events that end in uproar, accounts of what exactly transpired vary widely. In the most popular version, Wilberforce, when properly in flow, turned to Huxley with a dry smile and demanded of him whether he claimed attachment to the apes by way of his grandmother or grandfather. The remark was doubtless intended as a quip, but it came across as an icy challenge. According to his own account, Huxley turned to his neighbour and whispered, 'The Lord hath delivered him into my hands,' then rose with a certain relish.

Others, however, recalled a Huxley trembling with fury and indignation. At all events, Huxley declared that he would rather claim kinship to an ape than to someone who used his eminence to propound uninformed twaddle in what was supposed to be a serious scientific forum. Such a riposte was a scandalous impertinence, as well as an insult to Wilberforce's office, and the proceedings instantly collapsed into tumult. A Lady Brewster fainted. Robert FitzRoy, Darwin's companion on the *Beagle* twenty-five years before, wandered through the hall with a Bible held aloft, shouting, 'The Book, the Book!' (He was at the conference to present a paper on storms in his capacity as head of the newly created Meteorological Department.) Interestingly, each side afterwards claimed to have routed the other.

Darwin did eventually make his belief in our kinship with the apes explicit in *The Descent of Man* in 1871. The conclusion was a bold one, since nothing in the fossil record supported such a notion. The only known early human remains of that time were the famous Neandertal bones from Germany and a few uncertain fragments of jawbones, and many respected authorities refused to believe even in their antiquity. *The Descent of Man* was altogether a more controversial book than the *Origin*, but by the time of its appearance the world had grown less excitable and its arguments caused much less of a stir.

For the most part, however, Darwin passed his twilight years on other projects, most of which touched only tangentially on questions of natural selection. He spent amazingly long periods picking through bird droppings, scrutinizing the contents in an attempt to understand how seeds spread between continents, and spent years more studying the behaviour of worms. One of his experiments was to play the piano to them – not to amuse them but to study the effects on them of sound and vibration. He was the first to realize how vitally important worms are to soil fertility. 'It may be doubted whether there are many other animals which have played so important a part in the history of the world,' he wrote in his masterwork on the subject, *The Formation of Vegetable Mould Through the Action of Worms* (1881), which was actually more popular than *On the Origin of Species* had ever been. Among his other books were *On the Various Contrivances by which British and Foreign Orchids Are Fertilised by Insects* (1862), *Expressions of the Emotions in Man and Animals* (1872), which sold almost 5,300 copies on its first day, *The Effects of Cross and Self Fertilization in the Vegetable Kingdom* (1876) – a subject that came improbably close to Mendel's own work, without attaining anything like the same insights – and *The Power of Movement in Plants*. Finally, but not least, he devoted much effort to studying the consequences of inbreeding – a matter of private interest to him. Having married his own cousin, Darwin glumly suspected that certain physical and mental frailties among his children arose from a lack of diversity in his family tree.

Darwin was often honoured in his lifetime, but never for *On the Origin of Species* or *The Descent of Man*. When the Royal Society bestowed on him the prestigious Copley Medal it was for his geology, zoology and botany, not evolutionary theories, and the Linnaean Society was similarly pleased to honour Darwin without embracing his radical notions. He was never knighted, though he was buried in Westminster Abbey – next to Newton. He died at Down in April 1882. Mendel died two years later.

Darwin's theory didn't really gain widespread acceptance until the 1930s and 1940s, with the advance of a refined theory called, with a certain hauteur, the Modern Synthesis, combining Darwin's ideas with those of Mendel and others. For Mendel, appreciation was also posthumous, though it came somewhat sooner. In 1900, three scientists working separately in Europe rediscovered Mendel's work more or less simultaneously. It was only because one of them, a Dutchman named Hugo de Vries, seemed set to claim Mendel's insights as his own that a rival made it noisily clear that the credit really lay with the forgotten monk.

The world was almost – but not quite – ready to begin to understand how we got here: how we made each other. It is fairly amazing to reflect that at the beginning of the twentieth century, and for some years beyond, the best scientific minds in the world couldn't actually tell you, in any meaningful way, where babies came from.

And these, you may recall, were men who thought science was nearly at an end.

26

THE STUFF OF LIFE

If your two parents hadn't bonded just when they did – possibly to the second, possibly to the nanosecond – you wouldn't be here. And if their parents hadn't bonded in a precisely timely manner, you wouldn't be here either. And if their parents hadn't done likewise, and their parents before them, and so on, obviously and indefinitely, you wouldn't be here.

Push backwards through time and these ancestral debts begin to add up. Go back just eight generations to about the time that Charles Darwin and Abraham Lincoln were born, and already there are over 250 people on whose timely couplings your existence depends. Continue further, to the time of Shakespeare and the Mayflower pilgrims, and you have no fewer than 16,384 ancestors earnestly exchanging genetic material in a way that would, eventually and miraculously, result in you.

At twenty generations ago, the number of people procreating on your behalf has risen to 1,048,576. Five generations before that, and there are no fewer than 33,554,432 men and women on whose devoted couplings your existence depends. By thirty generations ago, your total number of forebears – remember, these aren't cousins and aunts and other incidental relatives, but only parents and parents of parents in a line leading ineluctably to you – is over one billion (1,073,741,824, to be precise). If you go back sixty-four generations, to the time of the Romans, the number of people on whose co-operative efforts your eventual existence depends has risen to approximately one million trillion, which is several thousand times the total number of people who have ever lived.

Clearly something has gone wrong with our maths here. The answer, it may interest you to learn, is that your line is not pure. You couldn't be here without a little incest – actually quite a lot of incest – albeit at a genetically discreet remove. With so many millions of ancestors in your background, there will have been many occasions when a relative from your mother's side of the family procreated with some distant cousin from your father's side of the ledger. In fact, if you are in a partnership now with someone from your own race and country, the chances are excellent that you are at some level related. Indeed, if you look around you on a bus or in a park or café or any crowded place, *most* of the people you see are very probably relatives. When someone boasts to you that he is descended from Shakespeare or William the Conqueror, you should answer at once: 'Me, too!' In the most literal and fundamental sense we are all family.

We are also uncannily alike. Compare your genes with any other human being's and on average they will be about 99.9 per cent the same. That is what makes us a species. The tiny differences in that remaining 0.1 per cent – 'roughly one nucleotide base in every thousand', to quote the British geneticist and recent Nobel laureate John Sulston – are what endow us with our individuality. Much has been made in recent years of the piecing together of the human genome. In fact, there is no such thing as 'the' human genome. Every human genome is different. Otherwise we would all be identical. It is the endless recombinations of our genomes – each nearly identical to all the others, but not quite – that make us what we are, both as individuals and as a species.

But what exactly is this thing we call the genome? And what, come to that, are genes? Well, start with a cell again. Inside the cell is a nucleus and inside each nucleus are the chromosomes – forty-six little bundles of complexity, of which twenty-three come from your mother and twenty-three from your father. With a very few exceptions, every cell in your body – 99.999 per cent of them, say – carries the same complement of chromosomes. (The exceptions are red blood cells, some immune system cells and egg and sperm cells, which for various organizational reasons don't carry the full genetic package.) Chromosomes constitute the complete set of instructions necessary to make and maintain you and are made of long strands of the little wonder chemical called deoxyribonucleic acid or DNA – 'the most extraordinary molecule on Earth', as it has been called.

DNA exists for just one reason – to create more DNA – and you have a lot of it inside you: nearly 2 metres of it squeezed into almost every cell. Each length of DNA comprises some 3.2 billion letters of coding, enough to provide $10^{3,480,000,000}$ possible combinations, 'guaranteed to be unique against all conceivable odds', in the words of Christian de Duve. That's a lot of possibility – a one followed by more than three billion zeroes. 'It would take more than five thousand average-size books just to print that figure,' notes de Duve. Look at yourself in the mirror and reflect upon the fact that you are beholding ten thousand trillion cells, and that almost every one of them holds two yards of densely compacted DNA, and you begin to appreciate just how much of this stuff you carry around with you. If all your DNA were woven into a single fine strand, there would be enough of it to stretch from the Earth to the Moon and back, not once or twice but again and again. Altogether, according to one calculation, you may have as much as 20 million kilometres of DNA bundled up inside you.

Your body, in short, loves to make DNA, and without it you couldn't live. Yet DNA is not itself alive. No molecule is, but DNA is, as it were, especially unalive. It is 'among the most nonreactive, chemically inert molecules in the living world', in the words of the geneticist Richard Lewontin. That is why it can be recovered from patches of long-dried blood or semen in murder investigations and coaxed from the bones of ancient Neandertals. It also explains why it took scientists so long to work out how a substance so mystifyingly low-key – so, in a word, lifeless – could be at the very heart of life itself.

As a known entity, DNA has been around longer than you might think. It was discovered as far back as 1869 by Johann Friedrich Miescher, a Swiss scientist working at the University of Tübingen in Germany. While delving microscopically through the pus in surgical bandages, Miescher found a substance he didn't recognize and called it nuclein (because it resided in the nuclei of cells). At the time, Miescher did little more than note its existence, but nuclein clearly remained on his mind, for twenty-three years later, in a letter to his uncle, he raised the possibility that such molecules could be the agents behind heredity. This was an extraordinary insight, but one so far in advance of the day's scientific requirements that it attracted no attention at all.

For most of the next half-century the common assumption was that the material – now called deoxyribonucleic acid, or DNA – had at most a subsidiary role in matters of heredity. It was too simple. It had just four basic components, called nucleotides, which was like having an alphabet of just four letters. How could you possibly write the story of life with such a rudimentary alphabet? (The answer is that you do it in much the way that you create complex messages with the simple dots and dashes of Morse code – by combining them.) DNA didn't do anything at all, as far as anyone could tell. It just sat there in the nucleus, possibly binding the chromosome in some way or adding a splash of acidity on command or fulfilling some other trivial task that no one had yet thought of. The necessary complexity, it was thought, had to exist in proteins in the nucleus.

There were, however, two problems with dismissing DNA. First, there was so much of it – nearly 2 metres in nearly every nucleus – so clearly the cells esteemed it in some important way. On top of this, it kept turning up, like the suspect in a murder mystery, in experiments. In two studies in particular, one involving the *Pneumonococcus* bacterium and another involving bacteriophages (viruses that infect bacteria), DNA betrayed an importance that could be explained only if its role were more central than prevailing thought allowed. The evidence suggested that DNA was somehow involved in the making of proteins, a process vital to life, yet it was also clear that proteins were being made *outside* the nucleus, well away from the DNA that was supposedly directing their assembly.

No-one could understand how DNA could possibly be getting messages to the proteins. The answer, we now know, was RNA, or ribonucleic acid, which acts as an interpreter between the two. It is a notable oddity of biology that DNA and proteins don't speak the same language. For almost four billion years they have been the living world's great double act, and yet they answer to mutually incompatible codes, as if one spoke Spanish and the other Hindi. To communicate they need a mediator in the form of RNA. Working with a kind of chemical clerk called a ribosome, RNA translates information from a cell's DNA into terms proteins can understand and act upon.

However, by the early 1900s, where we resume our story,

we were still a very long way from understanding that, or indeed almost anything else to do with the confused business of heredity.

Clearly there was a need for some inspired and clever experimentation, and happily the age produced a young person with the diligence and aptitude to undertake it. His name was Thomas Hunt Morgan and in 1904, just four years after the timely rediscovery of Mendel's experiments with pea plants and still almost a decade before *gene* would even become a word, he began to do remarkably dedicated things with chromosomes.

Chromosomes had been discovered by chance in 1888 and were so called because they readily absorbed dye and thus were easy to see under the microscope. By the turn of the century it was strongly suspected that they were involved in the passing on of traits, but no-one knew how, or even really whether, they did this.

Morgan chose as his subject of study a tiny, delicate fly formally called *Drosophila melanogaster*, but more commonly known as the fruit fly (or vinegar fly, banana fly or garbage fly). *Drosophila* is familiar to most of us as that frail, colourless insect that seems to have a compulsive urge to drown in our drinks. As laboratory specimens fruit flies had certain very attractive advantages: they cost almost nothing to house and feed, could be bred by the millions in milk bottles, went from egg to productive parenthood in ten days or less and had just four chromosomes, which kept things conveniently simple.

Working out of a small lab (which became known, inevitably, as the Fly Room) in Schermerhorn Hall at Columbia University in New York, Morgan and his team embarked on a programme of meticulous breeding and cross-breeding involving millions of flies (one biographer says billions, though that is probably an exaggeration), each of which had to be captured with tweezers and examined under a jeweller's glass for any tiny variations in inheritance. For six years they tried to produce mutations by any means they could think of – zapping the flies with radiation and X-rays, rearing them in bright light and darkness, baking them gently in ovens, spinning them crazily in centrifuges – but nothing worked. Morgan was on the brink of giving up when there occurred a sudden and repeatable mutation – a fly that had white eyes rather than the usual red ones. With this breakthrough, Morgan and his assistants were able to generate useful deformities, allowing them to track a trait through successive generations. By such means they could work out the correlations between particular characteristics and individual chromosomes, eventually proving to more or less everyone's satisfaction that chromosomes were at the heart of inheritance.

The problem, however, remained the next level of biological intricacy: the enigmatic genes and the DNA that composed them. These were much trickier to isolate and understand. As late as 1933, when Morgan was awarded a Nobel Prize for his work, many researchers still weren't convinced that genes even existed. As Morgan noted at the time, there was no consensus 'as to what the genes are – whether they are real or purely fictitious'. It may seem surprising that scientists could struggle to accept the physical reality of something so fundamental to cellular activity, but, as Wallace, King and Sanders point out in *Biology: The Science of Life* (that rarest thing: a readable textbook), we are in much the same position today in respect of mental processes such as thought and memory. We know that we have them, of course, but we don't know what, if any, physical form they take. So it was for a very long time with genes. The idea that you could pluck one from your body and take it away for study was as absurd to many of Morgan's peers as the idea that scientists today might capture a stray thought and examine it under a microscope.

What was certainly true was that *something* associated with chromosomes was directing cell replication. Finally, in 1944, after fifteen years of effort, a team at the Rockefeller Institute in Manhattan, led by a brilliant but diffident Canadian named Oswald Avery, succeeded with an exceedingly tricky experiment in which an innocuous strain of bacteria was made permanently infectious by crossing it with alien DNA, proving that DNA was far more than a passive molecule and almost certainly was the active agent in heredity. The Austrian-born biochemist Erwin Chargaff later quite seriously suggested that Avery's discovery was worth two Nobel Prizes.

Unfortunately, Avery was opposed by one of his own colleagues at the institute, a strong-willed and disagreeable protein enthusiast named Alfred Mirsky, who did everything in his power to discredit Avery's work – including, it has been said, lobbying the authorities at the Karolinska Institute in Stockholm not to give Avery a Nobel Prize. Avery by this time was sixty-six years old and tired. Unable to deal with the stress and controversy, he resigned his position and never went near a lab again. But other experiments elsewhere overwhelmingly supported his conclusions, and soon the race was on to find the structure of DNA.

Had you been a betting person in the early 1950s, your money would almost certainly have been on Linus Pauling of Caltech, America's leading chemist, to crack the structure of DNA. Pauling was unrivalled in determining the architecture of molecules and had been a pioneer in the field of X-ray crystallography, a technique that would prove crucial to peering into the heart of DNA. In an exceedingly distinguished career he would win two Nobel Prizes (for chemistry in 1954 and peace in 1962), but with DNA he became convinced that the structure was a triple helix, not a double one, and never quite got on the right track. Instead, victory fell to an unlikely quartet of scientists in England who didn't work as a team, often weren't on speaking terms and were for the most part novices in the field.

Of the four, the nearest to a conventional boffin was Maurice Wilkins, who had spent much of the Second World War helping to design the atomic bomb. Two of the others, Rosalind Franklin and Francis Crick, had passed their war years working for the British government on mines – Crick on the type that blow up, Franklin on the type that produce coal.

The most unconventional of the foursome was James Watson, an American prodigy who as a boy had distinguished himself as a member of a highly popular radio programme called *The Quiz Kids* (and thus could claim to be at least part of the inspiration for some of the members of the Glass family in *Frannie and Zooey* and other works by J. D. Salinger) and who had entered the University of Chicago aged just fifteen. He had earned his PhD by the age of twenty-two and was now attached to the famous Cavendish Laboratory in Cambridge. In 1951, he was a gawky 23-year-old with a strikingly lively head of hair that appears in photographs to be straining to attach itself to some powerful magnet just out of frame.

Crick, twelve years older and still without a doctorate, was less memorably hirsute and slightly more tweedy. In Watson's account he is presented as blustery, nosy, cheerfully argumentative, impatient with anyone slow to share a notion, and constantly in danger of being asked to go elsewhere. Neither was formally trained in biochemistry.

They assumed – correctly, as it turned out – that if you could determine the shape of a DNA molecule you would be able to see how it did what it did. They hoped to achieve this, it would appear, by doing as little work as possible beyond thinking, and no more of that than was absolutely necessary. As Watson cheerfully (if a touch disingenuously) remarked in his autobiographical book *The Double Helix*, 'it was my hope that the gene might be solved without my learning any chemistry.' They weren't actually assigned to work on DNA, and at one point were ordered to stop doing it. Watson was ostensibly mastering the art of crystallography; Crick was supposed to be completing a thesis on the X-ray diffraction of large molecules.

Although Crick and Watson enjoy nearly all the credit in popular accounts for solving the mystery of DNA, their breakthrough was crucially dependent on experimental work done by their competitors, the results of which were obtained 'fortuitously', in the tactful words of the historian Lisa Jardine. Far ahead of them, at least at the beginning, were two academics at King's College in London, Wilkins and Franklin.

The New Zealand-born Wilkins was a retiring figure, almost to the point of invisibility. A 1998 PBS documentary on the discovery of the structure of DNA – a feat for which he shared the 1962 Nobel Prize with Crick and Watson – managed to overlook him entirely.

Franklin was the most enigmatic character of them all. In a severely unflattering portrait, Watson in *The Double Helix* depicted Franklin as a woman who was unreasonable, secretive, chronically uncooperative and – this seemed especially to irritate him – almost wilfully unsexy. He allowed that she 'was not unattractive and might have been quite stunning had she taken even a mild interest in clothes', but in this she disappointed all expectations. She didn't even use lipstick, he noted in wonder, while her dress sense 'showed all the imagination of English blue-stocking adolescents'.*

*In 1968, Harvard University Press cancelled publication of *The Double Helix* after Crick and Wilkins complained about its characterizations, which Lisa Jardine has described as 'gratuitously hurtful'. The descriptions quoted above are as worded after Watson had softened his comments.

However, she did have the best images in existence of the possible structure of DNA, achieved by means of X-ray crystallography, the technique perfected by Linus Pauling. Crystallography had been used successfully to map atoms in crystals (hence 'crystallography'), but DNA molecules were a much more finicky proposition. Only Franklin was managing to get good results from the process, but to Wilkins's perennial exasperation she refused to share her findings.

If Franklin was not warmly forthcoming with her findings, she cannot be altogether blamed. Female academics at King's in the 1950s were treated with a formalized disdain that dazzles modern sensibilities (actually, any sensibilities). However senior or accomplished, they were not allowed into the college's senior common room but instead had to take their meals in a more utilitarian chamber that even Watson conceded was 'dingily pokey'. On top of this she was being constantly pressed – at times actively harassed – to share her results with a trio of men whose desperation to get a peek at them was seldom matched by more engaging qualities, like respect. 'I'm afraid we always used to adopt – let's say a patronizing attitude towards her,' Crick later recalled. Two of these men were from a competing institution and the third was more or less openly siding with them. It should hardly come as a surprise that she kept her results locked away.

That Wilkins and Franklin did not get along was a fact that Watson and Crick seem to have exploited to their benefit. Although the two of them were trespassing rather unashamedly on Wilkins's territory, it was with them that he increasingly sided – not altogether surprisingly, since Franklin herself was beginning to act in a decidedly queer way. Although her results showed that DNA definitely was helical in shape, she insisted to all that it was not. To Wilkins's presumed dismay and embarrassment, in the summer of 1952 she posted a mock notice around the King's physics department that said: 'It is with great regret that we have to announce the death, on Friday 18th July 1952 of D.N.A. helix . . . It is hoped that Dr. M. H. F. Wilkins will speak in memory of the late helix.'

The outcome of all this was that in January 1953 Wilkins showed Watson Franklin's images, 'apparently without her knowledge or consent'. It would be an understatement to call it a significant help to him. Years later, Watson conceded that it 'was the key event . . . it mobilized us.' Armed with the knowledge of the DNA molecule's basic shape and some important elements of its dimensions, Watson and Crick redoubled their efforts. Everything now seemed to go their way. At one point Pauling was en route to a conference in England at which he would in all likelihood have met Wilkins and learned enough to correct the misconceptions that had put him on the wrong line of enquiry; but this was the McCarthy era and Pauling found himself detained at Idlewild Airport in New York, his passport confiscated, on the grounds that he was too liberal of temperament to be allowed to travel abroad. Crick and Watson also had the no less convenient good fortune that Pauling's son was working at the Cavendish and innocently kept them abreast of any news of developments and setbacks at home.

Still facing the possibility of being trumped at any moment, Watson and Crick applied themselves feverishly to the problem. It was known that DNA had four chemical components – called adenine, guanine, cytosine and thiamine – and that these paired up in particular ways. By playing with pieces of cardboard cut into the shapes of molecules, Watson and Crick were able to work out how the pieces fit together. From this they made a Meccano-like model – perhaps the most famous in modern science – consisting of metal plates bolted together in a spiral, and invited Wilkins, Franklin and the rest of the world to have a look.

Any informed person could see at once that they had solved the problem. It was without question a brilliant piece of detective work, with or without the boost of Franklin's picture.

The 25 April 1953 edition of *Nature* carried a 900-word article by Watson and Crick titled 'A Structure for Deoxyribose Nucleic Acid'. Accompanying it were separate articles by Wilkins and Franklin. It was an eventful time in the world – Edmund Hillary was just about to clamber to the top of Everest, while Elizabeth II was shortly to be crowned Queen – so the discovery of the secret of life was mostly overlooked. It received a small mention in the *News Chronicle* and was ignored elsewhere.

Rosalind Franklin did not share in the Nobel Prize. She died of ovarian cancer at the age of just thirty-seven in 1958, four years before the award was granted. Nobel Prizes are not awarded posthumously. The cancer almost certainly arose as a result of chronic over-exposure to X-rays through her work and could have been avoided. In her much praised recent biography, Brenda Maddox noted that Franklin rarely wore a lead apron and often stepped carelessly in front of a beam. Oswald Avery never won a Nobel Prize either and was also largely overlooked by posterity, though he did at least have the satisfaction of living just long enough to see his findings vindicated. He died in 1955.

Watson and Crick's discovery wasn't actually confirmed until the 1980s. As Crick said in one of his books: 'It took over twenty-five years for our model of DNA to go from being only rather plausible, to being very plausible . . . and from there to being virtually certainly correct.'

Even so, with the structure of DNA understood, progress in genetics was swift, and by 1968 the journal *Science* could run an article entitled 'That Was the Molecular Biology That Was', suggesting – it hardly seems possible, but it is so – that the work of genetics was nearly at an end.

In fact, of course, it was only just beginning. Even now there is a great deal about DNA that we scarcely understand, not least why so much of it doesn't actually seem to *do* anything. Ninety-seven per cent of your DNA consists of nothing but long stretches of meaningless garble – 'junk' or 'non-coding DNA' as biochemists prefer to put it. Only here and there along each strand do you find sections that control and organize vital functions. These are the curious and long-elusive genes.

Genes are nothing more (nor less) than instructions to make proteins. This they do with a certain dull fidelity. In this sense, they are rather like the keys of a piano, each playing a single note and nothing else, which is obviously a trifle monotonous. But combine the genes, as you would combine piano keys, and you can create chords and melodies of infinite variety. Put all these genes together and you have (to continue the metaphor) the great symphony of existence known as the human genome.

An alternative and more common way to regard the genome is as a kind of instruction manual for the body. Viewed this way, the chromosomes can be imagined as the book's chapters and the genes as individual instructions for making proteins. The words in which the instructions are written are called codons and the letters are known as bases. The bases – the letters of the genetic alphabet – consist of the four nucleotides mentioned a page or two back: adenine, thymine, guanine and cytosine. Despite the importance of what they do, these substances are not made of anything exotic. Guanine, for instance, is the same stuff that abounds in, and gives its name to, guano.

The shape of a DNA molecule, as everyone knows, is rather like a spiral staircase or twisted rope ladder: the famous double helix. The uprights of this structure are made of a type of sugar called deoxyribose and the whole of the helix is a nucleic acid – hence the name 'deoxyribonucleic acid'. The rungs (or steps) are formed by two bases joining across the space between, and they can combine in only two ways: guanine is always paired with cytosine and thymine always with adenine. The order in which these letters appear as you move up or down the ladder constitutes the DNA code; logging it has been the job of the Human Genome Project.

Now, the particular brilliance of DNA lies in its manner of replication. When it is time to produce a new DNA molecule, the two strands part down the middle, like the zip on a jacket, and each half goes off to form a new partnership. Because each nucleotide along a strand pairs up with a specific other nucleotide, each strand serves as a template for the creation of a new matching strand. If you possessed just one strand of your own DNA, you could easily enough reconstruct the matching side by working out the necessary partnerships: if the topmost rung on one strand was made of guanine, then you would know that the topmost rung on the matching strand must be cytosine. Work your way down the ladder through all the nucleotide pairings and eventually you would have the code for a new molecule. That is just what happens in nature, except that nature does it really quickly – in only a matter of seconds, which is quite a feat.

Most of the time our DNA replicates with dutiful accuracy, but just occasionally – about one time in a million – a letter gets into the wrong place. This is known as a single nucleotide polymorphism, or SNP, familiarly known to biochemists as a 'Snip'. Generally these Snips are buried in stretches of non-coding DNA and have no detectable consequence for the body. But occasionally they make a difference. They might leave you predisposed to some disease, but equally they might confer some slight advantage – more protective pigmentation, for instance, or increased production of red blood cells for someone living at altitude.

Over time, these slight modifications accumulate both in individuals and in populations, contributing to the distinctiveness of both.

The balance between accuracy and errors in replication is a fine one. Too many errors and the organism can't function, but too few and it sacrifices adaptability. A similar balance must exist between stability and innovation in an organism. An increase in red blood cells can help a person or group living at high elevations to move and breathe more easily, because more red cells can carry more oxygen. But additional red cells also thicken the blood. Add too many 'and it's like pumping oil', in the words of Temple University anthropologist Charles Weitz. That's hard on the heart. Thus, those designed to live at high altitude get increased breathing efficiency, but pay for it with higher-risk hearts. By such means does Darwinian natural selection look after us. It also helps to explain why we are all so similar. Evolution simply won't let you become too different – not without becoming a new species, anyway.

The 0.1 per cent difference between your genes and mine is accounted for by our Snips. Now, if you compared your DNA with a third person's, there would also be 99.9 per cent correspondence, but the Snips would, for the most part, be in different places. Add more people to the comparison and you will get yet more Snips in yet more places. For every one of your 3.2 billion bases, somewhere on the planet there will be a person, or group of persons, with different coding in that position. So not only is it wrong to refer to 'the' human genome, but in a sense we don't even have 'a' human genome. We have 6 billion of them. We are all 99.9 per cent the same, but equally, in the words of the biochemist David Cox, 'you could say all humans share nothing, and that would be correct, too.'

But we have still to explain why so little of that DNA has any discernible purpose. It starts to get a little unnerving, but it does really seem that the purpose of life is to perpetuate DNA. The 97 per cent of our DNA commonly called junk is largely made up of clumps of letters that, in Matt Ridley's words, 'exist for the pure and simple reason that they are good at getting themselves duplicated'.* Most of your DNA, in other words, is devoted not to you but to itself: you are a machine for the benefit of it, not it for you. Life, you will recall, just wants to be, and DNA is what makes it so.

Even when DNA includes instructions for making genes – when it codes for them, as scientists put it – it is not necessarily with the smooth functioning of the organism in mind. One of the commonest genes we have is for a protein called reverse transcriptase, which has no known beneficial function in human beings at all. The one thing it *does* do is make it possible for retroviruses, such as HIV, to slip unnoticed into the human system.

In other words, our bodies devote considerable energies to producing a protein that does nothing that is beneficial and sometimes clobbers us. Our bodies have no choice but to make it because the genes order it. We are vessels for their whims. Altogether, almost half of human genes – the largest proportion known in any organism – don't do anything at all, as far as we can tell, except reproduce themselves.

All organisms are in some sense slaves to their genes. That's why salmon and spiders and other types of creature more or less beyond counting are prepared to die in the

*Junk DNA does have a use. It is the portion employed in DNA fingerprinting. Its practicality for this purpose was discovered accidentally by Alec Jeffreys, a scientist at the University of Leicester. In 1986 Jeffreys was studying DNA sequences for genetic markers associated with heritable diseases when he was approached by the police and asked if he could help connect a suspect to two murders. He realized his technique ought to work perfectly for solving criminal cases – and so it proved. A young baker with the improbable name of Colin Pitchfork was sentenced to two life terms in prison for the murders.

process of mating. The desire to breed, to disperse one's genes, is the most powerful impulse in nature. As Sherwin B. Nuland has put it: 'Empires fall, ids explode, great symphonies are written, and behind all of it is a single instinct that demands satisfaction.' From an evolutionary point of view, sex is really just a reward mechanism to encourage us to pass on our genetic material.

Scientists had only barely absorbed the surprising news that most of our DNA doesn't do anything when even more unexpected findings began to turn up. First in Germany and then in Switzerland, researchers performed some rather bizarre experiments that produced curiously unbizarre outcomes. In one, they took the gene that controlled the development of a mouse's eye and inserted it into the larva of a fruit fly. The thought was that it might produce something interestingly grotesque. In fact, the mouse-eye gene not only made a viable eye in the fruit fly, it made a *fly's* eye. Here were two creatures that hadn't shared a common ancestor for 500 million years, yet could swap genetic material as if they were sisters.

The story was the same wherever researchers looked. They found that they could insert human DNA into certain cells of flies and the flies would accept it as if it were their own. Over 60 per cent of human genes, it turns out, are fundamentally the same as those found in fruit flies. At least 90 per cent correlate at some level with those found in mice. (We even have the same genes for making a tail, if only they would switch on.) In field after field, researchers found that whatever organism they were working on – whether nematode worms or human beings – they were often studying essentially the same genes. Life, it appeared, was drawn up from a single set of blueprints.

Further probings revealed the existence of a clutch of master control genes, each directing the development of a section of body, which were dubbed homeotic (from a Greek word meaning 'similar') or hox genes. Hox genes answered the long-bewildering question of how billions of embryonic cells, all arising from a single fertilized egg and carrying identical DNA, know where to go and what to do – that this one should become a liver cell, this one a stretchy neuron, this one a bubble of blood, this one part of the shimmer on a beating wing. It is the hox genes that instruct them, and they do it for all organisms in much the same way.

Interestingly, the amount of genetic material and how it is organized doesn't necessarily, or even generally, reflect the level of sophistication of the creature that contains it. We have forty-six chromosomes, but some ferns have more than six hundred. The lungfish, one of the least evolved of all complex animals, has forty times as much DNA as we have. Even the common newt is more genetically splendorous than we are, by a factor of five.

Clearly it is not the number of genes you have that matters, so much as what you do with them. This is a very good thing, because the number of genes in humans has taken a big hit lately. Until recently it was thought that humans had at least one hundred thousand genes, possibly a good many more, but that number was drastically reduced by the first results of the Human Genome Project, which suggested a figure more like thirty-five thousand or forty thousand genes – about the same number as are found in grass. That came as both a surprise and a disappointment.

It won't have escaped your attention that genes have been commonly implicated in any number of human frailties. Exultant scientists have at various times declared themselves to have found the genes responsible for obesity, schizophrenia, homosexuality, criminality, violence, alcoholism, even shoplifting and homelessness. Perhaps the apogee (or nadir) of this faith in biodeterminism was a study published in the journal *Science* in 1980 contending that women are genetically inferior at mathematics. In fact, we now know, almost nothing about you is so accommodatingly simple.

This is clearly a pity in one important sense, for if you had individual genes that determined height or propensity to diabetes or to baldness or any other distinguishing trait, then it would be easy – comparatively easy, anyway – to isolate and tinker with them. Unfortunately, thirty-five thousand genes functioning independently is not nearly enough to produce the kind of physical complexity that makes a satisfactory human being. Genes clearly, therefore, must co-operate. A few disorders – haemophilia, Parkinson's disease, Huntington's disease and cystic fibrosis, for example – are caused by lone dysfunctional genes, but as a rule disruptive genes are weeded out by natural selection long before they can become permanently troublesome to a species or population. For the most part our fate and comfort – and even our eye colour – are determined not by individual genes but by complexes of genes working in alliance. That's why it is so hard to work out how it all fits together and why we won't be producing designer babies any time soon.

In fact, the more we have learned in recent years the more complicated matters have tended to become. Even thinking, it turns out, affects the ways genes work. How fast a man's beard grows, for instance, is partly a function of how much he thinks about sex (because thinking about sex produces a testosterone surge). In the early 1990s, scientists made an even more profound discovery when they found they could knock out supposedly vital genes from embryonic mice, and still see the mice often not only born healthy, but sometimes actually fitter than their brothers and sisters who had not been tampered with. When certain important genes were destroyed, it turned out, others were stepping in to fill the breach. This was excellent news for us as organisms, but not so good for our understanding of how cells work, since it introduced an extra layer of complexity to something that we had barely begun to understand anyway.

It is largely because of these complicating factors that cracking the human genome came to be seen almost at once as only a beginning. The genome, as Eric Lander of MIT has put it, is like a parts list for the human body: it tells us what we are made of, but says nothing about how we work. What's needed now is the operating manual – instructions for how to make it go. We are not close to that point yet.

So now the quest is to crack the human proteome – a concept so novel that the term *proteome* didn't even exist a decade ago. The proteome is the library of information that creates proteins. 'Unfortunately,' observed *Scientific American* in the spring of 2002, 'the proteome is much more complicated than the genome.'

That's putting it mildly. Proteins, you will remember, are the workhorses of all living systems; as many as a hundred million of them may be busy in any cell at any moment. That's a lot of activity to try to figure out. Worse, proteins' behaviour and functions are based not simply on their chemistry, as with genes, but also on their shapes. To function, a protein not only must have the necessary chemical components, properly assembled, but then must also be folded into an extremely specific shape. 'Folding' is the term that's used, but it's a misleading one as it suggests a geometrical tidiness that doesn't in fact apply. Proteins loop and coil and crinkle into shapes that are at once extravagant and complex. They are more like furiously mangled coat hangers than folded towels.

Moreover, proteins are (if I may be permitted to use a handy archaism) the swingers of the biological world. Depending on mood and metabolic circumstance, they will allow themselves to be phosphorylated, glycosylated, acetylated, ubiquitinated, farneysylated, sulphated and linked to glycophosphatidylinositol anchors, among rather a lot else. Often it takes relatively little to get them going, it appears. Drink a glass of wine, as *Scientific American* notes, and you materially alter the number and types of proteins at large in your system. This is pleasant for drinkers, but not nearly so helpful for geneticists who are trying to understand what is going on.

It can all begin to seem impossibly complicated, and in some ways it *is* impossibly complicated. But there is also an underlying simplicity in all this, too, owing to an equally elemental underlying unity in the way life works. All the tiny, deft chemical processes that animate cells – the co-operative efforts of nucleotides, the transcription of DNA into RNA – evolved just once and have stayed pretty well fixed ever since across the whole of nature. As the late French geneticist Jacques Monod put it, only half in jest: 'Anything that is true of E. coli must be true of elephants, except more so.'

Every living thing is an elaboration on a single original plan. As humans we are mere increments – each of us a musty archive of adjustments, adaptations, modifications and providential tinkerings stretching back 3.8 billion years. Remarkably, we are even quite closely related to fruit and vegetables. About half the chemical functions that take place in a banana are fundamentally the same as the chemical functions that take place in you.

It cannot be said too often: all life is one. That is, and I suspect will for ever prove to be, the most profound true statement there is.

VI

THE ROAD TO US

Descended from the apes! My dear, let us hope that it is not true, but if it is, let us pray that it will not become generally known.

Remark attributed to the wife of the Bishop of Worcester after Darwin's theory of evolution was explained to her

27

ICE TIME

I had a dream, which was not all a dream.
The bright sun was extinguish'd, and the stars
Did wander . . .

Byron, 'Darkness'

In 1815, on the island of Sumbawa in Indonesia, a handsome and long quiescent mountain named Tambora exploded spectacularly, killing a hundred thousand people with its blast and associated tsunamis. No-one living now has ever seen such fury. Tambora was far bigger than anything any living human has experienced. It was the biggest volcanic explosion in ten thousand years – 150 times the size of Mount St Helens, equivalent to sixty thousand Hiroshima-sized atom bombs.

News didn't travel terribly fast in those days. In London, *The Times* ran a small story – actually a letter from a merchant – seven months after the event. But by this time Tambora's effects were already being felt. Two hundred and forty cubic kilometres of smoky ash, dust and grit had diffused through the atmosphere, obscuring the Sun's rays and causing the Earth to cool. Sunsets were unusually but blearily colourful, an effect memorably captured by the artist

J. M. W. Turner, who could not have been happier, but mostly the world existed under an oppressive, dusky pall. It was this deathly dimness that inspired Bryon to write the lines quoted above.

Spring never came and summer never warmed: 1816 became known as the year without summer. Crops everywhere failed to grow. In Ireland a famine and associated typhoid epidemic killed sixty-five thousand people. In New England, the year became popularly known as Eighteen Hundred and Froze to Death. Morning frosts continued until June and almost no planted seed would grow. Short of fodder, livestock died or had to be prematurely slaughtered. In every way it was a dreadful year – almost certainly the worst for farmers in modern times. Yet globally the temperature fell by less than 1 degree Celsius. The Earth's natural thermostat, as scientists would learn, is an exceedingly delicate instrument.

The nineteenth century was already a chilly time. For two hundred years Europe and North America had been experiencing a Little Ice Age, as it has become known, which permitted all kinds of wintry events – frost fairs on the Thames, ice-skating races along Dutch canals – that are mostly impossible now. It was a period, in other words, when frigidity was much on people's minds. So we may perhaps excuse nineteenth-century geologists for being slow to realize that the world they lived in was in fact balmy compared with former epochs, and that much of the land around them had been shaped by crushing glaciers and cold that would wreck even a frost fair.

They knew there was something odd about the past. The European landscape was littered with inexplicable anomalies – the bones of Arctic reindeer in the warm south of France, huge rocks stranded in improbable places – and they often came up with inventive but not terribly plausible explanations. One French naturalist named de Luc, trying to explain how granite boulders had come to rest high up on the limestone flanks of the Jura Mountains, suggested that perhaps they had been shot there by compressed air in caverns, like corks out of a popgun. The term for a displaced boulder is an erratic, but in the nineteenth century the expression seemed to apply more often to the theories than to the rocks.

The great British geologist Arthur Hallam has suggested that if James Hutton, the eighteenth-century father of geology, had visited Switzerland, he would have seen at once the significance of the carved valleys, the polished striations, the telltale strand lines where rocks had been dumped, and the other abundant clues that point to passing ice sheets. Unfortunately, Hutton was not a traveller. But even with nothing better at his disposal than secondhand accounts, Hutton rejected out of hand the idea that huge boulders had been carried 1,000 metres up mountainsides by floods – all the water in the world won't make a boulder float, he pointed out – and became one of the first to argue for widespread glaciation. Unfortunately his ideas escaped notice, and for another half-century most naturalists continued to insist that the gouges on rocks could be attributed to passing carts or even the scrape of hobnailed boots.

Local peasants, uncontaminated by scientific orthodoxy, knew better, however. The naturalist Jean de Charpentier told the story of how in 1834 he was walking along a country lane with a Swiss woodcutter when they got to talking about the rocks along the roadside. The woodcutter matter-of-factly told him that the boulders had come from the Grimsel, a zone of granite some distance away. 'When I asked him how he thought that these stones had reached their location, he answered without hesitation: "The Grimsel glacier transported them on both sides of the valley, because that glacier extended in the past as far as the town of Bern."'

Charpentier was delighted, for he had come to such a view himself; but when he raised the notion at scientific gatherings, it was dismissed. One of Charpentier's closest friends was another Swiss naturalist, Louis Agassiz, who after some initial scepticism came to embrace, and eventually all but appropriate, the theory.

Agassiz had studied under Cuvier in Paris and now held the post of Professor of Natural History at the College of Neuchâtel in Switzerland. Another friend of Agassiz's, a botanist named Karl Schimper, was actually the first to coin the term 'ice age' (in German, *Eiszeit*), in 1837, and to propose that there was good evidence to show that ice had once lain heavily not just across the Swiss Alps, but over much of Europe, Asia and North America. It was a radical notion. He lent Agassiz his notes – then came very much to regret it as Agassiz increasingly got the credit for what Schimper felt, with some legitimacy, was his theory. Charpentier likewise ended up a bitter enemy of his old friend. Alexander von Humboldt, yet another friend, may have had Agassiz at least partly in mind when he observed that there are three stages in scientific discovery: first, people deny that it is true; then they deny that it is important; finally they credit the wrong person.

At all events, Agassiz made the field his own. In his quest to understand the dynamics of glaciation, he went everywhere – deep into dangerous crevasses and up to the summits of the craggiest Alpine peaks, often apparently unaware that he and his team were the first to climb them. Nearly everywhere Agassiz encountered an unyielding reluctance to accept his theories. Humboldt urged him to return to his area of real expertise, fossil fish, and give up this mad obsession with ice, but Agassiz was a man possessed by an idea.

Agassiz's theory found even less support in Britain, where most naturalists had never seen a glacier and often couldn't grasp the crushing forces that ice in bulk exerts. 'Could scratches and polish just be due to *ice*?' asked Roderick Murchison in a mocking tone at one meeting, evidently imagining the rocks as covered in a kind of light and glassy rime. To his dying day, he expressed the frankest incredulity at those 'ice-mad' geologists who believed that glaciers could account for so much. William Hopkins, a Cambridge professor and leading member of the Geological Society, endorsed this view, arguing that the notion that ice could transport boulders presented 'such obvious mechanical absurdities' as to make it unworthy of the society's attention.

Undaunted, Agassiz travelled tirelessly to promote his theory. In 1840 he read a paper to a meeting of the British Association for the Advancement of Science in Glasgow, at which he was openly criticized by the great Charles Lyell. The following year the Geological Society of Edinburgh passed a resolution conceding that there might be some general merit in the theory but that certainly none of it applied to Scotland.

Lyell did eventually come round. His moment of epiphany came when he realized that a moraine, or line of rocks, near his family estate in Scotland, which he had passed hundreds of times, could be understood only if one accepted that a glacier had dropped them there. But, having become converted, Lyell then lost his nerve and backed off public support of the ice age idea. It was a frustrating time for Agassiz. His marriage was breaking up, Schimper was hotly accusing him of the theft of his ideas, Charpentier wouldn't speak to him and the greatest living geologist offered support of only the most tepid and vacillating kind.

In 1846 Agassiz travelled to America to give a series of lectures, and there at last found the esteem he craved. Harvard gave him a professorship and built him a first-rate museum, the Museum of Comparative Zoology. Doubtless it helped that he had settled in New England, where the long winters encouraged a certain sympathy for the idea of interminable periods of cold. It also helped that six years after his arrival the first scientific expedition to Greenland reported that nearly the whole of that semi-continent was covered in an ice sheet just like the ancient one imagined in Agassiz's theory. At long last, his ideas began to find a real following. The one central defect of Agassiz's theory was that his ice ages had no cause. But assistance was about to come from an unlikely quarter.

In the 1860s, journals and other learned publications in Britain began to receive papers on hydrostatics, electricity and other scientific subjects from a James Croll of Anderson's University in Glasgow. One of the papers, on how variations in the Earth's orbit might have precipitated ice ages, was published in the *Philosophical Magazine* in 1864 and was recognized at once as a work of the highest standard. So there was some surprise, and perhaps just a touch of embarrassment, when it turned out that Croll was not an academic at the university, but a janitor.

Born in 1821, Croll grew up poor and his formal education lasted only to the age of thirteen. He worked at a variety of jobs – as a carpenter, insurance salesman, keeper of a temperance hotel – before taking a position as a janitor at Anderson's (now the University of Strathclyde) in Glasgow. By somehow inducing his brother to do much of his work, he was able to pass many quiet evenings in the university library teaching himself physics, mechanics, astronomy, hydrostatics and the other fashionable sciences of the day, and gradually began to produce a string of papers, with a particular emphasis on the motions of the Earth and their effect on climate.

Croll was the first to suggest that cyclical changes in the shape of the Earth's orbit, from elliptical (which is to say, slightly oval) to nearly circular to elliptical again, might explain the onset and retreat of ice ages. No-one had ever thought before to consider an astronomical explanation for variations in the Earth's weather. Thanks almost entirely to Croll's persuasive theory, people in Britain began to become more responsive to the notion that at some former time parts of the Earth had been in the grip of ice. When his ingenuity and aptitude were recognized, Croll was given a job at the Geological Survey of Scotland and widely honoured: he was made a fellow of the Royal Society in London and of the New York Academy of Science, and given an honorary degree from the University of St Andrews, among much else.

Unfortunately, just as Agassiz's theory was at last beginning to find converts in Europe, he was busy taking it into ever more exotic territory in America. He began to find evidence for glaciers practically everywhere he looked, including near the equator. Eventually he became convinced that ice had once covered the whole Earth, extinguishing all life, which God had then recreated. None of the evidence Agassiz cited supported such a view. Nonetheless, in his adopted country his stature grew and grew until he was regarded as only slightly below a deity. When he died in 1873 Harvard felt it necessary to appoint three professors to take his place.

Yet, as sometimes happens, his theories fell swiftly out of fashion. Less than a decade after his death his successor in the chair of geology at Harvard wrote that the 'so-called glacial epoch . . . so popular a few years ago among glacial geologists may now be rejected without hesitation.'

Part of the problem was that Croll's computations suggested that the most recent ice age occurred eighty thousand years ago, whereas the geological evidence increasingly indicated that the Earth had undergone some sort of dramatic perturbation much more recently than that. Without a plausible explanation for what might have provoked an ice age, the whole theory fell into abeyance. There it might have remained for some time had it not been for a Serbian academic named Milutin Milankovitch, who had no background in celestial motions at all – he was a mechanical engineer by training – but who in the early 1900s developed an unexpected interest in the matter. Milankovitch realized that the problem with Croll's theory was not that it was incorrect but that it was too simple.

As the Earth moves through space, it is subject not just to variations in the length and shape of its orbit, but also to rhythmic shifts in its angle of orientation to the Sun – its tilt and pitch and wobble – all affecting the duration and intensity of sunlight falling on any patch of land. In particular it is subject to three changes in position, known formally as its obliquity, precession and eccentricity, over long periods of time. Milankovitch wondered if there might be a relationship between these complex cycles and the comings and goings of ice ages. The difficulty was that the cycles were of widely different lengths – of approximately twenty thousand, forty thousand and a hundred thousand years respectively, but varying in each case by up to a few thousand years – which meant that determining their points of intersection over long spans of time involved a nearly endless amount of exceedingly devoted computation. Essentially, Milankovitch had to work out the angle and duration of incoming solar radiation at every latitude on Earth, in every season, for a million years, adjusted for three ever-changing variables.

Happily, this was precisely the sort of repetitive toil that suited Milankovitch's temperament. For the next twenty years, even while on holiday, he worked ceaselessly with pencil and slide rule computing the tables of his cycles – work that now could be completed in a day or two with a computer. The calculations all had to be made in his spare time, but in 1914 Milankovitch suddenly got a great deal of that when the First World War broke out and he was arrested owing to his position as a reservist in the Serbian army. He spent most of the next four years under loose house arrest in Budapest, required only to report to the police once a week. The rest of his time was spent working in the library of the Hungarian Academy of Sciences. He was possibly the happiest prisoner of war in history.

The eventual outcome of his diligent scribblings was the 1930 book *Mathematical Climatology and the Astronomical Theory of Climatic Changes*. Milankovitch was right that there was a relationship between ice ages and planetary wobble, though like most people he assumed that it was a gradual increase in harsh winters that led to these long spells of coldness. It was a Russian–German meteorologist, Wladimir Köppen – father-in-law of our tectonic friend Alfred Wegener – who saw that the process was more subtle, and rather more unnerving, than that.

The cause of ice ages, Köppen decided, is to be found in cool summers, not brutal winters. If summers are too cool to melt all the snow that falls on a given area, more incoming sunlight is bounced back by the reflective surface, exacerbating the cooling effect and encouraging yet more snow to fall. The consequence would tend to be self-perpetuating. As snow accumulated into an ice sheet, the region would grow cooler, prompting more ice to accumulate. As the glaciologist Gwen Schultz has noted: 'It is not necessarily the *amount* of snow that causes ice sheets but the fact that snow, however little, lasts.' It is thought that an ice age could start from a single unseasonal summer. The leftover snow reflects heat and exacerbates the chilling effect. 'The process is self-enlarging, unstoppable, and once the ice is really growing it moves,' says McPhee. You have advancing glaciers and an ice age.

In the 1950s, because of imperfect dating technology, scientists were unable to correlate Milankovitch's carefully worked-out cycles with the supposed dates of ice ages as then perceived, and so Milankovitch and his calculations increasingly fell out of favour. He died in 1958, unable to prove that his cycles were correct. By this time, in the words of one history of the period, 'you would have been hard pressed to find a geologist or meteorologist who regarded the model as being anything more than an historical curiosity.' Not until the 1970s and the refinement of a potassium–argon method for dating ancient sea-floor sediments were his theories finally vindicated.

The Milankovitch cycles alone are not enough to explain cycles of ice ages. Many other factors are involved – not least the disposition of the continents, in particular the presence of land masses over the poles – but the specifics of these are imperfectly understood. It has been suggested, however, that if you hauled North America, Eurasia and Greenland just 500 kilometres north we would have permanent and inescapable ice ages. We are very lucky, it appears, to get any good weather at all. Even less well understood are the cycles of comparative balminess within ice ages, known as interglacials. It is mildly disconcerting to reflect that the whole of meaningful human history – the development of farming, the creation of towns, the rise of mathematics and writing and science and all the rest – has taken place within an atypical patch of fair weather. Previous interglacials have lasted as little as eight thousand years. Our own has already passed its ten-thousandth anniversary.

The fact is, we are still very much in an ice age; it's just a somewhat shrunken one – though less shrunken than many people realize. At the height of the last period of glaciation, around twenty thousand years ago, about 30 per cent of the Earth's land surface was under ice. Ten per cent still is. (And a further 14 per cent is in a state of permafrost.) Three-quarters of all the fresh water on Earth is locked up in ice even now, and we have ice caps at both poles – a situation that may be unique in the Earth's history. That there are snowy winters through much of the world and permanent glaciers even in temperate places such as New Zealand may seem quite natural, but in fact it is a most unusual situation for the planet.

For most of its history until fairly recent times the general pattern for the Earth was to be hot, with no permanent ice anywhere. The current ice age – ice epoch, really – started about forty million years ago, and has ranged from murderously bad to not bad at all. We live in one of the few spells of the latter. Ice ages tend to wipe out evidence of earlier ice ages, so the further back you go the more sketchy the picture grows, but it appears that we have had at least seventeen severe glacial episodes in the last 2.5 million years or so – the period that coincides with the rise of *Homo erectus* in Africa followed by modern humans. Two commonly cited culprits for the present epoch are the rise of the Himalayas and the formation of the Isthmus of Panama, the first disrupting air flows, the second ocean currents. India, once an island, has pushed 2,000 kilometres into the Asian land mass over the past 45 million years, raising not only the Himalayas, but also the vast Tibetan plateau behind it. The hypothesis is that the higher landscape was not only cooler, but diverted winds in a way that made them flow north and towards North America, making it more susceptible to long-term chills. Then, beginning about five million years ago, Panama rose from the sea, closing the gap between North and South America, disrupting the flows of warming currents between the Pacific and the Atlantic, and changing patterns of precipitation across at least half the world. One consequence was a drying out of Africa, which caused apes to climb down out of trees and go looking for a new way of living on the emerging savannas.

At all events, with the oceans and continents arranged as they are now, it appears that ice will be a long-term part of our future. According to John McPhee, about fifty more glacial episodes can be expected, each lasting a hundred thousand years or so, before we can hope for a really long thaw.

Before 50 million years ago the Earth had no regular ice ages, but when we did have them they tended to be colossal. A massive freezing occurred about 2.2 billion years ago, followed by a billion years or so of warmth. Then there was another ice age even larger than the first – so large that some scientists are now referring to the period in which it occurred as the Cryogenian, or super ice age. The condition is more popularly known as Snowball Earth.

Snowball, however, barely captures the murderousness of conditions. The theory is that because of a fall in solar radiation of about 6 per cent and a dropoff in the production (or retention) of greenhouse gases, the Earth essentially lost its ability to hold on to its heat. It became a kind of all-over Antarctica. Temperatures plunged by as much as 45 degrees Celsius. The entire surface of the planet may have frozen solid, with ocean ice up to 800 metres thick at higher latitudes and tens of metres thick even in the tropics.

There is a serious problem in all this in that the geological evidence indicates ice everywhere, including around the equator, while the biological evidence suggests just as firmly that there must have been open water somewhere. For one thing, cyanobacteria survived the experience and they photosynthesize. For that they needed sunlight, but as you will know if you have ever tried to peer through it, ice very quickly becomes opaque and after only a few yards would pass on no light at all. Two possibilities have been suggested. One is that a little ocean water did remain exposed (perhaps because of some kind of localized warming at a hot spot); the other is that maybe the ice formed in such a way that it remained translucent – a condition that does sometimes happen in nature.

If Earth did freeze over, then there is the very difficult

question of how it ever got warm again. An icy planet should reflect so much heat that it would stay frozen for ever. It appears that rescue may have come from our molten interior. Once again we may be indebted to tectonics for allowing us to be here. The idea is that we were saved by volcanoes, which pushed through the buried surface, pumping out lots of heat and gases that melted the snows and re-formed the atmosphere. Interestingly, the end of this hyper-frigid episode is marked by the Cambrian outburst – the springtime event of life's history. In fact, it may not have been as tranquil as all that. As Earth warmed, it probably had the wildest weather it has ever experienced, with hurricanes powerful enough to raise waves to the heights of skyscrapers and rainfalls of indescribable intensity.

Throughout all this the tubeworms and clams and other life forms adhering to deep ocean vents undoubtedly went on as if nothing were amiss, but all other life on Earth probably came as close as it ever has to checking out entirely. It was all a long time ago and at this stage we just don't know.

Compared with a Cryogenian outburst, the ice ages of more recent times seem pretty small-scale, but of course they were immensely grand by the standards of anything to be found on Earth today. The Wisconsian ice sheet, which covered much of Europe and North America, was over 3 kilometres thick in places and marched forward at a rate of about 120 metres a year. What a thing it must have been to behold. Even at their leading edge, the ice sheets could be nearly 800 metres thick. Imagine standing at the base of a wall of ice that high. Behind this edge, over an area measuring in the millions of square kilometres, would be nothing but more ice, with only a few of the tallest mountain summits poking through here and there. Whole continents sagged under the weight of so much ice and even now, twelve thousand years after the glaciers' withdrawal, are still rising back into place. The ice sheets didn't just dribble out boulders and long lines of gravelly moraines, but dumped entire land masses – Long Island and Cape Cod and Nantucket, among others – as they slowly swept along. It's little wonder that geologists before Agassiz had trouble grasping their monumental capacity to rework landscapes.

If ice sheets advanced again, we have nothing in our armoury that could deflect them. In 1964, at Prince William Sound in Alaska, one of the largest glacial fields in North America was hit by the strongest earthquake ever recorded on the continent. It measured 9.2 on the Richter scale. Along the fault line, the land rose by as much as 6 metres. The quake was so violent, in fact, that it made water slosh out of pools in Texas. And what effect did this unparalleled outburst have on the glaciers of Prince William Sound? None at all. They just soaked it up and kept on moving.

For a long time it was thought that we moved into and out of ice ages gradually, over hundreds or thousands of years, but we now know that that has not been the case. Thanks to ice cores from Greenland we have a detailed record of climate for something over a hundred thousand years, and what is found there is not comforting. It shows that for most of its recent history the Earth has been nothing like the stable and tranquil place that civilization has known, but rather has lurched violently between periods of warmth and brutal chill.

Towards the end of the last big glaciation, some twelve thousand years ago, Earth began to warm, and quite rapidly, but then abruptly plunged back into bitter cold for a thousand years or so in an event known to science as the Younger Dryas. (The name comes from the Arctic plant the dryas, which is one of the first to recolonize land after an ice sheet withdraws. There was also an Older Dryas period, but it wasn't so sharp.) At the end of this thousand-year onslaught average temperatures leaped again, by as much as

4 degrees Celsius in twenty years, which doesn't sound terribly dramatic but is equivalent to exchanging the climate of Scandinavia for that of the Mediterranean in just two decades. Locally, changes have been even more dramatic. Greenland ice cores show the temperatures there changing by as much as 8 degrees Celsius in ten years, drastically altering rainfall patterns and growing conditions. This must have been unsettling enough on a thinly populated planet. Today the consequences would be pretty well unimaginable.

What is most alarming is that we have no idea – none – what natural phenomena could so swiftly rattle the Earth's thermometer. As Elizabeth Kolbert, writing in the *New Yorker*, has observed: 'No known external force, or even any that has been hypothesized, seems capable of yanking the temperature back and forth as violently, and as often, as these cores have shown to be the case.' There seems to be, she adds, 'some vast and terrible feedback loop', probably involving the oceans and disruptions of the normal patterns of ocean circulation, but all this is a long way from being understood.

One theory is that the heavy inflow of meltwater to the seas at the beginning of the Younger Dryas reduced the saltiness (and thus density) of northern oceans, causing the Gulf Stream to swerve to the south, like a driver trying to avoid a collision. Deprived of the Gulf Stream's warmth, the northern latitudes returned to chilly conditions. But this doesn't begin to explain why a thousand years later, when the Earth warmed once again, the Gulf Stream didn't veer as before. Instead, we were given the period of unusual tranquillity known as the Holocene, the time in which we live now.

There is no reason to suppose that this stretch of climatic stability should last much longer. In fact, some authorities believe that we are in for even worse. It is natural to suppose that global warming would act as a useful counterweight to the Earth's tendency to plunge back into glacial conditions. However, as Kolbert has pointed out, when you are confronted with a fluctuating and unpredictable climate, 'the last thing you'd want to do is conduct a vast unsupervised experiment on it'. It has even been suggested, with more plausibility than would at first seem evident, that an ice age might actually be induced by a rise in temperatures. The idea is that a slight warming would enhance evaporation rates and increase cloud cover, leading in the higher latitudes to more persistent accumulations of snow. In fact, global warming could plausibly, if paradoxically, lead to powerful localized cooling in North America and northern Europe.

Climate is the product of so many variables – rising and falling carbon dioxide levels, the shifts of continents, solar activity, the stately wobbles of the Milankovitch cycles – that it is as difficult to comprehend the events of the past as it is to predict those of the future. Much is simply beyond us. Take Antarctica. For at least 20 million years after it settled over the South Pole Antarctica remained covered in plants and free of ice. That simply shouldn't have been possible.

No less intriguing are the known ranges of some late dinosaurs. The British geologist Stephen Drury notes that forests within 10 degrees latitude of the North Pole were home to great beasts, including tyrannosaurus rex. 'That is bizarre,' he writes, 'for such a high latitude is continually dark for three months of the year.' Moreover, there is now evidence that these high latitudes suffered severe winters. Oxygen isotope studies suggest that the climate around Fairbanks, Alaska, was about the same in the late Cretaceous period as it is now. So what was tyrannosaurus doing there? Either it migrated seasonally over enormous distances or it spent much of the year in snowdrifts in the dark. In Australia – which at that time was more polar in its orientation – a retreat to warmer climes wasn't possible. How dinosaurs managed to survive in such conditions can only be guessed.

One thought to bear in mind is that if the ice sheets did start to form again, for whatever reason, there is a lot more water for them to draw on this time. The Great Lakes, Hudson Bay, the countless lakes of Canada – these weren't there to fuel the last ice age. They were created by it.

On the other hand, the next phase of our history could see us melting a lot of ice rather than making it. If all the ice sheets melted, sea levels would rise by 60 metres – the height of a twenty-storey building – and every coastal city in the world would be inundated. More likely, at least in the short term, is the collapse of the West Antarctic ice sheet. In the past fifty years the waters around it have warmed by 2.5 degrees Celsius and collapses have increased dramatically. Because of the underlying geology of the area, a large-scale collapse is all the more possible. If so, sea levels globally would rise – and pretty quickly – by between 4.5 and 6 metres on average.

The extraordinary fact is that we don't know which is more likely: a future offering us aeons of perishing frigidity or one giving us equal expanses of steamy heat. Only one thing is certain: we live on a knife edge.

In the long run, incidentally, ice ages are by no means altogether bad news for the planet. They grind up rocks, leaving behind new soils of sumptuous richness, and gouge out freshwater lakes that provide abundant nutritive possibilities for hundreds of species of being. They act as a spur to migration and keep the planet dynamic. As Tim Flannery has remarked: 'There is only one question you need ask of a continent in order to determine the fate of its people: "Did you have a good ice age?"' And with that in mind, it's time to look at a species of ape that truly did.

28

THE MYSTERIOUS BIPED

Just before Christmas 1887, a young Dutch doctor with an un-Dutch name, Marie Eugène François Thomas Dubois,* arrived in Sumatra, in the Dutch East Indies, with the intention of finding the earliest human remains on Earth.

Several things were extraordinary about this. To begin with, no-one had ever gone looking for ancient human bones before. Everything that had been found to this point had been found accidentally, and nothing in Dubois' background suggested that he was the ideal candidate to make the process intentional. He was an anatomist by training with no background in palaeontology. Nor was there any special reason to suppose that the East Indies would hold early human remains. Logic dictated that if ancient people were to be found at all, it would be on a large and long-populated land mass, not in the comparative fastness of an archipelago. Dubois was driven to the East Indies on nothing stronger than a hunch, the availability of employment and the knowledge that Sumatra was full of caves, the environment in which most of the important hominid fossils

*Though Dutch, Dubois was from Eijsden, a town bordering the French-speaking part of Belgium.

had so far been found.* What is most extraordinary in all this – nearly miraculous, really – is that he found what he was looking for.

At the time Dubois conceived his plan to search for a missing link, the human fossil record consisted of very little: five incomplete Neandertal skeletons, one partial jawbone of uncertain provenance and half a dozen ice-age humans recently found by railway workers in a cave at a cliff called Cro-Magnon near Les Eyzies, France. Of the Neandertal specimens, the best preserved was sitting unremarked on a shelf in London. It had been found by workers blasting rock from a quarry in Gibraltar in 1848, so its preservation was a wonder, but unfortunately no-one yet appreciated what it was. After being briefly described at a meeting of the Gibraltar Scientific Society, it had been sent to the Hunterian Museum, where it remained undisturbed but for an occasional light dusting for over half a century. The first formal description of it wasn't written until 1907, and then by a geologist named William Sollas 'with only a passing competency in anatomy'.

So instead the name and credit for the discovery of the first early humans went to the Neander valley in Germany – not unfittingly, as it happens, for by uncanny coincidence *Neander* in Greek means 'new man'. There, in 1856, workmen at another quarry, in a cliff face overlooking the Düssel River, found some curious-looking bones, which they passed to a local schoolteacher, knowing he had an interest

*Humans are put in the family Hominidae. Its members, traditionally called hominids, include any creatures (including extinct ones) that are more closely related to us than to any surviving chimpanzees. The apes, meanwhile, are lumped together in a family called Pongidae. Many authorities believe that chimps, gorillas and orangutans should also be included in the family Hominidae, with humans and chimps in a subfamily called Homininae. The upshot is that the creatures traditionally called hominids become, under this arrangement, hominins. (Leakey and others insist on that designation.) Hominoidea is the name of the ape superfamily, which includes us.

in all things natural. To his great credit the teacher, Johann Karl Fuhlrott, saw that he had some new type of human, though quite what it was, and how special, would be matters of dispute for some time.

Many people refused to accept that the Neandertal bones were ancient at all. August Mayer, a professor at the University of Bonn and a man of influence, insisted that the bones were merely those of a Mongolian Cossack soldier who had been wounded while fighting in Germany in 1814 and had crawled into the cave to die. Hearing of this, T. H. Huxley in England drily observed how remarkable it was that the soldier, though mortally wounded, had climbed 60 feet up a cliff, divested himself of his clothing and personal effects, sealed the cave opening and buried himself under 2 feet of soil. Another anthropologist, puzzling over the Neandertal's heavy brow ridge, suggested that it was the result of long-term frowning arising from a poorly healed forearm fracture. (In their eagerness to reject the idea of earlier humans, authorities were often willing to embrace the most singular possibilities. At about the time that Dubois was setting out for Sumatra, a skeleton found in Périgueux was confidently declared to be that of an Eskimo. Quite what an ancient Eskimo was doing in southwest France was never comfortably explained. It was actually an early Cro-Magnon.)

It was against this background that Dubois began his search for ancient human bones. He did no digging himself, but instead used fifty convicts lent by the Dutch authorities. For a year they worked on Sumatra, then transferred to Java. And there in 1891, Dubois – or rather his team, for Dubois himself seldom visited the sites – found a section of ancient human cranium now known as the Trinil skullcap. Though only part of a skull, it showed that the owner had had distinctly non-human features but a much larger brain than any ape. Dubois called it *Anthropithecus erectus* (later changed for technical reasons to *Pithecanthropus erectus*) and declared it the missing link between apes and humans. It quickly became popularized as 'Java Man'. Today we know it as *Homo erectus*.

The next year Dubois' workers found a virtually complete thighbone that looked surprisingly modern. In fact, many anthropologists think it *is* modern, and has nothing to do with Java Man. If it is an erectus bone, it is unlike any other found since. Nonetheless Dubois used the thighbone to deduce – correctly, as it turned out – that *Pithecanthropus* walked upright. He also produced, with nothing but a scrap of cranium and one tooth, a model of the complete skull, which also proved uncannily accurate.

In 1895 Dubois returned to Europe, expecting a triumphal reception. In fact, he met nearly the opposite reaction. Most scientists disliked both his conclusions and the arrogant manner in which he presented them. The skullcap, they said, was that of an ape, probably a gibbon, and not of any early human. Hoping to bolster his case, in 1897 Dubois allowed a respected anatomist from the University of Strasbourg, Gustav Schwalbe, to make a cast of the skullcap. To Dubois' dismay, Schwalbe thereupon produced a monograph that received far more sympathetic attention than anything Dubois had written, and followed it with a lecture tour in which he was celebrated nearly as warmly as if he had dug up the skull himself. Appalled and embittered, Dubois withdrew into an undistinguished position as a professor of geology at the University of Amsterdam and for the next two decades refused to let anyone examine his precious fossils again. He died in 1940, an unhappy man.

Meanwhile, and half a world away, in late 1924 Raymond Dart, the Australian-born head of anatomy at the University of the Witwatersrand in Johannesburg, was sent a small but remarkably complete skull of a child, with an intact face, a lower jaw and what is known as an endocast – a natural cast of the brain – from a limestone quarry on the edge of the Kalahari Desert at a dusty spot called Taung. Dart could see at once that the Taung skull was not of a *Homo erectus* like Dubois' Java Man, but from an earlier, more apelike creature. He placed its age at two million years and dubbed it *Australopithecus africanus*, or 'southern ape man of Africa'. In a report to *Nature*, Dart called the Taung remains 'amazingly human' and suggested the need for an entirely new family, *Homo simiadae* ('the man-apes'), to accommodate the find.

The authorities were even less favourably disposed towards Dart than they had been to Dubois. Nearly everything about his theory – indeed, nearly everything about Dart, it appears – annoyed them. To start with, he had proved himself lamentably presumptuous by conducting the analysis himself rather than calling on the help of more worldly experts. Even his chosen name, *Australopithecus*, showed a lack of scholarly application, combining as it did Greek and Latin roots. Above all, his conclusions flew in the face of accepted wisdom. Humans and apes, it was agreed, had split apart at least 15 million years ago in Asia. If humans had arisen in Africa, why, that would make us *negroid*, for goodness' sake. It was rather as if someone working today were to announce that he had found ancestral bones of humans in, say, Missouri. It just didn't fit with what was known.

Dart's sole supporter of note was Robert Broom, a Scottish-born physician and palaeontologist of considerable intellect and cherishably eccentric nature. It was Broom's habit, for instance, to do his fieldwork naked when the weather was warm, which was often. He was also known for conducting dubious anatomical experiments on his poorer and more tractable patients. When the patients died, which was also often, he would sometimes bury their bodies in his back garden to dig up for study later.

Broom was an accomplished palaeontologist and since he was also resident in South Africa he was able to examine the Taung skull at first hand. He could see at once that it was as important as Dart supposed and spoke out vigorously on Dart's behalf, but to no effect. For the next fifty years the received wisdom was that the Taung child was an ape and nothing more. Most textbooks didn't even mention it. Dart spent five years working up a monograph, but could find no-one to publish it. Eventually he gave up the quest to publish altogether (though he did continue hunting for fossils). For years, the skull – today recognized as one of the supreme treasures of anthropology – sat as a paperweight on a colleague's desk.

At the time Dart made his announcement in 1924, only four categories of ancient hominid were known – *Homo heidelbergensis*, *Homo rhodesiensis*, Neandertals and Dubois' Java Man – but all that was about to change in a very big way.

First, in China, a gifted Canadian amateur named Davidson Black began to poke around at a place called Dragon Bone Hill, which was locally famous as a hunting ground for old bones. Unfortunately, rather than preserving the bones for study, the Chinese ground them up to make medicines. We can only guess how many priceless *Homo erectus* bones ended up as a sort of Chinese equivalent of Beecham's powder. The site had been much denuded by the time Black arrived, but he found a single fossilized molar and on the basis of that alone quite brilliantly announced the discovery of *Sinanthropus pekinensis*, which quickly became known as Peking Man.

At Black's urging, more determined excavations were undertaken and many other bones found. Unfortunately all were lost the day after the Japanese attack on Pearl Harbor in 1941, when a contingent of US marines, trying to spirit the bones (and themselves) out of the country, was intercepted by the Japanese and imprisoned. Seeing that their crates held nothing but bones, the Japanese soldiers left them at the roadside. It was the last that was ever seen of them.

In the meantime, back on Dubois' old turf of Java, a team led by Ralph von Koenigswald had found another group of early humans which became known as the Solo People, from the site of their discovery on the Solo River at Ngandong. Koenigswald's discoveries might have been more impressive still but for a tactical error that was realized too late. He had offered locals 10 cents for every piece of hominid bone they could come up with, then discovered to his horror that they had been enthusiastically smashing large pieces into small ones to maximize their income.

In the following years, as more bones were found and identified, there came a flood of new names – *Homo aurignacensis*, *Australopithecus transvaalensis*, *Paranthropus crassidens*, *Zinjanthropus boisei* and scores of others, nearly all involving a new genus type as well as a new species. By the 1950s, the number of named hominid types had risen to comfortably over a hundred. To add to the confusion, individual forms often went by a succession of different names as palaeoanthropologists refined, reworked and squabbled over classifications. The Solo People were known variously as *Homo soloensis*, *Homo primigenius asiaticus*, *Homo neanderthalensis soloensis*, *Homo sapiens soloensis*, *Homo erectus erectus* and, finally, plain *Homo erectus*.

In an attempt to introduce some order, in 1960 F. Clark Howell of the University of Chicago, following the suggestions of Ernst Mayr and others the previous decade, proposed cutting the number of genera to just two – *Australopithecus* and *Homo* – and rationalizing many of the species. The Java and Peking men both became *Homo erectus*. For a time order prevailed in the world of the hominids. It didn't last.

After about a decade of comparative calm, palaeoanthropology embarked on another period of swift and numerous discovery, which hasn't abated yet. The 1960s produced *Homo habilis*, thought by some to be the missing link between apes and humans, but thought by others not to be a separate species at all. Then came (among many others) *Homo ergaster*, *Homo louisleakeyi*, *Homo rudolfensis*, *Homo microcranus* and *Homo antecessor*, as well as a raft of australopithecines: *A. afarensis*, *A. praegens*, *A. ramidus*, *A. walkeri*, *A. anamensis* and still others. Altogether, some twenty types of hominid are recognized in the literature today. Unfortunately, almost no two experts recognize the same twenty.

Some continue to observe the two hominid genera suggested by Howell in 1960, but others place some of the australopithecines in a separate genus called *Paranthropus*, and still others add an earlier group called *Ardipithecus*. Some put *praegens* into *Australopithecus* and some into a new classification, *Homo antiquus*, but most don't recognize *praegens* as a separate species at all. There is no central authority that rules on these things. The only way a name becomes accepted is by consensus, and there is often very little of that.

A big part of the problem, paradoxically, is a shortage of evidence. Since the dawn of time, several billion human (or humanlike) beings have lived, each contributing a little genetic variability to the total human stock. Out of this vast number, the whole of our understanding of human prehistory is based on the remains, often exceedingly fragmentary, of perhaps five thousand individuals. 'You could fit it all into the back of a pickup truck if you didn't mind how much you jumbled everything up,' Ian Tattersall, the bearded and friendly curator of anthropology at the American Museum of Natural History in New York, replied when I asked him the size of the total world archive of hominid and early human bones.

The shortage wouldn't be so bad if the bones were distributed evenly through time and space, but of course they are not. They appear randomly, often in the most tantalizing fashion. *Homo erectus* walked the Earth for well over a million years and inhabited territory from the Atlantic edge of Europe to the Pacific side of China, yet if you brought back to life every *Homo erectus* individual whose existence we can vouch for, they wouldn't fill a school bus. *Homo habilis* consists of even less: just two partial skeletons and a number of isolated limb bones. Something as short-lived as our own civilization would almost certainly not be known from the fossil record at all.

'In Europe,' Tattersall offers by way of illustration, 'you've got hominid skulls in Georgia dated to about 1.7 million years ago, but then you have a gap of almost a million years before the next remains turn up in Spain, right on the other side of the continent, and then you've got another three-hundred-thousand-year gap before you get a *Homo heidelbergensis* in Germany – and none of them looks terribly much like any of the others.' He smiled. 'It's from these kinds of fragmentary pieces that you're trying to work out the histories of entire species. It's quite a tall order. We really have very little idea of the relationships between many ancient species – which led to us and which were evolutionary dead ends. Some probably don't deserve to be regarded as separate species at all.'

It is the patchiness of the record that makes each new find look so sudden and distinct from all the others. If we had tens of thousands of skeletons distributed at regular intervals through the historical record, there would be appreciably more degrees of shading. Whole new species don't emerge instantaneously, as the fossil record implies, but gradually out of other, existing species. The closer you go back to a point of divergence, the closer the similarities are, so that it becomes exceedingly difficult, and sometimes impossible, to distinguish a late *Homo erectus* from an early *Homo sapiens*, since it is likely to be both and neither. Similar disagreements can often arise over questions of identification from fragmentary remains – deciding, for instance, whether a particular bone represents a female *Australopithecus boisei* or a male *Homo habilis*.

With so little to be certain about, scientists often have to make assumptions based on other objects found nearby, and these may be little more than valiant guesses. As Alan Walker and Pat Shipman have drily observed, if you correlate tool discovery with the species of creature most often found nearby, you would have to conclude that early hand tools were mostly made by antelopes.

Perhaps nothing better typifies the confusion than the fragmentary bundle of contradictions that was *Homo habilis*. Simply put, *habilis* bones make no sense. When arranged in sequence, they show males and females evolving at different rates and in different directions – the males becoming less apelike and more human with time, while females from the same period appear to be moving *away* from humanness towards greater apeness. Some authorities don't believe *habilis* is a valid category at all. Tattersall and his colleague Jeffrey Schwartz dismiss it as a mere 'wastebasket species' – one into which unrelated fossils 'could be conveniently swept'. Even those who see *habilis* as an independent species don't agree on whether it is of the same genus as us or is from a side branch that never came to anything.

Finally, but perhaps above all, human nature is a factor in all this. Scientists have a natural tendency to interpret finds in the way that most flatters their stature. It is a rare palaeontologist indeed who announces that he has found a cache of bones but that they are nothing to get excited about. As John Reader understatedly observes in the book *Missing Links*, 'It is remarkable how often the first interpretations of new evidence have confirmed the preconceptions of its discoverer.'

All this leaves ample room for arguments, of course, and

no bunch of people likes to argue more than palaeoanthropologists. 'And of all the disciplines in science, paleoanthropology boasts perhaps the largest share of egos,' say the authors of the recent *Java Man* – a book, it may be noted, that itself devotes long, wonderfully unselfconscious passages to attacks on the inadequacies of others, in particular the authors' former close colleague Donald Johanson.

So, bearing in mind that there is little you can say about human prehistory that won't be disputed by someone somewhere, other than that we most certainly had one, what we think we know about who we are and where we come from is roughly this.

For the first 99.99999 per cent of our history as organisms, we were in the same ancestral line as chimpanzees. Virtually nothing is known about the prehistory of chimpanzees, but whatever they were, we were. Then, about seven million years ago, something major happened. A group of new beings emerged from the tropical forests of Africa and began to move about on the open savanna.

These were the australopithecines, and for the next five million years they would be the world's dominant hominid species. (*Austral* is from the Latin for 'southern' and has no connection in this context with Australia.) Australopithecines came in several varieties, some slender and gracile, like Raymond Dart's Taung child, others more sturdy and robust, but all were capable of walking upright. Some of these species existed for well over a million years, others for a more modest few hundred thousand, but it is worth bearing in mind that even the least successful had histories many times longer than we have yet achieved.

The most famous hominid remains in the world are those of a 3.18-million-year-old australopithecine found at Hadar in Ethiopia in 1974 by a team led by Donald Johanson. Formally known as A.L. (for 'Afar Locality') 288-1, the skeleton became more familiarly known as Lucy, after the Beatles song 'Lucy in the Sky with Diamonds'. Johanson has never doubted her importance. 'She is our earliest ancestor, the missing link between ape and human,' he has said.

Lucy was tiny – just three and a half feet tall. She could walk, though how well is a matter of some dispute. She was evidently a good climber too. Much else is unknown. Her skull was almost entirely missing, so little could be said with confidence about her brain size, though skull fragments suggested it was small. Most books describe Lucy's skeleton as being 40 per cent complete, though some put it closer to half and one produced by the American Museum of Natural History describes Lucy as two-thirds complete. The BBC television series *Ape Man* actually called it 'a complete skeleton', even while showing that it was anything but.

A human body has 206 bones, but many of these are repeated. If you have the left femur from a specimen, you don't need the right to know its dimensions. Strip out all the redundant bones and the total you are left with is 120 – what is called a half skeleton. Even by this fairly accommodating standard, and even counting the slightest fragment as a full bone, Lucy constituted only 28 per cent of a half skeleton (and only about 20 per cent of a full one).

In *The Wisdom of the Bones*, Alan Walker recounts how he once asked Johanson how he had come up with a figure of 40 per cent. Johanson breezily replied that he had discounted the 106 bones of the hands and feet – more than half the body's total, and a fairly important half, too, one would have thought, since Lucy's principal defining attribute was the use of those hands and feet to deal with a changing world. At all events, rather less is known about Lucy than is generally supposed. It isn't even actually known that she was a female. Her sex is merely presumed from her diminutive size.

* * *

Two years after Lucy's discovery, at Laetoli in Tanzania Mary Leakey found footprints left by two individuals from – it is thought – the same family of hominids. The prints had been made when two australopithecines had walked through muddy ash following a volcanic eruption. The ash had later hardened, preserving the impressions of their feet for a distance of over 23 metres.

The American Museum of Natural History in New York has an absorbing diorama that records the moment of their passing. It depicts life-sized recreations of a male and a female walking side by side across the ancient African plain. They are hairy and chimp-like in dimensions, but have a bearing and gait that suggest humanness. The most striking feature of the display is that the male holds his left arm protectively around the female's shoulder. It is a tender and affecting gesture, suggestive of close bonding.

The tableau is presented with such conviction that it is easy to overlook the consideration that virtually everything above the footprints is imaginary. Almost every external aspect of the two figures – degree of hairiness, facial appendages (whether they had human noses or chimp noses), expressions, skin colour, size and shape of the female's breasts – is necessarily suppositional. We can't even say that they were a couple. The female figure may in fact have been a child. Nor can we be certain that they were australopithecines. They are assumed to be australopithecines because there are no other known candidates.

I had been told that they were posed like that because during the building of the diorama the female figure kept toppling over, but Ian Tattersall insists with a laugh that the story is untrue. 'Obviously we don't know whether the male had his arm around the female or not, but we do know from the stride measurements that they were walking side by side and close together – close enough to be touching. It was quite an exposed area, so they were probably feeling vulnerable. That's why we tried to give them slightly worried expressions.'

I asked him if he was troubled about the amount of licence that was taken in reconstructing the figures. 'It's always a problem in making recreations,' he agreed readily enough. 'You wouldn't believe how much discussion can go into deciding details like whether Neandertals had eyebrows or not. It was just the same for the Laetoli figures. We simply can't know the details of what they looked like, but we *can* convey their size and posture and make some reasonable assumptions about their probable appearance. If I had it to do again, I think I might have made them just slightly more apelike and less human. These creatures weren't humans. They were bipedal apes.'

Until very recently it was assumed that we were descended from Lucy and the Laetoli creatures, but now many authorities aren't so sure. Although certain physical features (the teeth, for instance) suggest a possible link between us, other parts of the australopithecine anatomy are more troubling. In their book *Extinct Humans*, Tattersall and Schwartz point out that the upper portion of the human femur is very like that of the apes but not like that of the australopithecines; so if Lucy is in a direct line between apes and modern humans, it means we must have adopted an australopithecine femur for a million years or so, then gone back to an ape femur when we moved on to the next phase of our development. They believe, in fact, that not only was Lucy not our ancestor, she wasn't even much of a walker.

'Lucy and her kind did not locomote in anything like the modern human fashion,' insists Tattersall. 'Only when these hominids had to travel between arboreal habitats would they find themselves walking bipedally, "forced" to do so by their own anatomies.' Johanson doesn't accept this. 'Lucy's hips and the muscular arrangement of her pelvis', he has written, 'would have made it as hard for her to climb trees as it is for modern humans.'

Matters grew murkier still in 2001 and 2002 when four exceptional new specimens were found. One, discovered by Meave Leakey of the famous fossil-hunting family at Lake Turkana in Kenya and called *Kenyanthropus platyops* ('Kenyan flat-face'), is from about the same time as Lucy and raises the possibility that it was our ancestor and Lucy merely an unsuccessful side branch. Also found in 2001 were *Ardipithecus ramidus kadabba*, dated at between 5.2 million and 5.8 million years old, and *Orrorin tugenensis*, thought to be six million years old, making it the oldest hominid yet found – but only for a brief while. In the summer of 2002 a French team working in the Djurab Desert of Chad (an area that had never before yielded ancient bones) found a hominid almost seven million years old, which they labelled *Sahelanthropus tchadensis*. (Some critics believe that it was not human but an early ape, and therefore should be called *Sahelpithecus*.) All these were early creatures and quite primitive, but they walked upright, and they were doing it far earlier than previously thought.

Bipedalism is a demanding and risky strategy. It means refashioning the pelvis into a full load-bearing instrument. To preserve the required strength, the birth canal in the female must be comparatively narrow. This has two very significant immediate consequences and one longer-term one. First, it means a lot of pain for any birthing mother and a greatly increased danger of fatality to mother and baby both. Moreover, to get the baby's head through such a tight space it must be born while its brain is still small – and while the baby, therefore, is still helpless. This means long-term infant care, which in turn implies solid male–female bonding.

All this is problematic enough when you are the intellectual master of the planet, but when you are a small, vulnerable australopithecine, with a brain about the size of an orange,* the risk must have been enormous.

So why did Lucy and her kind come down from the trees and out of the forests? Probably they had no choice. The slow rise of the Isthmus of Panama had cut the flow of waters from the Pacific into the Atlantic, diverting warming currents away from the Arctic and leading to the onset of an exceedingly sharp ice age in northern latitudes. In Africa, this would have produced seasonal drying and cooling, gradually turning jungle into savanna. 'It was not so much that Lucy and her like left the forests,' John Gribbin has written, 'but that the forests left them.'

But stepping out onto the open savanna also clearly left the early hominids much more exposed. An upright hominid could see better, but could also be seen better. Even now, as a species we are almost preposterously vulnerable in the wild. Nearly every large animal you care to name is stronger, faster and toothier than us. Faced with attack, modern humans have only two advantages. We have a good brain, with which we can devise strategies; and we have hands, with which we can fling or brandish hurtful objects. We are the only creature that can harm at a distance. We can thus afford to be physically vulnerable.

All the elements would appear to have been in place for the rapid evolution of a potent brain, and yet that seems not to have happened. For over three million years, Lucy and her fellow australopithecines scarcely changed at all. Their brain didn't grow and there is no sign that they used even

*Absolute brain size does not tell you everything – or possibly sometimes even much. Elephants and whales both have brains larger than ours, but you wouldn't have much trouble outwitting them in contract negotiations. It is relative size that matters, a point that is often overlooked. As Gould notes, *A. africanus* had a brain of only 450cc, smaller than that of a gorilla. But a typical *africanus* male weighed less than 45 kilos, and a female much less still, whereas gorillas can easily top out at over 150 kilos.

the simplest tools. What is stranger still is that we now know that for about a million years they lived alongside other early hominids who did use tools, yet the australopithecines never took advantage of this useful technology that was all around them.

At one point between three million and two million years ago, it appears there may have been as many as six hominid types co-existing in Africa. Only one, however, was fated to last: *Homo*, which emerged from the mists beginning about two million years ago. No-one knows quite what the relationship was between australopithecines and *Homo*, but what is known is that they co-existed for something over a million years before all the australopithecines, robust and gracile alike, vanished mysteriously, and possibly abruptly, over a million years ago. No-one knows why they disappeared. 'Perhaps,' suggests Matt Ridley, 'we ate them.'

Conventionally, the *Homo* line begins with *Homo habilis*, a creature about whom we know almost nothing, and concludes with us, *Homo sapiens* (literally 'man the thinker'). In between, and depending on which opinions you value, there have been half a dozen other *Homo* species: *Homo ergaster*, *Homo neanderthalensis*, *Homo rudolfensis*, *Homo heidelbergensis*, *Homo erectus* and *Homo antecessor*.

Homo habilis ('handy man') was named by Louis Leakey and colleagues in 1964 and was so called because it was the first hominid to use tools, albeit very simple ones. It was a fairly primitive creature, much more chimpanzee than human, but its brain was about 50 per cent larger than that of Lucy in gross terms and not much less large proportionally, so it was the Einstein of its day. No persuasive reason has ever been adduced for why hominid brains suddenly began to grow two million years ago. For a long time it was assumed that big brains and upright walking were directly related – that the movement out of the forests necessitated cunning new strategies that fed off or promoted braininess –

so it was something of a surprise, after the repeated discoveries of so many bipedal dullards, to realize that there was no apparent connection between them at all.

'There is simply no compelling reason we know of to explain why human brains got large,' says Tattersall. Huge brains are demanding organs: they make up only 2 per cent of the body's mass, but devour 20 per cent of its energy. They are also comparatively picky in what they use as fuel. If you never ate another morsel of fat, your brain would not complain because it won't touch the stuff. It wants glucose instead, and lots of it, even if it means short-changing other organs. As Guy Brown notes: 'The body is in constant danger of being depleted by a greedy brain, but cannot afford to let the brain go hungry as that would rapidly lead to death.' A big brain needs more food and more food means increased risk.

Tattersall thinks the rise of a big brain may simply have been an evolutionary accident. He believes with Stephen Jay Gould that if you replayed the tape of life – even if you ran it back only a relatively short way to the dawn of hominids – the chances are 'quite unlikely' that modern humans or anything like them would be here now.

'One of the hardest ideas for humans to accept,' he says, 'is that we are not the culmination of anything. There is nothing inevitable about our being here. It is part of our vanity as humans that we tend to think of evolution as a process that, in effect, was programmed to produce us. Even anthropologists tended to think this way right up until the 1970s.' Indeed, as recently as 1991, in the popular textbook *The Stages of Evolution*, C. Loring Brace stuck doggedly to the linear concept, acknowledging just one evolutionary dead end, the robust australopithecines. Everything else represented a straightforward progression – each species of hominid carrying the baton of development so far, then handing it on to a younger, fresher runner. Now, however, it seems certain that many of these early forms followed side trails that didn't come to anything.

Luckily for us, one did – a group of tool users who seemed to arise from out of nowhere and overlapped with the shadowy and much disputed *Homo habilis*. This is *Homo erectus*, the species discovered by Eugène Dubois in Java in 1891. Depending on which sources you consult, it existed from about 1.8 million years ago to possibly as recently as twenty thousand or so years ago.

According to the *Java Man* authors, *Homo erectus* is the dividing line: everything that came before him was apelike in character; everything that came after him was humanlike. *Homo erectus* was the first to hunt, the first to use fire, the first to fashion complex tools, the first to leave evidence of campsites, the first to look after the weak and frail. Compared with all that had gone before, the species was extremely human in form as well as behaviour, its members long-limbed and lean, very strong (much stronger than modern humans), and with the drive and intelligence to spread successfully over huge areas. To other hominids, *Homo erectus* must have seemed terrifyingly large, powerful, fleet and gifted. Their brains were vastly more sophisticated than anything the world had seen before.

Erectus was 'the velociraptor of its day', according to Alan Walker of Penn State University, one of the world's leading authorities. If you were to look one in the eyes, it might appear superficially to be human, but 'you wouldn't connect. You'd be prey.' According to Walker, it had the body of an adult human but the brain of a baby.

Although *erectus* had been known about for almost a century it was known only from scattered fragments – not enough to come even close to making one full skeleton. So it wasn't until an extraordinary discovery in Africa in the 1980s that its importance – or, at the very least, possible importance – as a precursor species for modern humans was fully appreciated. The remote valley of Lake Turkana (formerly Lake Rudolf) in Kenya is now one of the world's most productive sites for early human remains, but for a very long time no-one had thought to look there. It was only because Richard Leakey was on a flight that was diverted over the valley that he realized it might be more promising than had been thought. A team was dispatched to investigate, but at first found nothing. Then late one afternoon Kamoya Kimeu, Leakey's most renowned fossil hunter, found a small piece of hominid brow on a hill well away from the lake. Such a site was unlikely to yield much, but they dug anyway out of respect for Kimeu's instincts and to their astonishment found a nearly complete *Homo erectus* skeleton. It was from a boy aged between about nine and twelve who had died 1.54 million years ago. The skeleton had 'an entirely modern body structure', says Tattersall, in a way that was without precedent. The Turkana boy was 'very emphatically one of us'.

Also found at Lake Turkana by Kimeu was KNM-ER 1808, a female 1.7 million years old, which gave scientists their first clue that *Homo erectus* was more interesting and complex than previously thought. The woman's bones were deformed and covered in coarse growths, the result of an agonizing condition called hypervitaminosis A, which can only come from eating the liver of a carnivore. This told us, first of all, that *Homo erectus* was eating meat. Even more surprising was that the amount of growth showed that she had lived weeks or even months with the disease. Someone had looked after her. It was the first sign of tenderness in hominid evolution.

It was also discovered that *Homo erectus* skulls contained (or, in the view of some, possibly contained) a Broca's area, a region of the frontal lobe of the brain associated with speech. Chimps don't have such a feature. Alan Walker thinks the spinal canal didn't have the size and complexity to enable speech, that *erectus* probably would have communicated only about as well as modern chimps. Others, notably Richard Leakey, are convinced they could speak.

For a time, it appears, *Homo erectus* was the only hominid species. They were unprecedentedly adventurous and spread across the globe with what seems to have been breathtaking rapidity. The fossil evidence, if taken literally, suggests that some members of the species reached Java at about the same time as, or even slightly before, they left Africa. This has led some hopeful scientists to suggest that perhaps modern people arose not in Africa at all, but in Asia – which would be remarkable, not to say miraculous, as no possible precursor species has ever been found anywhere outside Africa. The Asian hominids would have had to appear, as it were, spontaneously. And anyway, an Asian beginning would merely reverse the problem of their spread; you would still have to explain how the Java people then got to Africa so quickly.

There are several more plausible alternative explanations for how *Homo erectus* managed to turn up in Asia so soon after its first appearance in Africa. First, a lot of plus-or-minusing goes into the dating of early human remains. If the actual age of the African bones is at the higher end of the range of estimates or the Javan ones at the lower end, or both, then there is plenty of time for African *erectus* to find their way to Asia. It is also entirely possible that older *erectus* bones await discovery in Africa. In addition, the Javan dates could be wrong altogether.

What is certain is that some time well over a million years ago, some new, comparatively modern, upright beings left Africa and boldly spread out across much of the globe. They possibly did so quite rapidly, increasing their range by as much as 40 kilometres a year on average, all while dealing with mountain ranges, rivers, deserts and other impediments and adapting to differences in climate and food sources. A particular mystery is how they passed along the west side of the Red Sea, an area of famously punishing aridity now, but even drier in the past. It is a curious irony that the conditions that prompted them to leave Africa would have made it much more difficult to do so. Yet somehow they managed to find their way around every barrier and to thrive in the lands beyond.

And that, I'm afraid, is where all agreement ends. What happened next in the history of human development is a matter of long and rancorous debate, as we shall see in the next chapter.

But it is worth remembering, before we move on, that all of these evolutionary jostlings over five million years, from distant, puzzled australopithecine to fully modern human, produced a creature that is still 98.4 per cent genetically indistinguishable from the modern chimpanzee. There is more difference between a zebra and a horse, or between a dolphin and a porpoise, than there is between you and the furry creatures your distant ancestors left behind when they set out to take over the world.

29

THE RESTLESS APE

Sometime about a million and a half years ago, some forgotten genius of the hominid world did an unexpected thing. He (or very possibly she) took one stone and carefully used it to shape another. The result was a simple teardrop-shaped hand-axe, but it was the world's first piece of advanced technology.

It was so superior to existing tools that soon others were following the inventor's lead and making hand-axes of their own. Eventually whole societies existed that seemed to do little else. 'They made them in their thousands,' says Ian Tattersall. 'There are some places in Africa where you literally can't move without stepping on them. It's strange because they are quite intensive objects to make. It was as if they made them for the sheer pleasure of it.'

From a shelf in his sunny workroom Tattersall took down an enormous cast, perhaps half a metre long and 20 centimetres wide at its widest point, and handed it to me. It was shaped like a spearhead, but one the size of a stepping stone. As a fibreglass cast it weighed only a few ounces, but the original, which was found in Tanzania, weighed 11 kilograms. 'It was completely useless as a tool,' Tattersall said. 'It would have taken two people to lift it adequately and even then it would have been exhausting to try to pound anything with it.'

'What was it used for, then?'

Tattersall gave a genial shrug, pleased at the mystery of it. 'No idea. It must have had some symbolic importance, but we can only guess what.'

The axes became known as Acheulean tools, after St Acheul, a suburb of Amiens in northern France, where the first examples were found in the nineteenth century, and contrast with the older, simpler tools known as Oldowan, originally found at Olduvai Gorge in Tanzania. In older textbooks, Oldowan tools are usually shown as blunt, rounded, hand-sized stones. In fact, palaeoanthropologists now tend to believe that the tool parts of Oldowan rocks were the pieces flaked off these larger stones, which could then be used for cutting.

Now here's the mystery. When early modern humans – the ones who would eventually become us – started to move out of Africa something over a hundred thousand years ago, Acheulean tools were the technology of choice. These early *Homo sapiens* loved their Acheulean tools, too. They carried them vast distances. Sometimes they even took unshaped rocks with them to make into tools later on. They were, in a word, devoted to the technology. But although Acheulean tools have been found throughout Africa, Europe and western and central Asia, they are almost never found in the Far East. This is deeply puzzling.

In the 1940s a Harvard palaeontologist named Hallum Movius drew something called the Movius line, dividing the side with Acheulean tools from the one without. The line runs in a southeasterly direction across Europe and the Middle East to the vicinity of modern-day Calcutta and Bangladesh. Beyond the Movius line, across the whole of southeast Asia and into China, only the older, simpler Oldowan tools have been found. We know that *Homo sapiens* went far beyond this point, so why would they carry an advanced and treasured stone technology to the edge of the Far East and then just abandon it?

'That troubled me for a long time,' recalls Alan Thorne of the Australian National University in Canberra. 'The whole of modern anthropology was built round the idea that humans came out of Africa in two waves – a first wave of *Homo erectus*, which became Java Man and Peking Man and the like, and a later, more advanced wave of *Homo sapiens*, which displaced the first lot. Yet to accept that you must believe that *Homo sapiens* got so far with their more modern technology and then, for whatever reason, gave it up. It was all very puzzling, to say the least.'

As it turned out, there would be a great deal else to be puzzled about, and one of the most puzzling findings of all would come from Thorne's own part of the world, in the outback of Australia. In 1968, a geologist named Jim Bowler was poking around on a long-dried lake bed called Mungo in a parched and lonely corner of western New South Wales when something very unexpected caught his eye. Sticking out of a crescent-shaped sand ridge of a type known as a lunette were some human bones. At the time, it was believed that humans had been in Australia for no more than eight thousand years, but Mungo had been dry for twelve thousand years. So what was anyone doing in such an inhospitable place?

The answer, provided by carbon dating, was that the bones' owner had lived there when Lake Mungo was a much more agreeable habitat, 20 kilometres long, full of water and fish, fringed by pleasant groves of casuarina trees. To everyone's astonishment, the bones turned out to be twenty-three thousand years old. Other bones found nearby proved to be as much as sixty thousand years old. This was unexpected to the point of seeming practically impossible. At no time since hominids first arose on Earth has Australia not been an island. Any human beings who arrived there must have come by sea, in large enough numbers to start a breeding population, after crossing 100 kilometres or more of open water without having any way of knowing that a convenient landfall awaited them. Having landed, the Mungo people had then found their way over 3,000 kilometres inland from Australia's north coast – the presumed point of entry – which suggests, according to a report in the *Proceedings of the National Academy of Sciences*, 'that people may have first arrived substantially earlier than 60,000 years ago'.

How they got there and why they came are questions that can't be answered. According to most anthropology texts, there's no evidence that people could even speak sixty thousand years ago, much less engage in the sorts of co-operative efforts necessary to build ocean-worthy craft and colonize island continents.

'There's just a whole lot we don't know about the movements of people before recorded history,' Alan Thorne told me when I met him in Canberra. 'Do you know that when nineteenth-century anthropologists first got to Papua New Guinea, they found people in the highlands of the interior, in some of the most inaccessible terrain on earth, growing sweet potatoes. Sweet potatoes are native to South America. So how did they get to Papua New Guinea? We don't know. Don't have the faintest idea. But what is certain is that people have been moving around with considerable assuredness for longer than traditionally thought, and almost certainly sharing genes as well as information.'

The problem, as ever, is the fossil record. 'Very few parts of the world are even vaguely amenable to the long-term preservation of human remains,' says Thorne, a sharp-eyed man with a white goatee and an intent but friendly manner. 'If it weren't for a few productive areas like Hadar and Olduvai in east Africa we'd know frighteningly little. And when you look elsewhere, often we *do* know frighteningly little. The whole of India has yielded just one ancient human fossil, from about three hundred thousand years ago. Between Iraq and Vietnam – that's a distance of some five thousand kilometres – there have been just two: the one in India and a Neandertal in Uzbekistan.' He grinned. 'That's not a whole hell of a lot to work with. You're left with the position that you've got a few productive areas for human fossils, like the Great Rift Valley in Africa and Mungo here in Australia, and very little in between. It's not surprising that palaeontologists have trouble connecting the dots.'

The traditional theory to explain human movements – and the one still accepted by the majority of people in the field – is that humans dispersed across Eurasia in two waves. The first wave consisted of *Homo erectus* who left Africa remarkably quickly – almost as soon as they emerged as a species – beginning nearly two million years ago. Over time, as they settled in different regions, these early erects further evolved into distinctive types – into Java Man and Peking Man in Asia, and into *Homo heidelbergensis* and finally *Homo neanderthalensis* in Europe.

Then, something over a hundred thousand years ago, a smarter, lither species of creature – the ancestors of every one of us alive today – arose on the African plains and began radiating outwards in a second wave. Wherever they went, according to this theory, these new *Homo sapiens* displaced their duller, less adept predecessors. Quite how they did this has always been a matter of disputation. No signs of slaughter have ever been found, so most authorities believe the newer hominids simply outcompeted the older ones, though other factors may also have contributed. 'Perhaps we gave them smallpox,' suggests Tattersall. 'There's no real way of telling. The one certainty is that we are here now and they aren't.'

These first modern humans are surprisingly shadowy. We

know less about ourselves, curiously enough, than about almost any other line of hominids. It is odd indeed, as Tattersall notes, 'that the most recent major event in human evolution – the emergence of our own species – is perhaps the most obscure of all'. Nobody can even quite agree where truly modern humans first appear in the fossil record. Many books place their debut at about a hundred and twenty thousand years ago in the form of remains found at the Klasies River mouth in South Africa, but not everyone accepts that these were fully modern people. Tattersall and Schwartz maintain that 'whether any or all of them actually represent our species still awaits definitive clarification'.

The first undisputed appearance of *Homo sapiens* is in the eastern Mediterranean, around modern-day Israel, where they begin to show up about a hundred thousand years ago – but even there they are described (by Trinkaus and Shipman) as 'odd, difficult-to-classify and poorly known'. Neandertals were already well established in the region and had a type of tool kit known as Mousterian, which the modern humans evidently found worthy enough to borrow. No Neandertal remains have ever been found in north Africa, but their tool kits turn up all over the place. Somebody must have taken them there: modern humans are the only candidate. It is also known that Neandertals and modern humans co-existed in some fashion for tens of thousands of years in the Middle East. 'We don't know if they time-shared the same space or actually lived side by side,' Tattersall says, but the moderns continued happily to use Neandertal tools – hardly convincing evidence of overwhelming superiority. No less curiously, Acheulean tools are found in the Middle East well over a million years ago, but scarcely exist in Europe until just three hundred thousand years ago. Again, why people who had the technology didn't take the tools with them is a mystery.

For a long time, it was believed that the Cro-Magnons, as

modern humans in Europe became known, drove the Neandertals before them as they advanced across the continent, eventually forcing them to the western margins of the continent where essentially they had no choice but to fall in the sea or go extinct. In fact, it is now known that Cro-Magnons were in the far west of Europe at about the same time they were also coming in from the east. 'Europe was a pretty empty place in those days,' Tattersall says. 'They may not have encountered each other all that often, even with all their comings and goings.' One curiosity of the Cro-Magnons' arrival is that it came at a time known to palaeoclimatology as the Boutellier interval, when Europe was plunging from a period of relative mildness into yet another long spell of punishing cold. Whatever it was that drew them to Europe, it wasn't the glorious weather.

In any case, the idea that Neandertals crumpled in the face of competition from newly arrived Cro-Magnons strains against the evidence at least a little. Neandertals were nothing if not tough. For tens of thousands of years they lived through conditions that no modern human outside a few polar scientists and explorers has experienced. During the worst of the ice ages, blizzards with hurricane-force winds were common. Temperatures routinely fell to minus 45 degrees Celsius. Polar bears padded across the snowy vales of southern England. Neandertals naturally retreated from the worst of it, but even so they will have experienced weather that was at least as bad as a modern Siberian winter. They suffered, to be sure – a Neandertal who lived much past thirty was lucky indeed – but as a species they were magnificently resilient and practically indestructible. They survived for at least a hundred thousand years, and perhaps twice that, over an area stretching from Gibraltar to Uzbekistan, which is a pretty successful run for any species of being.

Quite who they were and what they were like remain

matters of disagreement and uncertainty. Right up until the middle of the twentieth century the accepted anthropological view of the Neandertal was that he was dim, stooped, shuffling and simian – the quintessential caveman. It was only a painful accident that prodded scientists to reconsider this view. In 1947, while doing fieldwork in the Sahara, a Franco-Algerian palaeontologist named Camille Arambourg took refuge from the midday sun under the wing of his light aircraft. As he sat there, a tyre burst from the heat and the plane tipped suddenly, striking him a painful blow on the upper body. Later in Paris he went for an X-ray of his neck, and noticed that his own vertebrae were aligned exactly like those of the stooped and hulking Neandertal. Either he was physiologically primitive or the Neandertal's posture had been misdescribed. In fact, it was the latter. Neandertal vertebrae were not simian at all. It changed utterly how we viewed Neandertals – but only some of the time, it appears.

It is still commonly held that Neandertals lacked the intelligence or fibre to compete on equal terms with the continent's slender and more cerebrally nimble newcomers, *Homo sapiens*. Here is a typical comment from a recent book: 'Modern humans neutralized this advantage [the Neandertal's considerably heartier physique] with better clothing, better fires and better shelter; meanwhile the Neandertals were stuck with an oversize body that required more food to sustain.' In other words, the very factors that had allowed them to survive successfully for a hundred thousand years suddenly became an insuperable handicap.

Above all, the issue that is almost never addressed is that Neandertals had brains that were significantly larger than those of modern people – 1.8 litres for Neandertals versus 1.4 for modern people, according to one calculation. This is more than the difference between modern *Homo sapiens* and late *Homo erectus*, a species we are happy to regard as barely human. The argument put forward is that although our brains were smaller, they were somehow more efficient. I believe I speak the truth when I observe that nowhere else in human evolution is such an argument made.

So why then, you may well ask, if the Neandertals were so stout and adaptable and cerebrally well endowed, are they no longer with us? One possible (but much disputed) answer is that perhaps they are. Alan Thorne is one of the leading proponents of an alternative theory, known as the multiregional hypothesis, which holds that human evolution has been continuous – that just as australopithecines evolved into *Homo habilis* and *Homo heidelbergensis* became over time *Homo neanderthalensis*, so modern *Homo sapiens* simply emerged from more ancient *Homo* forms. *Homo erectus* is, on this view, not a separate species but just a transitional phase. Thus modern Chinese are descended from ancient *Homo erectus* forebears in China, modern Europeans from ancient European *Homo erectus*, and so on. 'Except that for me there are no *Homo erectus*,' says Thorne. 'I think it's a term which has outlived its usefulness. For me, *Homo erectus* is simply an earlier part of us. I believe only one species of humans has ever left Africa, and that species is *Homo sapiens*.'

Opponents of the multiregional theory reject it, in the first instance, on the grounds that it requires an improbable amount of parallel evolution by hominids throughout the Old World – in Africa, China, Europe, the most distant islands of Indonesia, wherever they appeared. Some also believe that multiregionalism encourages a racist view of which anthropology took a very long time to rid itself. In the early 1960s, a famous anthropologist named Carleton Coon of the University of Pennsylvania suggested that some modern races have different sources of origin, implying that some of us come from superior stock to others. This harkened back uncomfortably to earlier beliefs that some modern races such as the African 'Bushmen' (properly the Kalahari San) and Australian Aborigines were more primitive than others.

Whatever Coon may personally have felt, the implication for many people was that some races are inherently more advanced, and that some humans could essentially constitute different species. The view, so instinctively offensive now, was widely popularized in many respectable places until fairly recent times. I have before me a popular book published by Time-Life Publications in 1961 called *The Epic of Man*, based on a series of articles in *Life* magazine. In it you can find such comments as 'Rhodesian man . . . lived as recently as 25,000 years ago and may have been an ancestor of the African Negroes. His brain size was close to that of *Homo sapiens*.' In other words, black Africans were recently descended from creatures that were only 'close' to *Homo sapiens*.

Thorne emphatically (and I believe sincerely) dismisses the idea that his theory is in any measure racist, and accounts for the uniformity of human evolution by suggesting that there was a lot of movement back and forth between cultures and regions. 'There's no reason to suppose that people only went in one direction,' he says. 'People were moving all over the place, and where they met they almost certainly shared genetic material through interbreeding. New arrivals didn't replace the indigenous populations, they *joined* them. They became them.' He likens the situation to when explorers like Cook or Magellan encountered remote peoples for the first time. 'They weren't meetings of different species, but of the same species with some physical differences.'

What you actually see in the fossil record, Thorne insists, is a smooth, continuous transition. 'There's a famous skull from Petralona in Greece, dating from about three hundred thousand years ago, that has been a matter of contention among traditionalists because it seems in some ways *Homo erectus* but in other ways *Homo sapiens*. Well, what we say is that this is just what you would expect to find in species that were evolving rather than being displaced.'

One thing that would help to resolve matters would be evidence of interbreeding, but that is not at all easy to prove, or disprove, from fossils. In 1999, archaeologists in Portugal found the skeleton of a child about four years old who died 24,500 years ago. The skeleton was modern overall, but with certain archaic, possibly Neandertal, characteristics: unusually sturdy leg bones, teeth bearing a distinctive 'shovelling' pattern, and (though not everyone agrees on it) an indentation at the back of the skull called a suprainiac fossa, a feature exclusive to Neandertals. Erik Trinkaus of Washington University in St Louis, the leading authority on Neandertals, announced the child to be a hybrid: proof that modern humans and Neandertals interbred. Others, however, were troubled that the Neandertal and modern features weren't more blended. As one critic put it: 'If you look at a mule, you don't have the front end looking like a donkey and the back end looking like a horse.'

Ian Tattersall declared it to be nothing more than 'a chunky modern child'. He accepts that there may well have been some 'hanky-panky' between Neandertals and moderns, but doesn't believe it could have resulted in reproductively successful offspring.* 'I don't know of any two organisms from any realm of biology that are that different and still in the same species,' he says.

With the fossil record so unhelpful, scientists have turned increasingly to genetic studies, in particular the part known as mitochondrial DNA. Mitochondrial DNA was only

*One possibility is that Neandertals and Cro-Magnons had different numbers of chromosomes, a complication that commonly arises when species that are close but not quite identical conjoin. In the equine world, for example, horses have 64 chromosomes and donkeys 62. Mate the two and you get an offspring with a reproductively useless number of chromosomes, 63. You have, in short, a sterile mule.

discovered in 1964, but by the 1980s some ingenious souls at the University of California at Berkeley had realized that it has two features that lend it a particular convenience as a kind of molecular clock: it is passed on only through the female line, so it doesn't become scrambled with paternal DNA with each new generation, and it mutates about twenty times faster than normal nuclear DNA, making it easier to detect and follow genetic patterns over time. By tracking the rates of mutation they could work out the genetic histories and relationships of whole groups of people.

In 1987 the Berkeley team, led by the late Allan Wilson, did an analysis of mitochondrial DNA from 147 individuals and declared that the rise of anatomically modern humans occurred in Africa within the last 140,000 years and that 'all present-day humans are descended from that population.' It was a serious blow to the multiregionalists. But then people began to look a little more closely at the data. One of the most extraordinary points – almost too extraordinary to credit, really – was that the 'Africans' used in the study were actually African-Americans, whose genes had obviously been subjected to considerable mediation in the past few hundred years. Doubts also soon emerged about the assumed rates of mutations.

By 1992 the study was largely discredited. But the techniques of genetic analysis continued to be refined; in 1997 scientists from the University of Munich managed to extract and analyse some DNA from the arm bone of the original Neandertal Man, and this time the evidence stood up. The Munich study found that the Neandertal DNA was unlike any DNA found on Earth now, strongly indicating that there was no genetic connection between Neandertals and modern humans. Now this really *was* a blow to multiregionalism.

Then, in late 2000, *Nature* and other publications reported on a Swedish study of the mitochondrial DNA of fifty-three people which suggested that all modern humans emerged from Africa within the past hundred thousand years and came from a breeding stock of no more than ten thousand individuals. Soon afterwards, Eric Lander, director of the Whitehead Institute/Massachusetts Institute of Technology Center for Genome Research, announced that modern Europeans, and perhaps people further afield, are descended from 'no more than a few hundred Africans who left their homeland as recently as 25,000 years ago'.

As we have noted elsewhere in the book, modern human beings show remarkably little genetic variability – 'there's more diversity in one social group of fifty-five chimps than in the entire human population,' as one authority has put it – and this would explain why. Because we are recently descended from a small founding population, there hasn't been time enough or people enough to provide a source of great variability. It seemed a pretty severe blow to multiregionalism. 'After this,' a Penn State academic told the *Washington Post*, 'people won't be too concerned about the multiregional theory, which has very little evidence.'

But all of this overlooked the more or less infinite capacity for surprise offered by the ancient Mungo people of western New South Wales. In early 2001, Thorne and his colleagues at the Australian National University reported that they had recovered DNA from the oldest of the Mungo specimens – now dated at sixty-two thousand years – and that this DNA proved to be 'genetically distinct'.

The Mungo man, according to these findings, was anatomically modern – just like you and me – but carried an extinct genetic lineage. His mitochondrial DNA is no longer found in living humans, as it should be if, like all other modern people, he was descended from individuals who had left Africa in the recent past.

'It turned everything upside down again,' says Thorne with undisguised delight.

Then other, even more curious anomalies began to turn up. Rosalind Harding, a population geneticist at the Institute of Biological Anthropology in Oxford, while studying betaglobin genes in modern people, found two variants that are common among Asians and the indigenous people of Australia, but hardly exist in Africa. The variant genes, she is certain, arose more than two hundred thousand years ago not in Africa, but in east Asia – long before modern *Homo sapiens* reached the region. The only way to account for them is to say that the ancestors of people now living in Asia included archaic hominids – Java Man and the like. Interestingly, this same variant gene – the Java Man gene, so to speak – turns up in modern populations in Oxfordshire.

Confused, I went to see Ms Harding at the institute, which inhabits an old brick villa on the Banbury Road in Oxford. Harding is a small and chirpy Australian, from Brisbane originally, with the rare knack for being amused and earnest at the same time.

'Don't know,' she said at once, grinning, when I asked her how people in Oxfordshire came to harbour sequences of betaglobin that shouldn't be there. 'On the whole,' she went on more sombrely, 'the genetic record supports the out of Africa hypothesis. But then you find these anomalous clusters, which most geneticists prefer not to talk about. There's *huge* amounts of information that would be available to us if only we could understand it, but we don't yet. We've barely begun.' She refused to be drawn on what the existence of Asian-origin genes in Oxfordshire tells us other than that the situation is clearly complicated. 'All we can say at this stage is that it is very untidy and we don't really know why.'

At the time of our meeting, in early 2002, another Oxford scientist named Bryan Sykes had just produced a popular book called *The Seven Daughters of Eve* in which, using studies of mitochondrial DNA, he had claimed to be able to trace nearly all living Europeans back to a founding population of just seven women – the 'daughters of Eve' of the title – who lived between ten thousand and forty-five thousand years ago in the time known to science as the Palaeolithic. To each of these women Sykes had given a name – Ursula, Xenia, Jasmine and so on – and even a detailed personal history. ('Ursula was her mother's second child. The first had been taken by a leopard when he was only two . . .')

When I asked Harding about the book, she smiled broadly but carefully, as if not quite certain where to go with her answer. 'Well, I suppose you must give him some credit for helping to popularize a difficult subject,' she said and paused thoughtfully. 'And there remains the *remote* possibility that he's right.' She laughed, then went on more intently: 'Data from any single gene cannot really tell you anything so definitive. If you follow the mitochondrial DNA backwards, it will take you to a certain place – to an Ursula or Tara or whatever. But if you take any *other* bit of DNA, any gene at all, and trace *it* back, it will take you someplace else altogether.'

It was a little, I gathered, like following a road randomly out of London and finding that eventually it ends at John O'Groats, and concluding from this that anyone in London must therefore have come from the north of Scotland. They *might* have come from there, of course, but equally they could have arrived from any of hundreds of other places. In this sense, according to Harding, every gene is a different highway, and we have only barely begun to map the routes. 'No single gene is ever going to tell you the whole story,' she said.

So genetic studies aren't to be trusted?

'Oh, you can trust the studies well enough, generally speaking. What you can't trust are the sweeping conclusions that people often attach to them.'

She thinks out of Africa is 'probably ninety-five per cent correct', but adds: 'I think both sides have done a bit of a disservice to science by insisting that it must be one thing or the other. Things are likely to turn out to be not so straightforward as either camp would have you believe. The evidence is clearly starting to suggest that there were multiple migrations and dispersals in different parts of the world going in all kinds of directions and generally mixing up the gene pool. That's never going to be easy to sort out.'

Just at this time, there were also a number of reports questioning the reliability of claims concerning the recovery of very ancient DNA. An academic writing in *Nature* had noted how a palaeontologist, asked by a colleague whether he thought an old skull was varnished or not, had licked its top and announced that it was. 'In the process,' noted the *Nature* article, 'large amounts of modern human DNA would have been transferred to the skull', rendering it useless for future study. I asked Harding about this. 'Oh, it would almost certainly have been contaminated already,' she said. 'Just handling a bone will contaminate it. Breathing on it will contaminate it. Most of the water in our labs will contaminate it. We are all swimming in foreign DNA. In order to get a reliably clean specimen you have to excavate it in sterile conditions and do the tests on it at the site. It is the trickiest thing in the world not to contaminate a specimen.'

So should such claims be treated dubiously? I asked.

Ms Harding nodded solemnly. 'Very,' she said.

If you wish to understand at once why we know as little as we do about human origins, I have the place for you. It is to be found a little beyond the edge of the blue Ngong Hills in Kenya, to the south and west of Nairobi. Drive out of the city on the main highway to Uganda and there comes a moment of startling glory when the ground falls away and you are presented with a hang-glider's view of boundless, pale green African plain.

This is the Great Rift Valley, which arcs across 3,000 miles of east Africa, marking the tectonic rupture that is setting Africa adrift from Asia. Here, perhaps 65 kilometres out of Nairobi, along the baking valley floor, is an ancient site called Olorgesailie, which once stood beside a large and pleasant lake. In 1919, long after the lake had vanished, a geologist named J. W. Gregory was scouting the area for mineral prospects when he came across a stretch of open ground littered with anomalous dark stones that had clearly been shaped by human hand. He had found one of the great sites of Acheulean tool manufacture that Ian Tattersall had told me about.

Unexpectedly, in the autumn of 2002 I found myself a visitor to this extraordinary site. I was in Kenya for another purpose altogether, visiting some projects run by the charity CARE International, but my hosts, knowing of my interest in human origins for the present volume, had inserted a visit to Olorgesailie into the schedule.

After its discovery by the geologist Gregory, Olorgesailie lay undisturbed for over two decades before the famed husband and wife team of Louis and Mary Leakey began an excavation that isn't completed yet. What the Leakeys found was a site stretching to ten acres or so, where tools were made in incalculable numbers for roughly a million years, from about 1.2 million years ago to two hundred thousand years ago. Today the tool beds are sheltered from the worst of the elements beneath large tin lean-tos and fenced off with chicken wire to discourage opportunistic scavenging by visitors, but otherwise the tools are left just where their creators dropped them and where the Leakeys found them.

Jillani Ngalli, a keen young man from the Kenyan National Museum who had been dispatched to act as guide, told me that the quartz and obsidian rocks from which the axes were made were never found on the valley floor. 'They had to carry the stones from there,' he said, nodding at a pair of mountains in the hazy middle distance, in opposite directions from the site: Olorgesailie and Ol Esakut. Each was about 10 kilometres away – a long way to carry an armload of stone.

Why the early Olorgesailie people went to such trouble we can only guess, of course. Not only did they lug hefty stones considerable distances to the lakeside, but, perhaps even more remarkably, they then organized the site. The Leakeys' excavations revealed that there were areas where axes were fashioned and others where blunt axes were brought to be resharpened. Olorgesailie was, in short, a kind of factory; one that stayed in business for a million years.

Various replications have shown that the axes were tricky and labour-intensive objects to make – even with practice, an axe would take hours to fashion – and yet, curiously, they were not particularly good for cutting or chopping or scraping or any of the other tasks to which they were presumably put. So we are left with the position that for a million years – far, far longer than our own species has even been in existence, much less engaged in continuous co-operative efforts – early people came in considerable numbers to this particular site to make extravagantly large numbers of tools that appear to have been rather curiously pointless.

And who were these people? We have no idea, actually. We assume they were *Homo erectus* because there are no other known candidates, which means that at their peak – their *peak* – the Olorgesailie workers would have had the brains of a modern infant. But there is no physical evidence on which to base a conclusion. Despite over sixty years of searching, no human bone has ever been found in or around the vicinity of Olorgesailie. However much time they spent there shaping rocks, it appears they went elsewhere to die.

'It's all a mystery,' Jillani Ngalli told me, beaming happily.

The Olorgesailie people disappeared from the scene about two hundred thousand years ago when the lake dried up and the Rift Valley started to become the hot and challenging place it is today. But by this time their days as a species were already numbered. The world was about to get its first real master race, *Homo sapiens*. Things would never be the same again.

30

GOODBYE

In the early 1680s, at just about the time that Edmond Halley and his friends Christopher Wren and Robert Hooke were settling down in a London coffee house and embarking on the casual wager that would result eventually in Isaac Newton's *Principia*, Henry Cavendish's weighing of the Earth and many of the other inspired and commendable undertakings that have occupied us for the past four hundred or so pages, a rather less desirable milestone was being passed on the island of Mauritius, far out in the Indian Ocean some 1,300 kilometres off the east coast of Madagascar.

There, some forgotten sailor or sailor's pet was harrying to death the last of the dodos, the famously flightless bird whose dim but trusting nature and lack of leggy zip made it a rather irresistible target for bored young tars on shore leave. Millions of years of peaceful isolation had not prepared it for the erratic and deeply unnerving behaviour of human beings.

We don't know precisely the circumstances, or even the year, attending the last moments of the last dodo, so we don't know which arrived first, a world that contained a *Principia* or one that had no dodos, but we do know that they happened at more or less the same time. You would be hard pressed, I would submit, to find a better pairing of occurrences to illustrate the divine and felonious nature of the human being – a species of organism that is capable of unravelling the deepest secrets of the heavens while at the same time pounding into extinction, for no purpose at all, a creature that never did us any harm and wasn't even remotely capable of understanding what we were doing to it as we did it. Indeed, dodos were so spectacularly short on insight, it is reported, that if you wished to find all the dodos in a vicinity you had only to catch one and set it to squawking, and all the others would waddle along to see what was up.

The indignities to the poor dodo didn't end quite there. In 1755, some seventy years after the last dodo's death, the director of the Ashmolean Museum in Oxford decided that the institution's stuffed dodo was becoming unpleasantly musty and ordered it tossed on a bonfire. This was a surprising decision as it was by this time the only dodo in existence, stuffed or otherwise. A passing employee, aghast, tried to rescue the bird but could save only its head and part of one limb.

As a result of this and other departures from common sense, we are not now entirely sure what a living dodo was like. We possess much less information than most people suppose – a handful of crude descriptions by 'unscientific voyagers, three or four oil paintings, and a few scattered osseous fragments', in the somewhat aggrieved words of the nineteenth-century naturalist H. E. Strickland. As Strickland wistfully observed, we have more physical evidence of some ancient sea monsters and lumbering sauropods than we do of a bird that lived into modern times and required nothing of us to survive except our absence.

So what is known of the dodo is this: it lived on Mauritius, was plump but not tasty, and was the biggest-ever member of the pigeon family, though by quite what margin is unknown as its weight was never accurately recorded. Extrapolations from Strickland's 'osseous fragments' and the Ashmolean's modest remains show that it was a little over two and a half feet tall and about the same distance from beak-tip to backside. Being flightless, it nested on the ground, leaving its eggs and chicks tragically easy prey for pigs, dogs and monkeys brought to the island by outsiders. It was probably extinct by 1683 and was most certainly gone by 1693. Beyond that we know almost nothing except of course that we will not see its like again. We know nothing of its reproductive habits and diet, where it ranged, what sounds it made in tranquillity or alarm. We don't possess a single dodo egg.

From beginning to end, our acquaintance with animate dodos lasted just seventy years. That is a breathtakingly scanty period – though it must be said that by this point in our history we did have thousands of years of practice behind us in the matter of irreversible eliminations. Nobody knows quite how destructive human beings are, but it is a fact that over the last fifty thousand years or so, wherever we have gone animals have tended to vanish, often in astonishingly large numbers.

In America, thirty genera of large animals – some very large indeed – disappeared practically at a stroke after the arrival of modern humans on the continent between ten and twenty thousand years ago. Altogether North and South America between them lost about three-quarters of their big animals once man the hunter arrived with his flint-headed spears and keen organizational capabilities. Europe and Asia, where the animals had had longer to evolve a useful wariness of humans, lost between a third and a half of their big creatures. Australia, for exactly the opposite reasons, lost no less than 95 per cent.

Because the early hunter populations were comparatively small and the animal populations truly monumental – as many as ten million mammoth carcasses are thought to lie frozen in the tundra of northern Siberia alone – some authorities think there must be other explanations, possibly involving climate change or some kind of pandemic. As Ross MacPhee of the American Museum of Natural History put it: 'There's no material benefit to hunting dangerous animals more often than you need to – there are only so many mammoth steaks you can eat.' Others believe it may have been almost criminally easy to catch and clobber prey. 'In Australia and the Americas,' says Tim Flannery, 'the animals probably didn't know enough to run away.'

Some of the creatures that were lost were singularly spectacular and would take a little managing if they were still around. Imagine ground sloths that could look into an upstairs window, tortoises nearly the size of a small Fiat, monitor lizards 6 metres long basking beside desert highways in Western Australia. Alas, they are gone, and we live on a much diminished planet. Today, across the whole world, only four types of really hefty (a tonne or more) land animals survive: elephants, rhinos, hippos and giraffes. Not for tens of millions of years has life on Earth been so diminutive and tame.

The question that arises is whether the disappearances of the stone age and disappearances of more recent times are in effect part of a single extinction event – whether, in short, humans are inherently bad news for other living things. The sad likelihood is that we may well be. According to the University of Chicago palaeontologist David Raup, the background rate of extinction on Earth throughout biological history has been one species lost every four years on average. According to Richard Leakey and Roger Lewin in *The Sixth Extinction*, human-caused extinction now may be running at as much as 120,000 times that level.

In the mid-1990s, the Australian naturalist Tim Flannery,

now head of the South Australian Museum in Adelaide, became struck by how little we seemed to know about many extinctions, including relatively recent ones. 'Wherever you looked, there seemed to be gaps in the records – pieces missing, as with the dodo, or not recorded at all,' he told me in Melbourne in early 2002.

Flannery recruited his friend Peter Schouten, an artist and fellow Australian, and together they embarked on a slightly obsessive quest to scour the world's major collections to find out what was lost, what was left and what had never been known at all. They spent four years picking through old skins, musty specimens, old drawings and written descriptions – whatever was available. Schouten made life-sized paintings of every animal they could reasonably recreate and Flannery wrote the words. The result was an extraordinary book called *A Gap in Nature*, constituting the most complete – and, it must be said, moving – catalogue of animal extinctions from the last three hundred years.

For some animals, records were good, but nobody had done anything much with them, sometimes for years, sometimes for ever. Steller's sea cow, a walrus-like creature related to the dugong, was one of the last really big animals to go extinct. It was truly enormous – an adult could reach lengths of nearly 9 metres and weigh 10 tonnes – but we are acquainted with it only because in 1741 a Russian expedition happened to be shipwrecked on the sole place where the creatures still survived in any numbers, the remote and foggy Commander Islands in the Bering Sea.

Happily, the expedition had a naturalist, Georg Steller, who was fascinated by the animal. 'He took the most copious notes,' says Flannery. 'He even measured the diameter of its whiskers. The only thing he wouldn't describe was the male genitals – though, for some reason, he was happy enough to do the female's. He even saved a piece of skin, so we had a good idea of its texture. We weren't always so lucky.'

The one thing Steller couldn't do was save the sea cow itself. Already hunted to the brink of extinction, it would be gone altogether within twenty-seven years of Steller's discovery of it. Many other animals, however, couldn't be included because too little is known about them. The Darling Downs hopping mouse, Chatham Islands swan, Ascension Island flightless crake, at least five types of large turtle and many others are for ever lost to us except as names.

A great deal of extinction, Flannery and Schouten discovered, hasn't been cruel or wanton, but just kind of majestically foolish. In 1894, when a lighthouse was built on a lonely rock called Stephens Island in the tempestuous strait between the North and South Islands of New Zealand, the lighthouse keeper's cat kept bringing him strange little birds that it had caught. The keeper dutifully sent some specimens to the museum in Wellington. There a curator grew very excited because the bird was a relic species of flightless wrens – the only example of a flightless perching bird ever found anywhere. He set off at once for the island, but by the time he got there the cat had killed them all. Twelve stuffed museum species of the Stephens Island flightless wren are all that now exist.

At least we have those. All too often, it turns out, we are not much better at looking after species after they have gone than we were before they went. Take the case of the lovely Carolina parakeet. Emerald green, with a golden head, it was arguably the most striking and beautiful bird ever to live in North America – parrots don't usually venture so far north, as you may have noticed – and at its peak it existed in vast numbers, exceeded only by the passenger pigeon. But the Carolina parakeet was also considered a pest by farmers and easily hunted because it flocked tightly and had a peculiar habit of flying up at the sound of gunfire (as you would expect), but then returning almost at once to check on fallen comrades.

In his classic *American Ornithology*, written in the early nineteenth century, Charles Willson Peale describes an occasion in which he repeatedly empties a shotgun into a tree in which they roost:

> At each successive discharge, though showers of them fell, yet the affection of the survivors seemed rather to increase; for, after a few circuits around the place, they again alighted near me, looking down on their slaughtered companions with such manifest symptoms of sympathy and concern, as entirely disarmed me.

By the second decade of the twentieth century, the birds had been so relentlessly hunted that only a few remained alive in captivity. The last one, named Inca, died in Cincinnati Zoo in 1918 (not quite four years after the last passenger pigeon died in the same zoo) and was reverently stuffed. And where would you go to see poor Inca now? Nobody knows. The zoo lost it.

What is both most intriguing and puzzling about the story above is that Peale was a lover of birds, and yet did not hesitate to kill them in large numbers for no better reason than that it interested him to do so. It is a truly astounding fact that for a very long time the people who were most intensely interested in the world's living things were the ones most likely to extinguish it.

No-one represented this position on a larger scale (in every sense) than Lionel Walter Rothschild, the second Baron Rothschild. Scion of the great banking family, Rothschild was a strange and reclusive fellow. He lived his entire life, from 1868 to 1937, in the nursery wing of his home at Tring, in Buckinghamshire, using the furniture of his childhood – even sleeping in his childhood bed, though eventually he weighed 135 kilograms.

His passion was natural history and he became a devoted accumulator of objects. He sent hordes of trained men – as many as four hundred at a time – to every quarter of the globe to clamber over mountains and hack their way through jungles in the pursuit of new specimens – particularly things that flew. These were crated or boxed up and sent back to Rothschild's estate at Tring, where he and a battalion of assistants exhaustively logged and analysed everything that came before them, producing a constant stream of books, papers and monographs – some twelve hundred in all. Altogether, Rothschild's natural history factory processed well over two million specimens and added five thousand species of creature to the scientific archive.

Remarkably, Rothschild's collecting efforts were neither the most extensive nor the most generously funded of the nineteenth century. That title almost certainly belongs to a slightly earlier but also very wealthy British collector named Hugh Cuming, who became so preoccupied with accumulating objects that he built a large ocean-going ship and employed a crew to sail the world full-time picking up whatever they could find – birds, plants, animals of all types, and especially shells. It was his unrivalled collection of barnacles that passed to Darwin and served as the basis for his seminal study.

However, Rothschild was easily the most scientific collector of his age, though also the most regrettably lethal, for in the 1890s he became interested in Hawaii, perhaps the most temptingly vulnerable environment Earth has yet produced. Millions of years of isolation had allowed Hawaii to evolve 8,800 unique species of animals and plants. Of particular interest to Rothschild were the islands' colourful and distinctive birds, often consisting of very small populations inhabiting extremely specific ranges.

The tragedy for many Hawaiian birds was that they were not only distinctive, desirable and rare – a dangerous combination in the best of circumstances – but also often heartbreakingly easy to take. The greater koa finch, an innocuous member of the honeycreeper family, lurked shyly in the canopies of koa trees, but if someone imitated its song it would abandon its cover at once and fly down in a show of welcome. The last of the species vanished in 1896, killed by Rothschild's ace collector Harry Palmer, five years after the disappearance of its cousin the lesser koa finch, a bird so sublimely rare that only one has ever been seen: the one shot for Rothschild's collection. Altogether, during the decade or so of Rothschild's most intensive collecting, at least nine species of Hawaiian birds vanished, but it may have been more.

Rothschild was by no means alone in his zeal to capture birds at more or less any cost. Others in fact were more ruthless. In 1907, when a well-known collector named Alanson Bryan realized that he had shot the last three specimens of black mamos, a species of forest bird that had only been discovered the previous decade, he noted that the news filled him with 'joy'.

It was, in short, a difficult age to fathom – a time when almost any animal was persecuted if it was deemed the least bit intrusive. In 1890, New York State paid out over one hundred bounties for eastern mountain lions, even though it was clear that the much-harassed creatures were on the brink of extermination. Right up until the 1940s many states continued to pay bounties for almost any kind of predatory creature. West Virginia gave out an annual college scholarship to whoever brought in the greatest number of dead pests – and 'pests' was liberally interpreted to mean almost anything that wasn't grown on farms or kept as pets.

Perhaps nothing speaks more vividly for the strangeness of

the times than the fate of the lovely little Bachman's warbler. A native of the southern United States, the warbler was famous for its unusually lovely song, but its population numbers, never robust, gradually dwindled until by the 1930s the warbler vanished altogether and went unseen for many years. Then, in 1939, by happy coincidence two separate birding enthusiasts, in widely separated locations, came across lone survivors just two days apart. They both shot the birds.

The impulse to exterminate was by no means exclusively American. In Australia, bounties were paid on the Tasmanian tiger (properly the thylacine), a doglike creature with distinctive 'tiger' stripes across its back, until shortly before the last one died, forlorn and nameless, in a private Hobart zoo in 1936. Go to the Tasmanian Museum and Art Gallery today and ask to see the last of this species – the only large carnivorous marsupial to live into modern times – and all they can show you are photographs and 61 seconds of old film footage. Upon its death, the last surviving thylacine was thrown out with the weekly trash.

I mention all this to make the point that if you were designing an organism to look after life in our lonely cosmos, to monitor where it is going and keep a record of where it has been, you wouldn't choose human beings for the job.

But here's an extremely salient point: we have been chosen, by fate or providence or whatever you wish to call it. As far as we can tell, we are the best there is. We may be all there is. It's an unnerving thought that we may be the living universe's supreme achievement and its worst nightmare simultaneously.

Because we are so remarkably careless about looking after things, both when they are alive and when they are not, we have no idea – really none at all – about how many things have died off permanently, or may do so soon, or may never, and what role we have played in any part of the process. In 1979, in his book *The Sinking Ark*, Norman Myers suggested that human activities were causing about two extinctions a week on the planet. By the early 1990s he had raised the figure to some six hundred per week. (That's extinctions of all types – plants, insects and so on as well as animals.) Others have put the figure even higher – to well over a thousand a week. A United Nations report of 1995, on the other hand, put the total number of known extinctions in the last four hundred years at slightly under five hundred for animals and slightly over six hundred and fifty for plants – while allowing that this was 'almost certainly an underestimate', particularly with regard to tropical species. A few interpreters think most extinction figures are grossly inflated.

The fact is, we don't know. Don't have any idea. We don't know when we started doing many of the things we've done. We don't know what we are doing right now or how our present actions will affect the future. What we do know is that there is only one planet to do it on, and only one species of being capable of making a considered difference. Edward O. Wilson expressed it with unimprovable brevity in *The Diversity of Life*: 'One planet, one experiment.'

If this book has a lesson, it is that we are awfully lucky to be here – and by 'we' I mean every living thing. To attain any kind of life at all in this universe of ours appears to be quite an achievement. As humans we are doubly lucky, of course. We enjoy not only the privilege of existence, but also the singular ability to appreciate it and even, in a multitude of ways, to make it better. It is a trick we have only just begun to grasp.

We have arrived at this position of eminence in a stunningly short time. Behaviourally modern humans have been around for no more than about 0.0001 per cent of Earth's history – almost nothing, really – but even existing for that little while has required a nearly endless string of good fortune.

We really are at the beginning of it all. The trick, of course, is to make sure we never find the end. And that, almost certainly, will require a lot more than lucky breaks.

NOTES

Chapter 1: How to Build a Universse

p. 27 Protons are so small that: Bodanis, *E* = *mc*², p. 111.

p. 27 Now pack into that tiny, tiny space: Guth, *The Inflationary Universe*, p. 254.

p. 29 The consensus seems to be heading for a figure of about 13.7 billion years: *New York Times*, 'Cosmos Sits for Early Portrait, Gives Up Secrets', 12 Feb. 2003, p. 1; *US News and World Report*, 'How Old Is the Universe?', 18–25 Aug. 1997, pp. 34–6.

p. 29 there came the moment known to science as $t = 0$: Guth, *The Inflationary Universe*, p. 86.

p. 29 They climbed back into the dish with brooms and scrubbing brushes and carefully swept it clean: Lawrence M. Krauss, 'Rediscovering Creation', in Shore (ed.), *Mysteries of Life and the Universe*, p. 50.

p. 30 an instrument that might do the job: the Bell antenna: Overbye, *Lonely Hearts of the Cosmos*, p. 153.

p. 30 They had found the edge of the universe: *Scientific American*, 'Echoes from the Big Bang', Jan. 2001, pp. 38–43.

p. 30 Penzias and Wilson's finding pushed our acquaintance with the visible: Guth, *The Inflationary Universe*, p. 101.

p. 31 about 1 per cent of the dancing static you see: Gribbin, *In the Beginning*, p. 18.

p. 32 'These are very close to religious questions': *New York Times*, 'Before the Big Bang, There Was . . . What?', 22 May 2001, p. F1.

p. 32 or one ten million trillion trillion trillionths: Alan Lightman, 'First Birth', in Shore (ed.), *Mysteries of Life and the Universe*, p. 13.

p. 33 He was thirty-two years old and, by his own admission, had never: Overbye, *Lonely Hearts of the Cosmos*, p. 216.

p. 33 The lecture inspired Guth to take an interest: Guth, *The Inflationary Universe*, p. 89.

p. 33 doubling in size every 10^{-34} seconds: Overbye, *Lonely Hearts of the Cosmos*, p. 242.

p. 33 it changed the universe from something you could hold in your hand to something at least 10,000,000,000,000,000,000,000,000 times bigger: *New Scientist*, 'The First Split Second', 31 March 2001, pp. 27–30.

p. 34 perfectly arrayed for the creation of stars, galaxies and other complex systems: *Scientific American*, 'The First Stars in the Universe', Dec. 2001, pp. 64–71; *New York Times*, 'Listen Closely: From Tiny Hum Came Big Bang', 30 April 2001, p. 1.

p. 34 'Tryon emphasized that no one had counted the failed attempts': quoted by Guth, *The Inflationary Universe*, p. 14.

p. 34 He makes an analogy with a very large clothing store: *Discover*, 'Why Is There Life?', Nov. 2000, p.66.

p. 35 with the slightest tweaking of the numbers the universe: Rees, *Just Six Numbers*, p. 147.

p. 35 In the long term, gravity may turn out to be a little too strong: *Financial Times*, 'Riddle of the Flat Universe', 1–2 July 2000; *Economist*, 'The World is Flat after All', 20

May 2000, p. 97.

p. 36 the galaxies are rushing apart: Weinberg, *Dreams of a Final Theory*, p. 26.

p. 37 Scientists just assume that we can't really be the centre: Hawking, *A Brief History of Time*, p. 47.

p. 37 This visible universe – the universe we know and can talk about: Hawking, *A Brief History of Time*, p. 13.

p. 37 the number of light years to the edge of this larger, unseen universe: Rees, *Just Six Numbers*, p. 147.

Chapter 2: Welcome to the Solar System

p. 39 From the tiniest throbs and wobbles of distant stars: *New Yorker*, 'Among Planets', 9 Dec. 1996, p. 84.

p. 39 'less than the energy of a single snowflake striking the ground': Sagan, *Cosmos*, p. 261.

p. 39 In the summer of that year, a young astronomer named James Christy: US Naval Observatory press release, '20th Anniversary of the Discovery of Pluto's Moon Charon', 22 June 1998.

p. 40 Pluto was much smaller than anyone had supposed: *Atlantic Monthly*, 'When Is a Planet Not a Planet?', Feb. 1998, pp. 22–34.

p. 40 In the words of the astronomer Clark Chapman: quoted on PBS *Nova*, 'Doomsday Asteroid', first broadcast 29 April 1997.

p. 40 it took seven years for anyone to spot the moon again: US Naval Observatory press release, '20th Anniversary of the Discovery of Pluto's Moon Charon', 22 June 1998.

p. 41 after a year's patient searching he somehow spotted Pluto: Tombaugh paper, 'The Struggles to Find the Ninth Planet', from NASA website.

p. 42 A few astronomers continue to think there may yet be a Planet X out there: *Economist*, 'X marks the spot', 16 Oct. 1999, p. 83.

p. 42 The Kuiper belt was actually theorized by an astronomer named F. C. Leonard in 1930: *Nature*, 'Almost Planet X', 24 May 2001, p. 423.

p. 43 Only on 11 February 1999 did Pluto return to the outside lane: *Economist*, 'Pluto Out in the Cold,' 6 Feb. 1999, p. 85.

p. 43 as of early December 2002 had found over six hundred additional Trans-Neptunian Objects: *Nature*, 'Seeing Double in the Kuiper Belt', 12 Dec. 2002, p. 618.

p. 43 about the same as a lump of charcoal: *Nature*, 'Almost Planet X', 24 May 2001, p. 423.

p. 44 now flying away from us at about 56,000 kilometres an hour: PBS *NewsHour* transcript, 20 Aug. 2002.

p. 44 but all the visible stuff in it . . . fills less than a trillionth of the available space: *Natural History*, 'Between the Planets', Oct. 2001, p. 20.

p. 45 The total now is at least ninety: *New Scientist*, 'Many Moons', 17 March 2001, p. 39; *Economist*, 'A Roadmap for Planet-Hunting', 8 April 2000, p. 87.

p. 46 we won't reach the Oort cloud for another . . . ten thousand years: Sagan and Druyan, *Comet*, p. 198.

p. 46 and probably result in the deaths of all the crew: *New Yorker*, 'Medicine on Mars', 14 Feb. 2000, p. 39.

p. 47 so the comets drift in a stately manner, moving at only about 220 miles an hour: Sagan and Druyan, *Comet*, p. 195.

p. 47 The most perfect vacuum ever created by humans is not as empty as the emptiness of interstellar space: Ball, *H_2O*, p. 15.

p. 48 Our nearest neighbour in the cosmos, Proxima Centauri: Guth, *The Inflationary Universe*, p. 1; Hawking, *A Brief History of Time*, p. 39.

p. 48 The average distance between stars: Dyson, *Disturbing the Universe*, p. 251.

p. 49 'If we were randomly inserted into the universe,' Sagan wrote: Sagan, *Cosmos*, p. 5.

Chapter 3: The Reverend Evans's Universe

p. 52 releasing in an instant the energy of 100 billion suns: Ferris, *The Whole Shebang*, p. 37.

p. 52 'It's like a trillion hydrogen bombs going off at once': Robert Evans, interviewed Hazelbrook, Australia, 2 Sept. 2001.

p. 52 devotes a passage to him in a chapter on autistic savants: Sacks, *An Anthropologist on Mars*, p. 189.

p. 53 'an irritating buffoon': Thorne, *Black Holes and Time Warps*, p. 164.

p. 54 refused to be left alone with him: Ferris, *The Whole Shebang*, p. 125.

p. 54 On at least one occasion Zwicky threatened to kill Baade: Overbye, *Lonely Hearts of the Cosmos*, p. 18.

p. 54 Atoms would literally be crushed together: *Nature*, 'Twinkle, Twinkle, Neutron Star', 7 Nov. 2002, p. 31.

p. 54 enough to make the biggest bang in the universe: Thorne, *Black Holes and Time Warps*, p. 171.

p. 55 hasn't been verified yet: Thorne, *Black Holes and Time Warps*, p.174.

p. 55 'one of the most prescient documents in the history of physics and astronomy': Thorne, *Black Holes and Time Warps*, p. 174.

p. 55 'he did not understand the laws of physics': Thorne, *Black Holes and Time Warps*, p. 175.

p. 55 wouldn't attract serious attention for nearly four decades: Overbye, *Lonely Hearts of the Cosmos*, p. 18.

p. 56 Only about six thousand stars are visible to the naked eye: Harrison, *Darkness at Night*, p. 3.

p. 58 In 1987 Saul Perlmutter . . . set out to find a more systematic method of searching for them: BBC *Horizon* documentary, 'From Here to Infinity', transcript of programme first broadcast 28 Feb. 1999.

p. 59 'The news of such an event travels out at the speed of light, but so does the destructiveness': interview with John Thorstensen, Hanover, NH, 5 Dec. 2001.

p. 60 Only half a dozen times in recorded history have supernovae been close enough to be visible to the naked eye: note from Evans, 3 Dec. 2002.

p. 61 'cosmologist and controversialist': *Nature*, 'Fred Hoyle (1915–2001)', 17 Sept. 2001, p. 270.

p. 61 humans evolved projecting noses . . . as a way of keeping cosmic pathogens from falling into them: Gribbin and Cherfas, *The First Chimpanzee*, p. 190.

p. 61 continually creating new matter as it went: Rees, *Just Six Numbers*, p. 75.

p. 61 100 million degrees or more: Bodanis, $E = mc^2$, p. 187.

p. 62 99.9 per cent of the mass of the solar system: Asimov, *Atom*, p. 294.

p. 62 In just two hundred million years, possibly less: Stevens, *The Change in the Weather*, p. 6.

p. 62 Most of the lunar material, it is thought, came from the Earth's crust, not its core: *New Scientist* supplement, 'Firebirth', 7 Aug. 1999, n.p.

p. 63 in fact it was first proposed in the 1940s by Reginald Daly of Harvard: Powell, *Night Comes to the Cretaceous*, p. 38.

p. 63 the Earth might well have frozen over permanently: Drury, *Stepping Stones*, p. 144.

Chapter 4: The Measure of Things

p. 69 In the course of a long and productive career: Sagan, *Comet*, p. 52.

p. 69 'a very specific and precise curve': Feynman, *Six Easy Pieces*, p. 90.

p. 70 Hooke . . . claimed that he had solved the problem

already: Gjertsen, *The Classics of Science*, p. 219.

p. 70 and rubbed it around 'betwixt my eye and the bone': quoted by Ferris, *Coming of Age in the Milky Way*, p. 106.

p. 71 but then told no-one about it for twenty-seven years: Durant, *The Age of Louis XIV*, p. 538.

p. 72 Even the great German mathematician Gottfried von Leibniz: Durant, *The Age of Louis XIV*, p. 546.

p. 73 'one of the most inaccessible books ever written': Cropper, *Great Physicists*, p. 31.

p. 73 'proportional to the mass of each and varies inversely as the square of the distance between them': Feynman, *Six Easy Pieces*, p. 69.

p. 74 Newton, as was his custom, contributed nothing: Calder, *The Comet Is Coming!*, p. 39.

p. 74 He was to be paid instead in copies of *The History of Fishes*: Jardine, *Ingenious Pursuits*, p. 36.

p. 76 to 'within a scantling': Wilford, *The Mapmakers*, p. 98.

p. 78 The Earth was 43 kilometres stouter when measured equatorially than when measured from top to bottom around the poles: Asimov, *Exploring the Earth and the Cosmos*, p. 86.

p. 80 Unluckier still was Guillaume le Gentil: Ferris, *Coming of Age in the Milky Way*, p. 134.

p. 81 Mason and Dixon sent a note to the Royal Society: Jardine, *Ingenious Pursuits*, p. 141.

p. 83 'said to have been born in a coal mine': *Dictionary of National Biography*, vol. 12, p. 1302.

p. 83 We know that in 1772: *American Heritage*, 'Mason and Dixon: Their Line and its Legend', Feb. 1964, pp. 23–9.

p. 84 For convenience, Hutton had assumed: Jungnickel and McCormmach, *Cavendish*, p. 449.

p. 84 it was Michell to whom he turned for instruction in making telescopes: Calder, *The Comet Is Coming!*, p. 71.

p. 85 to a 'degree bordering on disease': Jungnickel and McCormmach, *Cavendish*, p. 306.

p. 86 'talk as it were into vacancy': Jungnickel and McCormmach, *Cavendish*, p. 305.

p. 87 he also foreshadowed 'the work of Kelvin and G. H. Darwin on the effect of tidal friction': Crowther, *Scientists of the Industrial Revolution*, pp. 214–15.

p. 87 At the heart of the machine were two 350-pound lead balls: *Dictionary of National Biography*, vol. 3, p. 1261.

p. 88 six billion trillion metric tons: *Economist*, 'G Whiz', 6 May 2000, p. 82.

Chapter 5: The Stone-Breakers

p. 90 Hutton was by all accounts a man of the keenest insights and liveliest conversation: *Dictionary of National Biography*, vol. 10, pp. 354–6.

p. 90 'almost entirely innocent of rhetorical accomplishments': Dean, *James Hutton and the History of Geology*, p. 18.

p. 91 He became a leading member of a society called the Oyster Club: McPhee, *Basin and Range*, p. 99.

p. 94 quotations from French sources, still in the original French: Gould, *Time's Arrow*, p. 66.

p. 94 A third volume was so unenticing that it wasn't published until 1899: Oldroyd, *Thinking About the Earth*, pp. 96–7.

p. 94 Even Charles Lyell . . . couldn't get through it: Schneer (ed.), *Toward a History of Geology*, p. 128.

p. 94 In the winter of 1807: Geological Society papers, *A Brief History of the Geological Society of London*.

p. 95 The members met twice a month from November until June: Rudwick, *The Great Devonian Controversy*, p. 25.

p. 95 As even a Murchison supporter conceded: Trinkaus and Shipman, *The Neandertals*, p. 28.

p. 96 In 1794 he was implicated in a faintly lunatic-sounding conspiracy: Cadbury, *Terrible Lizard*, p. 39.

p. 96 known ever since as Parkinson's disease: *Dictionary of National Biography*, vol. 15, pp. 314–15.

p. 96 because his mother was convinced that Scots were feckless drunks: Trinkaus and Shipman, *The Neandertals*, p. 26.

p. 97 Once Mrs Buckland found herself being shaken awake: Annan, *The Dons*, p. 27.

p. 98 His other slight peculiarity: Trinkaus and Shipman, *The Neandertals*, p. 30.

p. 98 Often when lost in thought: Desmond and Moore, *Darwin*, p. 202.

p. 99 but it was Lyell most people read: Schneer (ed.), *Toward a History of Geology*, p. 139.

p. 99 'and called for a new pack': Clark, *The Huxleys*, p. 48.

p. 99 'Never was there a dogma more calculated to foster indolence': quoted in Gould, *Dinosaur in a Haystack*, p. 167.

p. 99 He failed to explain . . . how mountain ranges were formed: Hallam, *Great Geological Controversies*, p. 135.

p. 99 'the refrigeration of the globe': Gould, *Ever since Darwin*, p. 151.

p. 99 He rejected the notion that animals and plants suffered sudden annihilations: Stanley, *Extinction*, p. 5.

p. 100 'one yet saw it partially through his eyes': quoted in Schneer (ed.), *Toward a History of Geology*, p. 288.

p. 100 'De la Beche is a dirty dog': quoted in Rudwick, *The Great Devonian Controversy*, p. 194.

p. 101 with the perky name of J. J. d'Omalius d'Halloy: McPhee, *In Suspect Terrain*, p. 190.

p. 101 Lyell originally intended to employ '-synchronous' for his endings: Gjertsen, *The Classics of Science*, p. 305.

p. 102 these number in the 'tens of dozens': McPhee, *In Suspect Terrain*, p. 50.

p. 103 Rocks are divided into quite separate units: Powell, *Night Comes to the Cretaceous*, p. 200.

p. 103 'I have seen grown men glow incandescent with rage': Fortey, *Trilobite!*, p. 238.

p. 103 When Buckland speculated: Cadbury, *Terrible Lizard*, p. 149.

p. 103 The most well known early attempt: Gould, *Eight Little Piggies*, p. 185.

p. 104 'most thinking people accepted the idea that the earth was young': cited in Gould, *Time's Arrow*, p. 114.

p. 104 'No geologist of any nationality whose work was taken seriously': Rudwick, *The Great Devonian Controversy*, p. 42.

p. 104 Even the Reverend Buckland: Cadbury, *Terrible Lizard*, p. 192.

p. 105 somewhere between 75,000 and 168,000 years old: Hallam, *Great Geological Controversies*, p.105 and Ferris, *Coming of Age in the Milky Way*, pp. 246–7.

p. 105 Darwin announced that the geological processes that created the Weald: Gjertsen, *The Classics of Science*, p. 335.

p. 106 The German scientist Hermann von Helmholtz: Cropper, *Great Physicists*, p. 78.

p. 106 and written (in French and English) a dozen papers in pure and applied mathematics of such dazzling originality that he had to publish them anonymously: Cropper, *Great Physicists*, p. 79.

p. 106 At the age of twenty-two he returned to Glasgow: *Dictionary of National Biography, Supplement 1901–1911*, p. 508.

Chapter 6: Science Red in Tooth and Claw

p. 109 who described it at a meeting of the American Philosophical Society: Colbert, *The Great Dinosaur*

Hunters and their Discoveries, p. 4.

p. 109 The cause of this froth was a strange assertion by the great French naturalist the Comte de Buffon: Kastner, *A Species of Eternity*, p. 123.

p. 110 A Dutchman named Corneille de Pauw: Kastner, *A Species of Eternity*, p. 124.

p. 112 in 1796 Cuvier wrote a landmark paper, *Note on the Species of Living and Fossil Elephants*: Trinkaus and Shipman, *The Neandertals*, p. 15.

p. 112 Jefferson for one couldn't abide the thought that whole species would ever be permitted to vanish: Simpson, *Fossils and the History of Life*, p. 7.

p. 113 On the evening of 5 January 1796, he was sitting in a coaching inn in Somerset: Harrington, *Dance of the Continents*, p. 175.

p. 113 'The whys and wherefores cannot come within the Province of a Mineral Surveyor': Lewis, *The Dating Game*, pp. 17–18.

p. 114 Cuvier resolved the matter to his own satisfaction: Barber, *The Heyday of Natural History*, p. 217.

p. 114 In 1806 the Lewis and Clark expedition passed through the Hell Creek formation: Colbert, *The Great Dinosaur Hunters and their Discoveries*, p. 5.

p. 115 She is commonly held to be the source for the famous tongue-twister: Cadbury, *Terrible Lizard*, p. 3.

p. 115 The plesiosaur alone took her ten years of patient excavation: Barber, *The Heyday of Natural History*, p. 127.

p. 116 Mantell could see at once it was a fossilized tooth: *New Zealand Geographic*, 'Holy Incisors! What a Treasure!', April–June 2000, p. 17.

p. 116 the name was actually suggested to Buckland by his friend Dr James Parkinson: Wilford, *The Riddle of the Dinosaur*, p. 31.

p. 117 Eventually he was forced to sell most of his collection to pay off his debts: Wilford, *The Riddle of the*

Dinosaur, p. 34.

p. 118 the world's first theme park: Fortey, *Life*, p. 214.

p. 119 he sometimes illicitly borrowed limbs, organs and other parts: Cadbury, *Terrible Lizard*, p. 133.

p. 119 Once his wife returned home to find a freshly deceased rhinoceros filling the front hallway: Cadbury, *Terrible Lizard*, p. 200.

p. 120 some were no bigger than rabbits: Wilford, *The Riddle of the Dinosaur*, p. 5.

p. 120 the one thing they most emphatically were not was lizards: Bakker, *The Dinosaur Heresies*, p. 22.

p. 120 dinosaurs constitute not one but two orders of reptiles: Colbert, *The Great Dinosaur Hunters and their Discoveries*, p. 33.

p. 120 He was the only person Charles Darwin was ever known to hate: *Nature*, 'Owen's Parthian shot', 12 July 2001, p. 123.

p. 120 referred to his father's 'lamentable coldness of heart': Cadbury, *Terrible Lizard*, p. 321.

p. 120 Huxley was leafing through a new edition of *Churchill's Medical Directory*: Clark, *The Huxleys*, p. 45.

p. 122 His deformed spine was removed and sent to the Royal College of Surgeons: Cadbury, *Terrible Lizard*, p. 291.

p. 122 'not quite as original as it appeared': Cadbury, *Terrible Lizard*, pp. 261–2.

p. 123 he became the driving force behind the creation of London's Natural History Museum: Colbert, *The Great Dinosaur Hunters and their Discoveries*, p. 30.

p. 123 Before Owen, museums were designed primarily for . . . the elite: Thackray and Press, *The Natural History Museum*, p. 24.

p. 123 He even proposed, very radically, to put informative labels on each display: Thackray and Press, *The Natural History Museum*, p. 98.

p. 125 'lying everywhere like logs': Wilford, *The Riddle of the Dinosaur*, p. 97.

p. 125 he managed to win them over by repeatedly taking out and replacing his false teeth: Wilford, *The Riddle of the Dinosaur*, p. 100.

p. 125 it was an affront that he would never forget: Colbert, *The Great Dinosaur Hunters and their Discoveries*, p. 73.

p. 126 increased the number of known dinosaur species in America from nine to almost one hundred and fifty: Colbert, *The Great Dinosaur Hunters and their Discoveries*, p. 93.

p. 126 Nearly every dinosaur that the average person can name: Wilford, *The Riddle of the Dinosaur*, p. 90.

p. 126 Between them they managed to 'discover' a species called *Uintatheres anceps* no fewer than twenty-two times: Psihoyos and Knoebber, *Hunting Dinosaurs*, p. 16.

p. 127 mercifully obliterated by a German bomb in the Blitz: Cadbury, *Terrible Lizard*, p. 325.

p. 127 much of it was taken to New Zealand by his son Walter: *Newsletter of the Geological Society of New Zealand*, 'Gideon Mantell – The New Zealand Connection', April 1992; *New Zealand Geographic*, 'Holy Incisors! What a Treasure!' April–June 2000, p. 17.

p. 128 hence the name: Colbert, *The Great Dinosaur Hunters and their Discoveries*, p. 151.

p. 128 He calculated that the Earth was 89 million years old: Lewis, *The Dating Game*, p. 37.

p. 129 Such was the confusion: Hallam, *Great Geological Controversies*, p. 173.

Chapter 7: Elemental Matters

p. 130 could make himself invisible: Ball, H_2O, p. 125.

p. 131 An ounce of phosphorus retailed for 6 guineas: Durant, *Age of Louis XIV*, p. 516.

p. 131 and got credit for none of them: Strathern,

Mendeleyev's Dream, p. 193.

p. 132 which is why we ended up with two branches of chemistry: Davies, *The Fifth Miracle*, p. 14.

p. 133 perhaps £12 million in today's money: White, *Rivals*, p. 63.

p. 133 the fourteen-year-old daughter of one of his bosses: Brock, *The Norton History of Chemistry*, p. 92.

p. 133 *jour de bonheur*: Gould, *Bully for Brontosaurus*, p. 366.

p. 133 It was in such a capacity in 1780 that Lavoisier made some dismissive remarks: Brock, *The Norton History of Chemistry*, pp. 95–6.

p. 134 failed to uncover a single one: Strathern, *Mendeleyev's Dream*, p. 239.

p. 135 taken away and melted down for scrap: Brock, *The Norton History of Chemistry*, p. 124.

p. 135 'a highly pleasurable thrilling': Cropper, *Great Physicists*, p. 139.

p. 136 Theatres put on 'laughing gas evenings': Hamblyn, *The Invention of Clouds*, p. 76.

p. 136 What Brown noticed: Silver, *The Ascent of Science*, p. 201.

p. 137 'lukewarmness in the cause of liberty': *Dictionary of National Biography*, vol. 19, p. 686.

p. 139 a diameter of 0.00000008 centimetres: Asimov, *The History of Physics*, p. 501.

p. 140 Later, for no special reason: Ball, *H_20*, p. 139.

p. 141 Luck was not always with the Mendeleyevs: Brock, *The Norton History of Chemistry*, p. 312.

p. 141 There he was a competent but not terribly outstanding chemist: Brock, *The Norton History of Chemistry*, p. 111.

p. 142 this was an idea whose time had not quite yet come: Carey (ed.), *The Faber Book of Science*, p. 155.

p. 143 chemistry really is just a matter of counting: Ball, *H_2O*, p. 139.

p. 143 'the most elegant organizational chart ever devised':

Krebs, *The History and Use of our Earth's Chemical Elements*, p. 23.

p. 143 '120 or so' known elements: from a review in *Nature*, 'Mind over Matter?', by Gautum R. Desiraju, 26 Sept. 2002.

p. 145 'purely speculative': Heiserman, *Exploring Chemical Elements and their Compounds*, p. 33.

p. 145 Marie Curie dubbed the effect 'radioactivity': Bodanis, $E = mc^2$, p. 75.

p. 147 He never accepted the revised figures: Lewis, *The Dating Game*, p. 55.

p. 148 'Appropriately . . . it is an unstable element': Strathern, *Mendeleyev's Dream*, p. 294.

p. 148 featured with pride the therapeutic effects of its 'Radio-active mineral springs': advertisement in *Time* magazine, 3 Jan. 1927, p. 24.

p. 148 It wasn't banned in consumer products until 1938: Biddle, *A Field Guide to the Invisible*, p. 133.

p. 148 Her lab books are kept in lead-lined boxes: *Science*, 'We Are Made of Starstuff', 4 May 2001, p. 863.

Chapter 8: Einstein's Universe

p. 154 his courses attracted an average of slightly over one student a semester: Cropper, *Great Physicists*, p. 106.

p. 154 a deck of cards: Ebbing, *General Chemistry*, p. 755.

p. 155 which dazzlingly elucidated the thermodynamic principles of, well, nearly everything: Cropper, *Great Physicists*, p. 109.

p. 155 In essence, what Gibbs did was show that thermodynamics didn't apply simply to heat and energy: Snow, *The Physicists*, p. 7.

p. 155 Gibbs's *Equilibrium* has been called 'the *Principia* of thermodynamics': Kevles, *The Physicists*, p. 33.

p. 156 he came to the United States with his family as an infant and grew up in a mining camp in California's gold rush country: Kevles, *The Physicists*, pp. 27–8.

p. 157 'The speed of light turned out to be the same in *all* directions and at *all* seasons': Thorne, *Black Holes and Time Warps*, p. 64.

p. 157 'probably the most famous negative result in the history of physics': Cropper, *Great Physicists*, p. 208.

p. 158 Michelson counted himself among those who believed that the work of science was nearly at an end: *Nature*, 'Physics from the Inside', 12 July 2001, p. 121.

p. 159 of which three, according to C. P. Snow, 'were among the greatest in the history of physics': Snow, *The Physicists*, p. 101.

p. 160 His very first paper, on the physics of fluids in drinking straws: Bodanis, $E = mc^2$, p. 6.

p. 160 only to discover that the quietly productive J. Willard Gibbs in Connecticut had done that work as well: Boorse et al., *The Atomic Scientists*, p. 142.

p. 160 is one of the most extraordinary scientific papers ever published: Ferris, *Coming of Age in the Milky Way*, p. 193.

p. 160 It was . . . as if Einstein 'had reached the conclusions by pure thought, unaided': Snow, *The Physicists*, p. 101.

p. 161 you will contain within your modest frame no less than 7×10^{18} joules of potential energy: Thorne, *Black Holes and Time Warps*, p. 172.

p. 161 Even a uranium bomb . . . releases less than 1 per cent of the energy it could release: Bodanis, $E = mc^2$, p. 77.

p. 162 'Oh, that's not necessary,' he replied. 'It's so seldom I have one': *Nature*, 'In the Eye of the Beholder', 21 March 2002, p. 264.

p. 162 'it is undoubtedly the highest intellectual achievement of humanity': Boorse et al., *The Atomic Scientists*, p. 53.

p. 162 According to Einstein himself, he was simply sitting in a chair when the problem of gravity occurred to him: Bodanis, $E = mc^2$, p. 204.

p. 163 the publication in early 1917 of a paper entitled 'Cosmological Considerations on the General Theory of Relativity': Guth, *The Inflationary Universe*, p. 36.

p. 163 'Without it,' wrote Snow in 1979: Snow, *The Physicists*, p. 21.

p. 163 Crouch was hopelessly out of his depth, and got nearly everything wrong: Bodanis, $E = mc^2$, p. 215.

p. 164 'I am trying to think who the third person is': quoted in Hawking, *A Brief History of Time*, p. 91; Aczel, *God's Equation*, p. 146.

p. 164 and the faster one moves the more pronounced these effects become: Guth, *The Inflationary Universe*, p. 37.

p. 165 a baseball thrown at 160 kilometres an hour will pick up 0.000000000002 grams of mass on its way to home plate: Brockman and Matson, *How Things Are*, p. 263.

p. 165 However, to turn to Bodanis again, we all commonly encounter other kinds of relativity: Bodanis, $E = mc^2$, p. 83.

p. 166 'the ultimate sagging mattress': Overbye, *Lonely Hearts of the Cosmos*, p. 55.

p. 166 'In some sense, gravity does not exist': Kaku, 'The Theory of the Universe?' in Shore (ed.), *Mysteries of Life and the Universe*, p. 161.

p. 168 and Edwin enjoyed a wealth of physical endowments, too: Cropper, *Great Physicists*, p. 423.

p. 168 At a single high-school track meeting: Christianson, *Edwin Hubble*, p. 33.

p. 170 One Harvard computer, Annie Jump Cannon, used her repetitive acquaintance with the stars to devise a system of stellar classifications: Ferris, *Coming of Age in the*

Milky Way, p. 258.

p. 171 they are elderly stars that have moved past their 'main sequence phase': Ferguson, *Measuring the Universe*, pp. 166–7.

p. 171 They could be used as standard candles: Ferguson, *Measuring the Universe*, p. 166.

p. 171 was developing *his* seminal theory that dark patches on the Moon were caused by swarms of seasonally migrating insects: Moore, *Fireside Astronomy*, p. 63.

p. 171 In 1923 he showed that a puff of distant gossamer in the Andromeda constellation known as M31 wasn't a gas cloud at all: Overbye, *Lonely Hearts of the Cosmos*, p. 45; *Natural History*, 'Delusions of Centrality', Dec. 2002–Jan. 2003, pp. 28–32.

p. 172 The wonder, as Stephen Hawking has noted, is that no-one had hit on the idea of the expanding universe before: Hawking, *The Universe in a Nutshell*, pp. 71–2.

p. 173 In 1936 Hubble produced a popular book called *The Realm of the Nebulae*: Overbye, *Lonely Hearts of the Cosmos*, p. 13.

p. 174 Half a century later the whereabouts of the century's greatest astronomer remain unknown: Overbye, *Lonely Hearts of the Cosmos*, p. 28.

Chapter 9: The Mighty Atom

p. 175 'All things are made of atoms': Feynman, *Six Easy Pieces*, p. 4.

p. 176 45 billion billion molecules: Gribbin, *Almost Everyone's Guide to Science*, p. 250.

p. 176 up to a billion for each of us, it has been suggested: Davies, *The Fifth Miracle*, p. 127.

p. 176 Atoms themselves, however, go on practically for ever: Rees, *Just Six Numbers*, p. 96.

p. 177 If you wanted to see . . . a paramecium swimming in a drop of water: Feynman, *Six Easy Pieces*, pp. 4–5.

p. 178 'We might as well attempt to introduce a new planet into the solar system': Boorstin, *The Discoverers*, p. 679.

p. 178 In 1826, the French chemist P. J. Pelletier travelled to Manchester: Gjertsen, *The Classics of Science*, p. 260.

p. 179 a confused Pelletier, upon beholding the great man, stammered: Holmyard, *Makers of Chemistry*, p. 222.

p. 179 forty thousand people viewed the coffin and the funeral cortège stretched for two miles: *Dictionary of National Biography*, vol. 5, p. 433.

p. 179 For a century after Dalton made his proposal: von Baeyer, *Taming the Atom*, p. 17.

p. 179 it was said to have played a part in the suicide of . . . Ludwig Boltzmann: Weinberg, *The Discovery of Subatomic Particles*, p. 3.

p. 180 to raise a little flax and a lot of children: Weinberg, *The Discovery of Subatomic Particles*, p. 104.

p. 180 'Had she taken a bullfighter': quoted in Cropper, *Great Physicists*, p. 259.

p. 180 It was a feeling Rutherford would have understood: Cropper, *Great Physicists*, p. 317.

p. 180 would give up halfway through and tell the students to work it out for themselves: Wilson, *Rutherford*, p. 174.

p. 181 'as far as he could see': Wilson, *Rutherford*, p. 208.

p. 181 He was one of the first . . . to see: Wilson, *Rutherford*, p. 208.

p. 181 'Why use radio?': quoted in Cropper, *Great Physicists*, p. 328.

p. 181 'Every day I grow in girth. And in mentality': Snow, *Variety of Men*, p. 47.

p. 182 gave it up when he was persuaded by a senior colleague that radio had little future: Cropper, *Great Physicists*, p. 94.

p. 182 Some physicists thought that atoms might be cube-shaped: Asimov, *The History of Physics*, p. 551.

p. 183 The number of protons is what gives an atom its chemical identity: Guth, *The Inflationary Universe*, p. 90.
p. 184 Add or subtract a neutron or two and you get an isotope: Atkins, *The Periodic Kingdom*, p. 106.
p. 184 only one-millionth of a billionth of the full volume of the atom: Gribbin, *Almost Everyone's Guide to Science*, p. 35.
p. 184 but a fly many thousands of times heavier than the cathedral: Cropper, *Great Physicists*, p. 245.
p. 184 'they could, like galaxies, pass right through each other unscathed': Ferris, *Coming of Age in the Milky Way*, p. 288.
p. 185 'Because atomic behaviour is so unlike ordinary experience': Feynman, *Six Easy Pieces*, p. 117.
p. 187 the delay in discovery was probably a very good thing: Boorse et al., *The Atomic Scientists*, p. 338.
p. 188 'I do not even know what a matrix *is*': Cropper, *Great Physicists*, p. 269.
p. 188 This isn't a matter of simply needing more precise instruments: Ferris, *Coming of Age in the Milky Way*, p. 288.
p. 188 until it is observed an electron must be regarded as being 'at once everywhere and nowhere': David H. Freedman, 'Quantum Liaisons', in Shore (ed.), *Mysteries of Life and the Universe*, p. 137.
p. 189 'a person who wasn't outraged on first hearing about quantum theory didn't understand what had been said': Overbye, *Lonely Hearts of the Cosmos*, p. 109.
p. 189 'Don't try': Von Baeyer, *Taming the Atom*, p. 43.
p. 189 The cloud itself is essentially just a zone of statistical probability: Ebbing, *General Chemistry*, p. 295.
p. 189 'an area of the universe that our brains just aren't wired to understand': Trefil, *101 Things You Don't Know About Science and No One Else Does Either*, p. 62.
p. 189 'things on a small scale behave *nothing* like things on

a large scale': Feynman, *Six Easy Pieces*, p. 33.

p. 189 where matter could pop into existence from nothing at all: Alan Lightman, 'First Birth', in Shore (ed.), *Mysteries of Life and the Universe*, p. 13.

p. 190 It is as if . . . you had two identical pool balls: Lawrence Joseph, 'Is Science Common Sense?' in Shore (ed.), *Mysteries of Life and the Universe*, pp. 42–3.

p. 190 Remarkably, the phenomenon was proved in 1997: *Christian Science Monitor*, 'Spooky Action at a Distance', 4 Oct. 2001.

p. 190 one cannot 'predict future events exactly': Hawking, *A Brief History of Time*, p. 61.

p. 191 Scientists have dealt with this problem . . . 'by not thinking about it': David H. Freedman, 'Quantum Liaisons', in Shore (ed.), *Mysteries of Life and the Universe*, p. 141.

p. 192 The weak nuclear force . . . is ten billion billion billion times stronger than gravity: Ferris, *The Whole Shebang*, p. 297.

p. 192 The grip of the strong force reaches out only to about one-hundred-thousandth of the diameter of an atom: Asimov, *Atom*, p. 258.

p. 192 'He devoted the rest of his life': Snow, *The Physicists*, p. 89.

Chapter 10: Getting the Lead Out

p. 194 Among the many symptoms associated with over-exposure are blindness, insomnia, kidney failure, hearing loss, cancer: McGrayne, *Prometheans in the Lab*, p. 88.

p. 195 'These men probably went insane because they worked too hard': McGrayne, *Prometheans in the Lab*, p. 92.

p. 195 In fact, Midgley knew only too well the perils of lead poisoning: McGrayne, *Prometheans in the Lab*, p. 92.

p. 195 One leak from a refrigerator at a hospital in

Cleveland, Ohio, in 1929 killed more than a hundred people: McGrayne, *Prometheans in the Lab*, p. 96.

p. 196 A single kilogram of CFCs can capture and annihilate 70,000 kilograms of atmospheric ozone: Biddle, *A Field Guide to the Invisible*, p. 62.

p. 196 A single CFC molecule is about ten thousand times more efficient at exacerbating greenhouse effects than a molecule of carbon dioxide: *Science*, 'The Ascent of Atmospheric Sciences', 13 Oct. 2000, p. 299.

p. 196 His death was itself memorably unusual: *Nature*, 27 Sept. 2001, p. 364.

p. 197 Up to this time, the oldest reliable dates went back no further than the First Dynasty in Egypt: Willard Libby, 'Radiocarbon Dating', from Nobel Lecture, 12 Dec. 1960.

p. 197 After eight half-lives, only 0.39 per cent of the original radioactive carbon remains: Gribbin and Gribbin, *Ice Age*, p. 58.

p. 197 'every raw radiocarbon date you read today is given as too young by around 3 per cent': Flannery, *The Eternal Frontier*, p. 174.

p. 198 it is like miscounting by a dollar when counting to a thousand: Flannery, *The Future Eaters*, p. 151.

p. 198 Among the more dubious are dates just around the time that people first came to the Americas: Flannery, *The Eternal Frontier*, pp. 174–5.

p. 198 the long-running debate over whether syphilis originated in the New World or the Old: *Science*, 'Can Genes Solve the Syphilis Mystery?', 11 May 2001, p. 109.

p. 200 Unfortunately, he now met yet another formidable impediment to acceptance: Lewis, *The Dating Game*, p. 204.

p. 202 It was this that eventually led him to create a sterile laboratory: Powell, *Mysteries of Terra Firma*, p. 58.

p. 202 'a figure that stands unchanged 50 years later':

McGrayne, *Prometheans in the Lab*, p. 173.

p. 202 In one such study, a doctor who had no specialized training in chemical pathology: McGrayne, *Prometheans in the Lab*, p. 94.

p. 203 about 90 per cent of it appeared to come from car exhaust pipes: *Nation*, 'The Secret History of Lead', 20 March 2000.

p. 203 The notion became the foundation of ice core studies, on which much modern climatological work is based: Powell, *Mysteries of Terra Firma*, p. 60.

p. 204 Ethyl executives allegedly offered to endow a chair at Caltech 'if Patterson was sent packing': *Nation*, 'The Secret History of Lead', 20 March 2000.

p. 204 Almost immediately lead levels in the blood of Americans fell by 80 per cent: McGrayne, *Prometheans in the Lab*, p. 169.

p. 204 Americans alive today each have about 625 times more lead in their blood than people did a century ago: *Nation*, 20 March 2000.

p. 204 The amount of lead in the atmosphere also continues to grow, quite legally, by about a hundred thousand tonnes a year: Green, *Water, Ice and Stone*, p. 258.

p. 204 '44 years after most of Europe': McGrayne, *Prometheans in the Lab*, p. 191.

p. 204 continued to contend 'that research has failed to show that leaded gasoline poses a threat to human health': McGrayne, *Prometheans in the Lab*, p. 191.

p. 205 will almost certainly be around and devouring ozone long after you and I have shuffled off: Biddle, *A Field Guide to the Invisible*, pp. 110–11.

p. 205 Worse, we are still introducing huge amounts of CFCs into the atmosphere every year: Biddle, *A Field Guide to the Invisible*, p. 63.

p. 205 Two recent popular books on the history of the dating of the Earth actually manage to misspell his name:

The books are *Mysteries of Terra Firma* and *The Dating Game*, both of which make his name 'Claire'. (Since this note first appeared, I have received a rather severe rebuke from the author of the latter book, Cherry Lewis, informing me that her choice of spelling was intentional and arose from correspondence she had had with Patterson's widow. Except for the other cited book, Lewis's choice of spelling accords with no other published sources I can find, including Patterson's many obituaries in leading journals – which were, after all, literally the last word on the man and his name. Nonetheless I am happy to accept that Ms Lewis's variant spelling of Patterson's name was done intentionally and I unreservedly apologize to her for any dismay caused.)

p. 205 made the additional, rather astounding error of thinking Patterson was a woman: *Nature*, 'The Rocky Road to Dating the Earth', 4 Jan. 2001, p. 20.

Chapter 11: Muster Mark's Quarks

p. 206 In 1911, a British scientist named C. T. R. Wilson: Cropper, *Great Physicists*, p. 325.

p. 207 'if I could remember the names of these particles, I would have been a botanist': quoted in Cropper, *Great Physicists*, p. 403.

p. 207 can do 47,000 laps around a 7-kilometre tunnel in under a second: *Discover*, 'Gluons', July 2000, p. 68.

p. 207 Even the most sluggish: Guth, *The Inflationary Universe*, p. 121.

p. 208 In 1998, Japanese observers reported that neutrinos do have mass: *Economist*, 'Heavy Stuff', 13 June 1998, p. 82; *National Geographic*, 'Unveiling the Universe', Oct. 1999, p. 36.

p. 208 Breaking up atoms . . . is easy: Trefil, *101 Things You Don't Know About Science and No One Else Does Either*, p. 48.

p. 209 CERN's new Large Hadron Collider . . . will achieve 14 trillion volts of energy: *Economist*, 'Cause for conCERN', 28 Oct. 2000, p. 75.

p. 209 'dotted along the circumference by a series of disappointed small towns': letter from Jeff Guinn.

p. 209 A proposed neutrino observatory at the old Homestake Mine in Lead, South Dakota: *Science*, 'U.S. Researchers Go for Scientific Gold Mine', 15 June 2001, p. 1979.

p. 210 A particle accelerator at Fermilab in Illinois . . . cost $260 million: *Science*, 8 Feb. 2002, p. 942.

p. 210 Today the particle count is well over 150: Guth, *The Inflationary Universe*, p. 120; Feynman, *Six Easy Pieces*, p. 39.

p. 210 Some people think there are particles called tachyons: *Nature*, 27 Sept. 2001, p. 354.

p. 210 'which are themselves universes at the next level and so on forever': Sagan, *Cosmos*, pp. 265–6.

p. 210 'The charged pion and antipion decay respectively': Weinberg, *The Discovery of Subatomic Particles*, p. 163.

p. 211 'to restore some economy to the multitude of hadrons': Weinberg, *The Discovery of Subatomic Particles*, p. 165.

p. 211 wanted to call these new basic particles *partons*: von Baeyer, *Taming the Atom*, p. 17.

p. 211 Eventually out of all this emerged what is called the Standard Model: *Economist*, 'New realities?', 7 Oct. 2000, p. 95; *Nature*, 'The Mass Question', 28 Feb. 2002, pp. 969–70.

p. 212 Bosons . . . are particles that produce and carry forces: *Scientific American*, 'Uncovering Supersymmetry', July 2002, p. 74.

p. 212 'It has too many arbitrary parameters': quoted on the PBS video *Creation of the Universe*, 1985; also quoted, with slightly different numbers, in Ferris, *Coming of Age in the*

Milky Way, pp. 298–9.

p. 212 the notional Higgs boson: CERN website document 'The Mass Mystery', undated.

p. 212 'So we are stuck with a theory': Feynman, *Six Easy Pieces*, p. 39.

p. 213 This postulates that all those little things like quarks: *Science News*, 22 Sept. 2001, p. 185.

p. 213 tiny enough to pass for point particles: Weinberg, *Dreams of a Final Theory*, p. 168.

p. 213 'The heterotic string consists of a closed string that has two types of vibrations': Kaku, *Hyperspace*, p. 158.

p. 213 String theory has further spawned something called M theory: *Scientific American*, 'The Universe's Unseen Dimensions', Aug. 2000, pp. 62–9; *Science News*, 'When Branes Collide', 22 Sept. 2001, pp. 184–5.

p. 214 'The ekpyrotic process begins far in the indefinite past': *New York Times*, 'Before the Big Bang, There Was . . . What?', 22 May 2001, p. F1.

p. 214 it is 'almost impossible for the non-scientist to discriminate between the legitimately weird and the outright crackpot': *Nature*, 27 Sept. 2001, p. 354.

p. 214 The question came interestingly to a head: *New York Times* website, 'Are They a) Geniuses or b) Jokers?; French Physicists' Cosmic Theory Creates a Big Bang of Its Own', 9 Nov. 2002; *Economist*, 'Publish and Perish', 16 Nov. 2002, p. 75.

p. 214 Karl Popper . . . once suggested that there may not in fact be an ultimate theory: Weinberg, *Dreams of a Final Theory*, p. 184.

p. 214 'we do not seem to be coming to the end of our intellectual resources': Weinberg, *Dreams of a Final Theory*, p. 187.

p. 215 Hubble calculated that the universe was about two billion years old: *US News and World Report*, 'How Old Is the Universe?', 25 Aug. 1997, p. 34.

p. 215 a new age for the universe of between seven billion and twenty billion years: Trefil, *101 Things You Don't Know About Science and No One Else Does Either*, p. 91.

p. 215 In the years that followed there erupted a dispute that would run and run: Overbye, *Lonely Hearts of the Cosmos*, p. 268.

p. 216 In February 2003, a team from NASA: *New York Times*, 'Cosmos Sits for Early Portrait, Gives up Secrets', 12 Feb. 2003, p. 1.

p. 217 'a mountain of theory built on a molehill of evidence': *Economist*, 'Queerer than we can suppose', 5 Jan. 2002, p. 58.

p. 217 'may reflect the paucity of the data rather than the excellence of the theory': *National Geographic*, 'Unveiling the Universe', Oct. 1999, p. 25.

p. 217 what they really mean: Goldsmith, *The Astronomers*, p. 82.

p. 218 'two-thirds of the universe is still missing from the balance sheet': *Economist*, 'Dark for Dark Business', 5 Jan. 2002, p. 51.

p. 219 The theory is that empty space isn't so empty at all: PBS *Nova*, 'Runaway Universe', transcript of programme first broadcast 21 Nov. 2000.

p. 219 the one thing that resolves all this is Einstein's cosmological constant: *Economist*, 'Dark for Dark Business', 5 Jan. 2002, p. 51.

Chapter 12: The Earth Moves

p. 220 In a tone that all but invited the reader to join him in a tolerant chuckle: Hapgood, *Earth's Shifting Crust*, p. 29.

p. 223 they posited ancient 'land bridges' wherever they were needed: Simpson, *Fossils and the History of Life*, p. 98.

p. 223 Even land bridges couldn't explain some things: Gould, *Ever since Darwin*, p. 163.

p. 223 full of 'numerous grave theoretical difficulties': *Encylopaedia Britannica*, vol. 6, p. 418.

p. 224 One reviewer there fretted . . . that students might actually come to believe them: Lewis, *The Dating Game*, p. 182.

p. 224 about half of those present now embraced the idea of continental drift: Hapgood, *Earth's Shifting Crust*, p. 31.

p. 224 'I feel the hypothesis is a fantastic one': Powell, *Mysteries of Terra Firma*, p. 147.

p. 225 Interestingly, oil company geologists had known for years: McPhee, *Basin and Range*, p. 175.

p. 225 Aboard this vessel was a fancy new depth sounder called a fathometer: McPhee, *Basin and Range*, p. 187.

p. 226 seamounts that he called guyots after an earlier Princeton geologist: Harrington, *Dance of the Continents*, p. 208.

p. 229 'probably the most significant paper in the earth sciences ever to be denied publication': Powell, *Mysteries of Terra Firma*, pp. 131–2.

p. 229 Well into the 1970s: Powell, *Mysteries of Terra Firma*, p. 141.

p. 229 one American geologist in eight still didn't believe in plate tectonics: McPhee, *Basin and Range*, p. 198.

p. 229 Today we know that the Earth's surface is made up of eight to twelve big plates: Simpson, *Fossils and the History of Life*, p. 113.

p. 230 The connections . . . were found to be infinitely more complex than anyone had imagined: McPhee, *Assembling California*, pp. 202–8.

p. 231 at about the speed a fingernail grows: Vogel, *Naked Earth*, p. 19.

p. 231 one-tenth of 1 per cent of the Earth's history: Margulis and Sagan, *Microcosmos*, p. 44.

p. 231 It is thought . . . that tectonics is an important part of the planet's organic well-being: Trefil, *Meditations at*

10,000 Feet, p. 181.

p. 231 suggesting that there may well be a relationship between the history of rocks and the history of life: *Science*, 'Inconstant Ancient Seas and Life's Path', 8 Nov. 2002, p. 1165.

p. 232 'the whole earth suddenly made sense': McPhee, *Rising from the Plains*, p. 158.

p. 232 a habit of appearing inconveniently where they shouldn't and failing to be where they ought: Simpson, *Fossils and the History of Life*, p. 115.

p. 233 There are also many surface features that tectonics can't explain: *Scientific American*, 'Sculpting the Earth from Inside Out', March 2001.

p. 233 Alfred Wegener never lived to see his ideas vindicated: Kunzig, *The Restless Sea*, p. 51.

p. 234 One of his students was a bright young fellow named Walter Alvarez: Powell, *Night Comes to the Cretaceous,* p. 7.

Chapter 13: Bang!

p. 237 In 1912, a man drilling a well for the town water supply reported bringing up a lot of strangely deformed rock: Raymond R. Anderson, Geological Society of America GSA Special Paper 302, '*The Manson Impact Structure: A Late Cretaceous Meteor Crater in the Iowa Subsurface*', Spring 1996.

p. 238 Virtually the whole town turned out: *Des Moines Register*, 30 June 1979.

p. 239 'Very occasionally we get people coming in and asking where they should go to see the crater': Interview with Schlapkohl, Manson, Iowa, 18 June 2001.

p. 239 The leading early investigator, G. K. Gilbert of Columbia University: Lewis, *Rain of Iron and Ice*, p. 38.

p. 239 Gilbert conducted these experiments not in a laboratory at Columbia but in a hotel room: Powell,

Night Comes to the Cretaceous, p. 37.

p. 240 'At the time we started, only slightly more than a dozen of these things had ever been discovered': transcript from BBC *Horizon* documentary, 'New Asteroid Danger', p. 4; programme first transmitted 18 March 1999.

p. 241 He called them asteroids – Latin for 'starlike': *Science News*, 'A Rocky Bicentennial', 28 July 2001, pp. 61–3.

p. 242 it was finally tracked down in 2000 after being missing for eighty-nine years: Ferris, *Seeing in the Dark*, p. 150.

p. 242 As of July 2001, 26,000 asteroids had been named and identified: *Science News*, 'A Rocky Bicentennial', 28 July 2001, pp. 61–3.

p. 242 down which we are cruising at over 100,000 kilometres an hour: Ferris, *Seeing in the Dark*, p. 147.

p. 243 'all of which are capable of colliding with the Earth and all of which are moving on slightly different courses through the sky at different rates': transcript from BBC *Horizon* documentary 'New Asteroid Danger', p. 5; first transmitted 18 March 1999.

p. 243 such near misses probably happen two or three times a week and go unnoticed: *New Yorker*, 'Is This the End?', 27 Jan. 1997, pp. 44–52.

p. 245 Every year the Earth accumulates some 30,000 tonnes of 'cosmic spherules': Vernon, *Beneath our Feet*, p. 191.

p. 246 'Well, they were very charming, very persuasive': telephone interview with Asaro, 10 March 2002.

p. 246 a Northwestern University astrophysicist named Ralph B. Baldwin had suggested such a possibility in an article in *Popular Astronomy* magazine: Powell, *Mysteries of Terra Firma*, p. 184.

p. 247 In 1956 a professor at Oregon State University,

M. W. de Laubenfels: Peebles, *Asteroids: A History*, p. 170.

p. 247 may have been the cause of an earlier event known as the Frasnian extinction: Lewis, *Rain of Iron and Ice*, p. 107.

p. 248 'They're more like stamp collectors': quoted by Officer and Page, *Tales of the Earth*, p. 142.

p. 248 even while conceding in a newspaper interview that he had no actual evidence of it: *Boston Globe*, 'Dinosaur Extinction Theory Backed', 16 Dec. 1985.

p. 248 continued to believe that the extinction of the dinosaurs was in no way related to an asteroid or cometary impact: Peebles, *Asteroids: A History*, p. 175.

p. 249 a big part of the work you do is to evaluate Manure Management Plans: Iowa Department of Natural Resources Publication, *Iowa Geology 1999*, Number 24.

p. 250 'Suddenly we were at the centre of things': interview with Anderson and Witzke, Iowa City, 15 June 2001.

p. 250 One of those moments came at the annual meeting of the American Geophysical Union in 1985: *Boston Globe*, 'Dinosaur Extinction Theory Backed', 16 Dec. 1985.

p. 251 The formation had been found by Pemex, the Mexican oil company, in 1952: Peebles, *Asteroids: A History*, pp. 177–8; *Washington Post*, 'Incoming', 19 April 1998.

p. 251 'I remember harboring some strong initial doubts about the efficacy of such an event': Gould, *Dinosaur in a Haystack*, p. 162.

p. 252 'Jupiter will swallow these comets up without so much as a burp': quoted by Peebles, *Asteroids: A History*, p. 196.

p. 252 One fragment, known as Nucleus G, struck with the force of about six million megatonnes: Peebles, *Asteroids: A History*, p. 202.

p. 252 Shoemaker was killed instantly, his wife injured: Peebles, *Asteroids: A History*, p. 204.

p. 254 nearly every standing thing would be flattened or on fire, and nearly every living thing would be dead: Anderson, Iowa Department of Natural Resources, *Iowa Geology 1999*, 'Iowa's Mansion Impact Structure'.

p. 255 fleeing would mean 'selecting a slow death over a quick one': Lewis, *Rain of Iron and Ice*, p. 209.

p. 255 analysed helium isotopes from sediments left from the later KT impact and concluded that it affected the Earth's climate for about ten thousand years: *Arizona Republic*, 'Impact Theory Gains New Supporters', 3 March 2001.

p. 256 First, as John S. Lewis notes, our missiles are not designed for space work: Lewis, *Rain of Iron and Ice*, p. 215.

p. 256 Tom Gehrels . . . thinks that even a year's warning would probably be insufficient: *New York Times* magazine, 'The Asteroids Are Coming! The Asteroids Are Coming!', 28 July 1996, pp. 17–19.

p. 256 Shoemaker–Levy 9 had been orbiting Jupiter in a fairly conspicuous manner since 1929, but it was over half a century before anyone noticed: Ferris, *Seeing in the Dark*, p. 168.

Chapter 14: The Fire Below

p. 259 'It was a dumb place to look for bones': interview with Mike Voorhies, Ashfall Fossil Beds State Park, Nebraska, 13 June 2001.

p. 259 At first they thought the animals were buried alive: *National Geographic*, 'Ancient Ashfall Creates Pompeii of Prehistoric Animals', Jan. 1981, p. 66.

p. 261 'far better than we understand the interior of the earth': Feynman, *Six Easy Pieces*, p. 60.

p. 261 The distance from the surface of Earth to the middle is 6,370 kilometres: Williams and Montaigne, *Surviving Galeras*, p. 78.

p. 262 A modest fellow, he never referred to the scale by his own name: Ozima, *The Earth*, p. 49.

p. 263 It rises exponentially: Officer and Page, *Tales of the Earth*, p. 33.

p. 264 sixty thousand people were dead: Officer and Page, *Tales of the Earth*, p. 52.

p. 265 'the city waiting to die': McGuire, *A Guide to the End of the World*, p. 21.

p. 265 the potential economic cost has been put as high as $7 trillion: McGuire, *A Guide to the End of the World*, p. 130.

p. 266 'collapsed scaffolding erected around the Capitol Building': Trefil, *101 Things You Don't Know About Science and No One Else Does Either*, p. 158.

p. 267 The project became known, all but inevitably, as the Mohole: Vogel, *Naked Earth*, p. 37.

p. 267 'like trying to drill a hole . . . using a strand of spaghetti': *Valley News*, 'Drilling the Ocean Floor for Earth's Deep Secrets', 21 August 1995.

p. 267 the crust of the Earth represents only about 0.3 per cent of the planet's volume: Schopf, *Cradle of Life*, p. 73.

p. 268 We know a little bit about the mantle from what are known as kimberlite pipes: McPhee, *In Suspect Terrain*, pp. 16–18.

p. 268 Scientists are generally agreed: *Scientific American*, 'Sculpting the Earth from Inside Out', March 2001, pp. 40–7, and *New Scientist*, 'Journey to the Centre of the Earth', supplement, 14 Oct. 2000, p. 1.

p. 269 By all the laws of geophysics: *Earth*, 'Mystery in the High Sierra', June 1996, p. 16.

p. 270 The rocks are viscous . . . in the same way that glass is: Vogel, *Naked Earth*, p. 31.

p. 270 The movements occur not just laterally: *Science*, 'Much About Motion in the Mantle', 1 Feb. 2002,

p. 982.

p. 270 an English vicar named Osmond Fisher presciently suggested: Tudge, *The Time Before History*, p. 43.

p. 271 'then had suddenly found out about wind': Vogel, *Naked Earth*, p. 53.

p. 271 'there are two sets of data, from two different disciplines': Trefil, *101 Things You Don't Know About Science and No One Else Does Either*, p. 146.

p. 271 82 per cent of the Earth's volume and 65 per cent of its mass: *Nature*, 'The Earth's Mantle', 2 Aug. 2001, pp. 501–6.

p. 271 something over three million times those found at the surface: Drury, *Stepping Stones*, p. 50.

p. 272 during the age of the dinosaurs, it was up to three times as strong as it is now: *New Scientist*, 'Dynamo Support', 10 March 2001, p. 27.

p. 272 37 million years appears to be the longest stretch: *New Scientist*, 'Dynamo Support', 10 March 2001, p. 27.

p. 272 'the greatest unanswered question in the geological sciences': Trefil, *101 Things You Don't Know About Science and No One Else Does Either*, p. 150.

p. 273 'Geologists and geophysicists rarely go to the same meetings': Vogel, *Naked Earth*, p. 139.

p. 274 The seismologists resolutely based their conclusions on the behaviour of Hawaiian volcanoes: Fisher et al., *Volcanoes*, p. 24.

p. 274 It was the biggest landslide in human history: Thompson, *Volcano Cowboys*, p. 118.

p. 274 with the force of five hundred Hiroshima-sized atomic bombs: Williams and Montaigne, *Surviving Galeras*, p. 7.

p. 275 Fifty-seven people were killed: Fisher et al., *Volcanoes*, p. 12.

p. 275 'only shake my head in wonder': Williams and Montaigne, *Surviving Galeras*, p. 151.

p. 276 An airliner . . . reported being pelted with rocks: Thompson, *Volcano Cowboys*, p. 123.

p. 276 Yet Yakima had no volcano emergency procedures: Fisher et al., *Volcanoes*, p. 16.

Chapter 15: Dangerous Beauty

p. 278 In 1943 at Parícutin in Mexico: Smith, *The Weather*, p. 112.

p. 280 'you wouldn't be able to get within a thousand kilometres of it': BBC *Horizon* documentary, 'Crater of Death', first broadcast 6 May 2001.

p. 281 a bang that reverberated around the world for nine days: Lewis, *Rain of Iron and Ice*, p. 152.

p. 282 The last supervolcano eruption on Earth was at Toba: McGuire, *A Guide to the End of the World*, p. 104.

p. 282 there is some evidence to suggest that for the next twenty thousand years the total number of people on Earth was never more than a few thousand: McGuire, *A Guide to the End of the World*, p. 107.

p. 283 'It may not feel like it, but you're standing on the largest active volcano in the world': interview with Paul Doss, Yellowstone National Park, Wyoming, 16 June 2001.

p. 287 as was made devastatingly evident on the night of 17 August 1959, at a place called Hebgen Lake: Smith and Siegel, *Windows into the Earth*, pp. 5–6.

p. 291 as little as a single molecule in ideal conditions: Sykes, *The Seven Daughters of Eve*, p. 12.

p. 291 Meanwhile, scientists were finding even hardier microbes: Ashcroft, *Life at the Extremes*, p. 275.

p. 291 As NASA scientist Jay Bergstralh has put it: PBC *NewsHour* transcript, 20 Aug. 2002.

Chapter 16: Lonely Planet

p. 295 no less than 99.5 per cent of the world's habitable

space by volume: *New York Times Book Review*, 'Where Leviathan Lives', 20 April 1997, p. 9.

p. 295 water is about 1,300 times heavier than air: Ashcroft, *Life at the Extremes*, p. 51.

p. 296 your veins would collapse and your lungs would compress to the approximate dimensions of a Coke can: *New Scientist*, 'Into the Abyss', 31 March 2001.

p. 296 the pressure is equivalent to being squashed beneath a stack of fourteen loaded cement trucks: *New Yorker*, 'The Pictures', 15 Feb. 2000, p. 47.

p. 297 Because we are made largely of water ourselves: Ashcroft, *Life at the Extremes*, p. 68.

p. 297 'humans may be more like whales and dolphins than had been expected': Ashcroft, *Life at the Extremes*, p. 69.

p. 297 'all that is left in the suit are his bones and some rags of flesh': Haldane, *What Is Life?*, p. 188.

p. 298 Ashcroft relates a story concerning the directors of a new tunnel under the Thames who held a celebratory banquet: Ashcroft, *Life at the Extremes*, p. 59.

p. 299 When roused, Haldane explained that he had found himself disrobing and assumed it was bedtime: Norton, *Stars beneath the Sea*, p. 111.

p. 299 Haldane's gift to diving was to work out the rest intervals necessary to manage an ascent from the depths without getting the bends: Haldane, *What Is Life?*, p. 202.

p. 300 his blood saturation level had reached 56 per cent: Norton, *Stars beneath the Sea*, p. 105.

p. 300 'But is it oxyhaemoglobin or carboxyhaemoglobin?': quoted in Norton, *Stars beneath the Sea*, p. 121.

p. 300 called him 'the cleverest man I ever knew': Gould, *The Lying Stones of Marrakech*, p. 305.

p. 300 the younger Haldane found the First World War 'a very enjoyable experience': Norton, *Stars beneath the Sea*, p. 124.

p. 301 'Almost every experiment . . . ended with someone having a seizure, bleeding or vomiting': Norton, *Stars beneath the Sea*, p. 133.

p. 301 Perforated eardrums were quite common: Haldane, *What Is Life?*, p. 192.

p. 302 left Haldane without feeling in his buttocks and lower spine for six years: Haldane, *What Is Life?*, p. 202.

p. 302 It also produced wild mood swings: Ashcroft, *Life at the Extremes*, p. 78.

p. 302 'the tester was usually as intoxicated as the testee': Haldane, *What Is Life?*, p. 197.

p. 302 The cause of the inebriation is even now a mystery: Ashcroft, *Life at the Extremes*, p. 79.

p. 303 Even in quite mild weather half the calories you burn go to keep your body warm: Attenborough, *The Living Planet*, p. 39.

p. 303 the portions of the Earth on which we are prepared or able to live are modest indeed: Smith, *The Weather*, p. 40.

p. 304 Had our sun been ten times as massive, it would have exhausted itself after ten million years instead of ten billion: Ferris, *The Whole Shebang*, p. 81.

p. 305 The Sun's warmth reaches it just two minutes before it touches us: Grinspoon, *Venus Revealed*, p. 9.

p. 305 It appears that during the early years of the solar system Venus was only slightly warmer than the Earth and probably had oceans: *National Geographic*, 'The Planets', Jan. 1985, p. 40.

p. 305 the atmospheric pressure at the surface is ninety times that of Earth: McSween, *Stardust to Planets*, p. 200.

p. 307 The Moon is slipping from our grasp at a rate of about 4 centimetres a year: Ward and Brownlee, *Rare Earth*, p. 33.

p. 308 The most elusive element of all, however, appears to be francium: Atkins, *The Periodic Kingdom*, p. 28.

p. 309 discarded the state silver dinner service and replaced it with an aluminium one: Bodanis, *The Secret House*, p. 13.

p. 309 accounting for a very modest 0.048 per cent of the Earth's crust: Krebs, *The History and Use of our Earth's Chemical Elements*, p. 148.

p. 309 'If it wasn't for carbon, life as we know it would be impossible': Davies, *The Fifth Miracle*, p. 126.

p. 309 Of every 200 atoms in your body, 126 are hydrogen, 51 are oxygen, and just 19 are carbon: Snyder, *The Extraordinary Chemistry of Ordinary Things*, p. 24.

p. 310 The degree to which organisms require or tolerate certain elements is a relic of their evolution: Parker, *Inscrutable Earth*, p. 100.

p. 310 Drop a small lump of pure sodium into ordinary water and it will explode with enough force to kill: Snyder, *The Extraordinary Chemistry of Ordinary Things*, p. 42.

p. 311 The Romans also flavoured their wine with lead: Parker, *Inscrutable Earth*, p. 103.

p. 312 The physicist Richard Feynman used to make a joke: Feynman, *Six Easy Pieces*, p. xix.

Chapter 17: Into the Troposphere

p. 313 Without it, Earth would be a lifeless ball of ice: Stevens, *The Change in the Weather*, p. 7.

p. 314 and was discovered in 1902 by a Frenchman in a balloon, Léon-Philippe Teisserenc de Bort: Stevens, *The Change in the Weather*, p. 56; *Nature*, '1902 and All That', 3 Jan. 2002, p. 15.

p. 314 it's from the same Greek root as *menopause*: Smith, *The Weather*, p. 52.

p. 314 would, at the very least, result in severe cerebral and pulmonary oedemas: Ashcroft, *Life at the Extremes*, p. 7.

p. 314 The temperature 10 kilometres up can be minus 57

degrees Celsius: Smith, *The Weather*, p. 25.

p. 315 about eight-millionths of a centimetre, to be precise: Allen, *Atmosphere*, p. 58.

p. 315 if an incoming vehicle hit the thermosphere at too shallow an angle, it could well bounce back into space: Allen, *Atmosphere*, p. 57.

p. 316 Dickinson records how Howard Somervell . . . 'found himself choking to death after a piece of infected flesh came loose and blocked his windpipe': Dickinson, *The Other Side of Everest*, p. 86.

p. 316 The absolute limit of human tolerance for continuous living appears to be about 5,500 metres: Ashcroft, *Life at the Extremes*, p. 8.

p. 316 above 5,500 metres even the most well-adapted women cannot provide a growing foetus with enough oxygen: Attenborough, *The Living Planet*, p. 18.

p. 317 'nearly half a ton has been quietly piled upon us during the night': quoted by Hamilton-Paterson, *The Great Deep*, p. 177.

p. 318 a typical weather front may consist of 750 million tonnes of cold air pinned beneath a billion tonnes of warmer air: Smith, *The Weather*, p. 50.

p. 318 an amount of energy equivalent to four days' use of electricity for the whole United States: Junger, *The Perfect Storm*, p. 102.

p. 318 At any one moment 1,800 thunderstorms are in progress around the globe: Stevens, *The Change in the Weather*, p. 55.

p. 319 Much of our knowledge of what goes on up there is surprisingly recent: Biddle, *A Field Guide to the Invisible*, p. 161.

p. 320 a wind blowing at 300 kilometres an hour is not simply ten times stronger than a wind blowing at 30 kilometres an hour, but a hundred times stronger: Bodanis, $E = mc^2$, p. 68.

p. 321 as much energy as a rich, medium-sized nation . . . uses in a year: Ball, *H_2O*, p. 51.

p. 321 The impulse of the atmosphere to seek equilibrium was first suspected by Edmond Halley: *Science*, 'The Ascent of Atmospheric Sciences', 13 Oct. 2000, p. 300.

p. 321 Coriolis's other distinction at the school was to introduce water coolers, which are still known there as Corios: Trefil, *The Unexpected Vista*, p. 24.

p. 322 gives weather systems their curl and sends hurricanes spinning off like tops: Drury, *Stepping Stones*, p. 25.

p. 323 Celsius made boiling point zero and freezing point 100: Trefil, *The Unexpected Vista*, p. 107.

p. 323 Howard is chiefly remembered now for giving cloud types their names in 1803: *Dictionary of National Biography*, vol. 10, pp. 51–2.

p. 323 Howard's system has been much added to over the years: Trefil, *Meditations at Sunset*, p. 62.

p. 324 That seems to have been the source of the expression 'to be on cloud nine': Hamblyn, *The Invention of Clouds*, p. 252.

p. 324 A fluffy summer cumulus several hundred metres to a side may contain no more than 100–150 litres of water: Trefil, *Meditations at Sunset*, p. 66.

p. 324 Only about 0.035 per cent of the Earth's fresh water is floating around above us at any moment: Ball, *H_2O*, p. 57.

p. 324 Depending on where it falls, the prognosis for a water molecule varies widely: Dennis, *The Bird in the Waterfall*, p. 8.

p. 325 Even something as large as the Mediterranean would dry out in a thousand years if it were not continually replenished: Gribbin and Gribbin, *Being Human*, p. 123.

p. 325 Such an event occurred a little under six million years ago: *New Scientist*, 'Vanished', 7 Aug. 1999.

p. 325 equivalent to the world's output of coal for ten years: Trefil, *Meditations at 10,000 Feet*, p. 122.

p. 326 For that reason there tends to be a lag in the official, astronomical start of a season and the actual feeling that that season has started: Stevens, *The Change in the Weather*, p. 111.

p. 327 As for the question of how anyone could possibly figure out how long it takes a drop of water to get from one ocean to another: *National Geographic*, 'New Eyes on the Oceans', Oct. 2000, p. 101.

p. 328 Altogether there is about twenty thousand times as much carbon locked away in the Earth's rocks as in the atmosphere: Stevens, *The Change in the Weather*, p. 7.

p. 329 the 'natural' level of carbon dioxide in the atmosphere: *Science*, 'The Ascent of Atmospheric Sciences', 13 Oct. 2000, p. 303.

Chapter 18: The Bounding Main

p. 330 Imagine trying to live in a world dominated by dihydrogen oxide: Margulis and Sagan, *Microcosmos*, p. 100.

p. 330 A potato is 80 per cent water, a cow 74 per cent, a bacterium 75 per cent: Schopf, *Cradle of Life*, p. 107.

p. 330 Almost nothing about it can be used to make reliable predictions about the properties of other liquids: Green, *Water, Ice and Stone*, p. 29; Gribbin, *In the Beginning*, p. 174.

p. 331 By the time it is solid, it is almost a tenth more voluminous than it was before: Trefil, *Meditations at 10,000 Feet*, p. 121.

p. 331 'an utterly bizarre property': Gribbin, *In the Beginning*, p. 174.

p. 331 like the ever-changing partners in a quadrille: Kunzig, *The Restless Sea*, p. 8.

p. 331 At any given moment only 15 per cent of them are

actually touching: Dennis, *The Bird in the Waterfall*, p. 152.

p. 332 Within days, the lips vanish 'as if amputated, the gums blacken, the nose withers to half its length': *Economist*, 13 May 2000, p. 4.

p. 332 A typical litre of sea water will contain only about 2.5 teaspoons of common salt: Dennis, *The Bird in the Waterfall*, p. 248.

p. 332 we sweat and cry sea water: Margulis and Sagan, *Microcosmos*, p. 184.

p. 333 There are 1.3 billion cubic kilometres of water on Earth and that is all we're ever going to get: Green, *Water, Ice and Stone*, p. 25.

p. 333 By 3.8 billion years ago, the oceans had (at least more or less) achieved their present volumes: Ward and Brownlee, *Rare Earth*, p. 360.

p. 333 Altogether the Pacific holds just over half of all the ocean water: Dennis, *The Bird in the Waterfall*, p. 226.

p. 333 we would better call our planet not Earth but Water: Ball, *H_2O*, p. 21.

p. 333 Of the 3 per cent of Earth's water that is fresh: Dennis, *The Bird in the Waterfall*, p. 6; *Scientific American*, 'On Thin Ice', Dec. 2002, pp. 100–5.

p. 333 Go to the South Pole and you will be standing on over 2 miles of ice, at the North Pole just 15 feet of it: Smith, *The Weather*, p. 62.

p. 333 enough to raise the oceans by a height of 200 feet if it all melted: Schultz, *Ice Age Lost*, p. 75.

p. 335 'driven to distraction by the mind-numbing routine of years of dredging': Weinberg, *A Fish Caught in Time*, p. 34.

p. 335 But they sailed across almost 70,000 nautical miles of sea: Hamilton-Paterson, *The Great Deep*, p. 178.

p. 335 female assistants whose jobs were inventively described as 'historian and technicist' or 'assistant in fish problems': Norton, *Stars beneath the Sea*, p. 57.

p. 335 Soon afterwards he teamed up with Barton, who came from an even wealthier family: Ballard, *The Eternal Darkness*, pp. 14–15.

p. 336 The sphere had no manoeuvrability . . . and only the most primitive breathing system: Weinberg, *A Fish Caught in Time*, p. 158; Ballard, *The Eternal Darkness*, p. 17.

p. 337 Whatever it was, nothing like it has been seen by anyone since: Weinberg, *A Fish Caught in Time*, p. 159.

p. 338 In 1958, they did a deal with the US Navy: Broad, *The Universe Below*, p. 54.

p. 339 'We didn't learn a hell of a lot from it, other than that we could do it': quoted in *Underwater* magazine, 'The Deepest Spot on Earth', Winter 1999.

p. 339 There was just one problem: the designers couldn't find anyone willing to build it: Broad, *The Universe Below*, p. 56.

p. 340 In 1994, 34,000 ice hockey gloves were swept overboard from a Korean cargo ship during a storm in the Pacific: *National Geographic*, 'New Eyes on the Oceans', Oct. 2000, p. 93.

p. 340 humans may have scrutinized 'perhaps a millionth or a billionth of the sea's darkness': Kunzig, *The Restless Sea*, p. 47.

p. 341 tube worms over 3 metres long, clams 30 centimetres wide, shrimps and mussels in profusion: Attenborough, *The Living Planet*, p. 30.

p. 341 Before this it had been thought that no complex organisms could survive in water warmer than about 54 degrees Celsius: *National Geographic*, 'Deep Sea Vents', Oct. 2000, p. 123.

p. 342 enough to bury every bit of land on the planet to a depth of about 150 metres: Dennis, *The Bird in the Waterfall*, p. 248.

p. 342 it can take up to ten million years to clean an ocean: Vogel, *Naked Earth*, p. 182.

p. 342 Perhaps nothing speaks more clearly of our psychological remoteness from the ocean depths: Engel, *The Sea*, p. 183.

p. 343 When they failed to sink, which was usually, navy gunners riddled them with bullets to let water in: Kunzig, *The Restless Sea*, pp. 294–305.

p. 343 Blue whales will sometimes break off a song, then pick it up again at exactly the same spot six months later: Sagan, *Cosmos*, p. 271.

p. 344 Consider . . . the fabled giant squid: *Good Weekend*, 'Armed and Dangerous', 15 July 2000, p. 35.

p. 344 there could be as many as 30 million species of animals living in the sea, most still undiscovered: *Time*, 'Call of the Sea', 5 Oct. 1998, p. 60.

p. 344 Even at a depth of nearly 5 kilometres, they found some 3,700 creatures: Kunzig, *The Restless Sea*, pp. 104–5.

p. 345 Altogether less than a tenth of the ocean is considered naturally productive: *Economist* survey, 'The Sea', 23 May 1998, p. 4.

p. 345 it doesn't even make it into the top fifty among fishing nations: Flannery, *The Future Eaters*, p. 104.

p. 346 Many fishermen 'fin' sharks: *Audubon*, May–June 1998, p. 54.

p. 346 and haul behind them nets big enough to hold a dozen jumbo jets: *Time*, 'The Fish Crisis', 11 Aug. 1997, p. 66.

p. 347 'We're still in the Dark Ages. We just drop a net down and see what comes up': *Economist*, 'Pollock Overboard', 6 Jan. 1996, p. 22.

p. 347 Perhaps as much as 22 million tonnes of such unwanted fish are dumped back in the sea each year, mostly in the form of corpses: *Economist* survey, 'The Sea',

23 May 1998, p. 12.

p. 347 Large areas of the North Sea floor are dragged clean by beam trawlers as many as seven times a year: *Outside*, Dec. 1997, p. 62.

p. 347 sailors scooped them up in baskets: *National Geographic*, Oct. 1993, p. 18.

p. 347 By 1990 this had sunk to 22,000 tonnes: *Economist* survey, 'The Sea', 23 May 1998, p. 8.

p. 347 'Fishermen . . . had caught them all': Kurlansky, *Cod*, p. 186.

p. 348 stocks had still not staged a comeback: *Nature*, 'How Many More Fish in the Sea?', 17 Oct. 2002, p. 662.

p. 348 These days, he notes drily, 'fish' is 'whatever is left': Kurlansky, *Cod*, p. 138.

p. 348 'Biologists . . . estimate that 90 per cent of lobsters are caught within a year after they reach the legal minimum size': *New York Times* magazine, 'A Tale of Two Fisheries', 27 Aug. 2000, p. 40.

p. 348 As many as 15 million of them may live on the pack ice around Antarctica: BBC *Horizon* transcript, 'Antarctica: The Ice Melts', p. 16.

Chapter 19: The Rise of Life

p. 350 After a few days, the water in the flasks had turned green and yellow in a hearty broth of amino acids: *Earth*, 'Life's Crucible', Feb. 1998, p. 34.

p. 350 Repeating Miller's experiments with these more challenging inputs has so far produced only one fairly primitive amino acid: Ball, *H_2O*, p. 209.

p. 351 there may be as many as a million types of protein in the human body: *Discover*, 'The Power of Proteins', Jan. 2002, p. 38.

p. 352 the odds against all 200 coming up in a prescribed sequence are 1 in 10^{260}: Crick, *Life Itself*, p. 51.

p. 352 Haemoglobin is only 146 amino acids long, a runt

by protein standards: Sulston and Ferry, *The Common Thread*, p. 14.

p. 352 DNA is a whiz at replicating – it can make a copy of itself in seconds: Margulis and Sagan, *Microcosmos*, p. 63.

p. 353 'If everything needs everything else, how did the community of molecules ever arise in the first place?': Davies, *The Fifth Miracle*, p. 71.

p. 353 there must have been some kind of cumulative selection process that allowed amino acids to assemble: Dawkins, *The Blind Watchmaker*, p. 45.

p. 354 Lots of molecules in nature get together to form long chains called polymers: Dawkins, *The Blind Watchmaker*, p. 115.

p. 354 'an obligatory manifestation of matter': quoted in Nuland, *How We Live*, p. 121.

p. 354 If you wished to create another living object . . . you would need really only four principal elements: Schopf, *Cradle of Life*, p. 107.

p. 354 'There is nothing special about the substances from which living things are made': Dawkins, *The Blind Watchmaker*, p. 112.

p. 355 As one leading biology text puts it, with perhaps just a tiny hint of discomfort: Wallace et al., *Biology*, p. 428.

p. 355 Well into the 1950s, it was thought that life was less than six hundred million years old: Margulis and Sagan, *Microcosmos*, p. 71.

p. 356 'We can only infer from this rapidity that it is not "difficult" for life of bacterial grade to evolve': *New York Times*, 'Life on Mars? So What?', 11 Aug. 1996.

p. 356 'life, arising as soon as it could, was chemically destined to be': Gould, *Eight Little Piggies*, p. 328.

p. 356 when tens of thousands of Australians were startled by a series of sonic booms and the sight of a fireball streaking from east to west across the sky: *Sydney Morning Herald*, 'Aerial Blast Rocks Towns', 29 Sept. 1969, and 'Farmer Finds "Meteor Soot"', 30 Sept. 1969.

p. 356 it was studded with amino acids – seventy-four types in all: Davies, *The Fifth Miracle*, pp. 209–10.

p. 357 A few other carbonaceous chondrites have strayed into the Earth's path since: *Nature*, Life's Sweet Beginnings?, 20–27 Dec. 2001, p. 857; *Earth*, 'Life's Crucible', Feb.1998, p. 37.

p. 357 'at the very fringe of scientific respectability': Gribbin, *In the Beginning*, p. 78.

p. 358 'Wherever you go in the world, whatever animal, plant, bug or blob you look at': Ridley, *Genome*, p. 21.

p. 358 'We can't be certain that what you are holding once contained living organisms': interview with Victoria Bennett, Australia National University, Canberra, 21 Aug. 2001.

p. 362 full of noxious vapours from hydrochloric and sulphuric acids powerful enough to eat through clothing and blister skin: Ferris, *Seeing in the Dark*, p. 200.

p. 362 'undoubtedly the most important single metabolic innovation in the history of life on the planet': Margulis and Sagan, *Microcosmos*, p. 78.

p. 362 Our white blood cells actually use oxygen to kill invading bacteria: note provided by Dr Laurence Smaje.

p. 363 But about 3.5 billion years ago something more emphatic became apparent: Wilson, *The Diversity of Life*, p. 186.

p. 364 'This is truly time travelling': Fortey, *Life*, p. 66.

p. 364 the cyanobacteria at Shark Bay are perhaps the most slowly evolving organisms on Earth: Schopf, *Cradle of Life*, p. 212.

p. 365 'Animals could not summon up the energy to work': Fortey, *Life*, p. 89.

p. 365 would be nothing more than a sludge of simple

microbes: Margulis and Sagan, *Microcosmos*, p. 128.

p. 365 you could pack a billion into the space occupied by a grain of sand: Brown, *The Energy of Life*, p. 101.

p. 366 Such fossils have been found just once and then no more are known for 500 million years: Ward and Brownlee, *Rare Earth*, p. 10.

p. 366 little more than 'bags of chemicals': Drury, *Stepping Stones*, p. 68.

p. 367 enough, as Carl Sagan noted, to fill eighty books of five hundred pages: Sagan, *Cosmos*, p. 273.

Chapter 20: Small World

p. 368 Louis Pasteur, the great French chemist and bacteriologist, became so preoccupied with his that he took to peering critically at every dish placed before him with a magnifying glass: Biddle, *A Field Guide to the Invisible*, p. 16.

p. 368 If you are in good health and averagely diligent about hygiene, you will have a herd of about one trillion bacteria grazing on your fleshy plains: Ashcroft, *Life at the Extremes*, p. 248; Sagan and Margulis, *Garden of Microbial Delights*, p. 4.

p. 369 Your digestive system alone is host to more than a hundred trillion microbes, of at least 400 types: Biddle, *A Field Guide to the Invisible*, p. 57.

p. 369 A surprising number . . . have no detectable function at all: *National Geographic*, 'Bacteria', Aug. 1993, p. 51.

p. 369 Every human body consists of about ten quadrillion cells, but is host to about a hundred quadrillion bacterial cells: Margulis and Sagan, *Microcosmos*, p. 67.

p. 369 We couldn't survive a day without them: *New York Times*, 'From Birth, Our Body Houses a Microbe Zoo', 15 Oct. 1996, p. C-3.

p. 370 Algae and other tiny organisms bubbling away in

the sea blow out about 150 billion kilograms of the stuff every year: Sagan and Margulis, *Garden of Microbial Delights*, p. 11.

p. 370 *Clostridium perfringens*, the disagreeable little organism that causes gangrene, can reproduce in nine minutes: *Outside*, July 1999, p. 88.

p. 370 At such a rate, a single bacterium could theoretically produce more offspring in two days than there are protons in the universe: Margulis and Sagan, *Microcosmos*, p. 75.

p. 370 'a single bacterial cell can generate 280,000 billion individuals in a single day': de Duve, *A Guided Tour of the Living Cell*, vol. 2, p. 320.

p. 370 Essentially . . . all bacteria swim in a single gene pool: Margulis and Sagan, *Microcosmos*, p. 16.

p. 371 Scientists in Australia found microbes known as *Thiobacillus concretivorans*: Davies, *The Fifth Miracle*, p. 145.

p. 371 Some bacteria break down chemical materials from which, as far as we can tell, they gain no benefit at all: *National Geographic*, 'Bacteria', August 1993, p. 39.

p. 371 'like the scuttling limbs of an undead creature from a horror movie': *Economist*, 'Human Genome Survey', 1 July 2000, p. 9.

p. 371 Perhaps the most extraordinary survival yet found was that of a *Streptococcus* bacterium that was recovered from the sealed lens of a camera that had stood on the Moon for two years: Davies, *The Fifth Miracle*, p. 146.

p. 372 It has even been suggested that their tireless nibblings created the Earth's crust: *New York Times*, 'Bugs Shape Landscape, Make Gold', 15 Oct. 1996, p. C-1.

p. 372 if you took all the bacteria out of the Earth's interior and dumped them on the surface, they would cover the planet to a depth of 15 metres: *Discover*, 'To Hell and Back', July 1999, p. 82.

p. 372 The liveliest of them may divide no more than

once a century: *Scientific American*, 'Microbes Deep Inside the Earth', Oct. 1996, p. 71.

p. 372 'The key to long life, it seems, is not to do too much': *Economist*, 'Earth's Hidden Life', 21 Dec. 1996, p. 112.

p. 372 Other micro-organisms have leaped back to life after being released from a 118-year-old can of meat and a 166-year-old bottle of beer: *Nature*, 'A Case of Bacterial Immortality?', 19 Oct. 2000, p. 844.

p. 372 claimed to have revived bacteria frozen in Siberian permafrost for three million years: *Economist*, 'Earth's hidden life', 21 Dec. 1996, p. 111.

p. 372 But the record claim for durability so far is one made by Russell Vreeland and colleagues at West Chester University: *New Scientist*, 'Sleeping Beauty', 21 Oct. 2000, p. 12.

p. 373 The more doubtful scientists suggested that the sample might have been contaminated: BBC News online, 'Row over Ancient Bacteria', 7 June 2001.

p. 373 Bacteria were usually lumped in with plants, too: Sagan and Margulis, *Garden of Microbial Delights*, p. 22.

p. 375 In 1969, in an attempt to bring some order to the growing inadequacies of classification: Sagan and Margulis, *Garden of Microbial Delights*, p. 23.

p. 376 By one calculation it contained as many as two hundred thousand species of organism: Sagan and Margulis, *Garden of Microbial Delights*, p. 24.

p. 376 At that time, according to Woese, only about five hundred species of bacteria were known: *New York Times*, 'Microbial Life's Steadfast Champion', 15 Oct. 1996, p. C-3.

p. 376 Only about 1 per cent will grow in culture: *Science*, 'Microbiologists Explore Life's Rich, Hidden Kingdoms', 21 March 1997, p. 1740.

p. 376 'like learning about animals from visiting zoos':

New York Times, 'Microbial Life's Steadfast Champion', 15 Oct. 1996, p. C-7.

p. 378 Woese . . . 'felt bitterly disappointed': Ashcroft, *Life at the Extremes*, pp. 274–5.

p. 378 'Biology, like physics before it . . . has moved to a level where the objects of interest and their interactions often cannot be perceived through direct observation': *Proceedings of the National Academy of Sciences*, 'Default Taxonomy: Ernst Mayr's View of the Microbial World', 15 Sept. 1998.

p. 378 'Woese was not trained as a biologist and quite naturally does not have an extensive familiarity with the principles of classification': *Proceedings of the National Academy of Sciences*, 'Two Empires or Three?', 18 Aug. 1998.

p. 379 Of the twenty-three main divisions of life only three . . . are large enough to be seen by the human eye: Schopf, *Cradle of Life*, p. 106.

p. 379 if you totalled up all the biomass of the planet . . . microbes would account for at least 80 per cent of all there is: *New York Times*, 'Microbial Life's Steadfast Champion', 15 Oct. 1996, p. C-7.

p. 380 the most rampantly infectious organism on Earth, a bacterium called Wolbachia: *Nature*, 'Wolbachia: a tale of sex and survival', 11 May 2001, p. 109.

p. 380 only about one microbe in a thousand is a pathogen for humans: *National Geographic*, 'Bacteria', Aug. 1993, p. 39.

p. 380 microbes are still the number three killer in the Western world: *Outside*, July 1999, p. 88.

p. 381 History . . . is full of diseases that 'once caused terrifying epidemics and then disappeared as mysteriously as they had come': Diamond, *Guns, Germs and Steel*, p. 208.

p. 382 a disease called necrotizing fasciitis in which

bacteria essentially eat the victim from the inside out: Gawande, *Complications*, p. 234.

p. 383 'The time has come to close the book on infectious diseases': *New Yorker*, 'No Profit, No Cure', 5 Nov. 2001, p. 46.

p. 383 Even as he spoke, however, some 90 per cent of those strains were in the process of developing immunity to penicillin: *Economist*, 'Disease Fights Back', 20 May 1995, p. 15.

p. 384 in 1997 a hospital in Tokyo reported the appearance of a strain that could resist even that: *Boston Globe*, 'Microbe Is Feared to Be Winning Battle Against Antibiotics', 30 May 1997, p. A-7.

p. 384 As James Surowiecki noted: *New Yorker*, 'No Profit, No Cure', 5 Nov. 2001, p. 46.

p. 384 America's National Institutes of Health . . . didn't officially endorse the idea until 1994: *Economist*, 'Bugged by Disease', 21 March 1998, p. 93.

p. 384 'Hundreds, even thousands of people must have died from ulcers who wouldn't have': *Forbes*, 'Do Germs Cause Cancer?', 15 Nov. 1999, p. 195.

p. 384 Since then, further research has shown that there is or may well be a bacterial component in all kinds of other disorders: *Science*, 'Do Chronic Diseases Have an Infectious Root?', 14 Sept. 2001, pp. 1974–6.

p. 385 'a piece of nucleic acid surrounded by bad news': quoted in Oldstone, *Viruses, Plagues and History*, p. 8.

p. 385 About five thousand types of virus are known: Biddle, *A Field Guide to the Invisible*, pp. 153–4.

p. 385 Smallpox in the twentieth century alone killed an estimated three hundred million people: Oldstone, *Viruses, Plagues and History*, p. 1.

p. 386 In ten years the disease killed some five million people and then quietly went away: Kolata, *Flu*, p. 292.

p. 386 The First World War killed 21 million people in

four years; swine flu did the same in its first four months: *American Heritage*, 'The Great Swine Flu Epidemic of 1918', June 1976, p. 82.

p. 387 In an attempt to devise a vaccine, medical authorities conducted experiments on volunteers at a military prison on Deer Island in Boston Harbor: *American Heritage*, 'The Great Swine Flu Epidemic of 1918', June 1976, p. 82.

p. 388 Researchers at the Manchester Royal Infirmary discovered that a sailor who had died of mysterious, untreatable causes in 1959 in fact had AIDS: *National Geographic*, 'The Disease Detectives', Jan. 1991, p. 132.

p. 388 In 1969, a doctor at a Yale University lab in New Haven, Connecticut, who was studying lassa fever: Oldstone, *Viruses, Plagues and History*, p. 126.

p. 389 In 1990, a Nigerian living in Chicago was exposed to lassa fever on a visit to his homeland: Oldstone, *Viruses, Plagues and History*, p. 128.

Chapter 21: Life Goes On

p. 390 The fate of nearly all living organisms: Schopf, *Cradle of Life*, p. 72.

p. 390 Only about 15 per cent of rocks can preserve fossils: Lewis, *The Dating Game*, p. 24.

p. 391 less than one species in ten thousand has made it into the fossil record: Trefil, *101 Things You Don't Know About Science*, p. 280.

p. 391 statement . . . that there are 250,000 species of creature in the fossil record: Leakey and Lewin, *The Sixth Extinction*, p. 45.

p. 391 About 95 per cent of all the fossils we possess are of animals that once lived under water: Leakey and Lewin, *The Sixth Extinction*, p. 45.

p. 392 'It seems like a big number,' he agreed: interview with Richard Fortey, Natural History Museum, London,

19 Feb. 2001.

p. 392 Humans . . . have survived so far for one-half of 1 per cent as long: Fortey, *Trilobite!*, p. 24.

p. 393 'a whole *Profallotaspis* or *Elenellus* as big as a crab': Fortey, *Trilobite!*, p. 121.

p. 394 and built up a collection of sufficient distinction that it was bought by Louis Agassiz: 'From Farmer-Laborer to Famous Leader: Charles D. Walcott (1850–1927)', *GSA Today*, Jan. 1996.

p. 394 In 1879 Walcott took a job as a field researcher: Gould, *Wonderful Life*, pp. 242–3.

p. 394 'His books fill a library shelf': Fortey, *Trilobite!*, p. 53.

p. 395 'our sole vista upon the inception of modern life': Gould, *Wonderful Life*, p. 56.

p. 395 Gould, ever scrupulous, discovered: Gould, *Wonderful Life*, p. 71.

p. 396 140 species, by one count: Leakey and Lewin, *The Sixth Extinction*, p. 27.

p. 396 'a range of disparity . . . never again equaled': Gould, *Wonderful Life*, p. 208.

p. 396 'Under such an interpretation,' Gould sighed: Gould, *Eight Little Piggies*, p. 225.

p. 397 Then in 1973 a graduate student from Cambridge: *National Geographic*, 'Explosion of Life', Oct. 1993, p. 126.

p. 397 There was so much unrecognized novelty . . . that at one point: Fortey, *Trilobite!*, p. 123.

p. 397 they all use architecture first created in the Cambrian party: *US News and World Report*, 'How Do Genes Switch On?', 18–25 Aug. 1997, p. 74.

p. 398 at least fifteen and perhaps as many as twenty: Gould, *Wonderful Life*, p. 25.

p. 398 'Wind back the tape of life': Gould, *Wonderful Life*, p. 14.

p. 399 In 1946 Sprigg, a young assistant government geologist: Corfield, *Architects of Eternity*, p. 287.

p. 399 but it failed to find favour with the association's head: Corfield, *Architects of Eternity*, p. 287.

p. 399 Nine years later, in 1957: Fortey, *Life*, p. 85.

p. 401 'There is nothing closely similar alive today': Fortey, *Life*, p. 88.

p. 401 'They are difficult to interpret': Fortey, *Trilobite!*, p 125.

p. 401 'If only Stephen Gould could think as clearly as he writes!': Dawkins, *Sunday Telegraph*, 25 Feb. 1990.

p. 402 One, writing in the *New York Times Book Review*: *New York Times Book Review*, 'Survival of the Luckiest', 22 Oct. 1989.

p. 402 Dawkins attacked Gould's assertions: review of *Full House* in *Evolution*, June 1997.

p. 403 who startled many in the palaeontological community by rounding abruptly on Gould in a book of his own, *The Crucible of Creation*: *New York Times Book Review*, 'Rock of Ages', 10 May 1998, p. 15.

p. 403 'I have never encountered such spleen in a book by a professional': Fortey, *Trilobite!*, p. 138.

p. 403 Fortey gives as an example the idea of comparing a shrew and an elephant: Fortey, *Trilobite!*, p. 132.

p. 404 'None was as strange as a present day barnacle': Fortey, *Life*, p. 111.

p. 404 'no less interesting, or odd, just more explicable': Fortey, 'Shock Lobsters', *London Review of Books*, 1 Oct. 1998.

p. 405 It is one thing to have one well-formed creature like a trilobite burst forth in isolation: Fortey, *Trilobite!*, p. 137.

Chapter 22: Goodbye to All That

p. 407 In areas of Antarctica where virtually nothing else

will grow: Attenborough, *The Living Planet*, p. 48.

p. 407 'Spontaneously, inorganic stone becomes living plant!': Marshall, *Mosses and Lichens*, p. 22.

p. 408 The world has more than twenty thousand species of lichens: Attenborough, *The Private Life of Plants*, p. 214.

p. 408 Those the size of dinner plates . . . are therefore 'likely to be hundreds if not thousands of years old': Attenborough, *The Living Planet*, p. 42.

p. 408 If you imagine the 4,500 million or so years of Earth's history compressed into a normal earthly day: adapted from Schopf, *Cradle of Life*, p. 13.

p. 409 stretch your arms to their fullest extent and imagine that width as the entire history of the Earth: McPhee, *Basin and Range*, p. 126.

p. 411 Oxygen levels . . . were as high as 35 per cent: Officer and Page, *Tales of the Earth*, p. 123.

p. 412 the isotopes accumulate at different rates depending on how much oxygen or carbon dioxide is in the atmosphere: Officer and Page, *Tales of the Earth*, p. 118.

p. 413 'The U.S. Air Force . . . has put them in wind tunnels to see how they do it, and despaired': Conniff, *Spineless Wonders*, p. 84.

p. 413 In Carboniferous forests dragonflies grew as big as ravens: Fortey, *Life*, p. 201.

p. 414 Luckily the team found just such a creature: BBC *Horizon*, 'The Missing Link', first broadcast 1 Feb. 2001.

p. 414 The names simply refer to the number and location of small holes found in the sides of their owners' skulls: Tudge, *The Variety of Life*, p. 411.

p. 415 but the number has been put as high as 4,000 billion: Tudge, *The Variety of Life*, p. 9.

p. 415 'To a first approximation . . . all species are extinct': quoted by Gould, *Eight Little Piggies*, p. 46.

p. 416 the average lifespan of a species is only about four million years: Leakey and Lewin, *The Sixth Extinction*,

p. 38.

p. 416 'The alternative to extinction is stagnation': interview with Ian Tattersall, American Museum of Natural History, New York, 6 May 2002.

p. 416 Crises in the Earth's history are invariably associated with dramatic leaps afterwards: Stanley, *Extinction*, p. 95; Stevens, *The Change in the Weather*, p. 12.

p. 416 In the Permian, at least 95 per cent of animals known from the fossil record checked out, never to return: *Harper's*, 'Planet of Weeds', Oct. 1998, p. 58.

p. 417 Even about a third of insect species went – the only occasion on which they were lost en masse: Stevens, *The Change in the Weather*, p. 12.

p. 417 'It was, truly, a mass extinction': Fortey, *Life*, p. 235.

p. 417 Estimates for the number of animal species alive at the end of the Permian range from as low as 45,000 to as high as 240,000: Gould, *Hen's Teeth and Horse's Toes*, p. 340.

p. 417 For individuals the death toll could be much higher – in many cases, practically total: Powell, *Night Comes to the Cretaceous*, p. 143.

p. 417 Grazing animals, including horses, were nearly wiped out in the Hemphillian event: Flannery, *The Eternal Frontier*, p. 100.

p. 418 At least two dozen potential culprits have been identified as causes or prime contributors: *Earth*, 'The Mystery of Selective Extinctions', Oct. 1996, p. 12.

p. 418 'tons of conjecture and very little evidence': *New Scientist*, 'Meltdown', 7 Aug. 1999.

p. 419 Such an outburst is not easily imagined: Powell, *Night Comes to the Cretaceous*, p. 19.

p. 419 The KT meteor had the additional advantage – advantage if you are a mammal, that is: Flannery, *The Eternal Frontier*, p. 17.

p. 420 'Why should these delicate creatures have emerged unscathed from such an unparalleled disaster?': Flannery, *The Eternal Frontier*, p. 43.

p. 420 In the seas it was much the same story: Gould, *Eight Little Piggies*, p. 304.

p. 420 'Somehow it does not seem satisfying just to call them "lucky ones" and leave it at that': Fortey, *Life*, p. 292.

p. 421 the period immediately after the dinosaur extinction could well be known as the Age of Turtles: Flannery, *The Eternal Frontier*, p. 39.

p. 422 'Evolution may abhor a vacuum . . . but it often takes a long time to fill it': Stanley, *Extinction*, p. 92.

p. 422 For perhaps as many as ten million years mammals remained cautiously small: Novacek, *Time Traveler*, p. 112

p. 422 For a time, there were guinea pigs the size of rhinos and rhinos the size of a two-storey house: Dawkins, *The Blind Watchmaker*, p. 102.

p. 422 For millions of years, a gigantic, flightless, carnivorous bird called *Titanis* was possibly the most ferocious creature in North America: Flannery, *The Eternal Frontier*, p. 138.

p. 422 built in 1903 in Pittsburgh and presented to the museum by Andrew Carnegie: Colbert, *The Great Dinosaur Hunters and their Discoveries*, p. 164.

p. 423 Until very recently, everything we know about the dinosaurs of this period came from only about three hundred specimens: Powell, *Night Comes to the Cretaceous*, pp. 168–9.

p. 424 'There is no reason to believe that the dinosaurs were dying out gradually': BBC *Horizon*, 'Crater of Death', first broadcast 6 May 2001.

p. 424 'Humans are here today because our particular line never fractured': Gould, *Eight Little Piggies*, p. 229.

Chapter 23: The Richness of Being

p. 426 The spirit room alone holds 15 miles of shelving: Thackray and Press, *The Natural History Museum*, p. 90.

p. 427 forty-four years after the expedition had concluded: Thackray and Press, *The Natural History Museum*, p. 74.

p. 428 published in 1956 and still to be found on many library shelves as almost the only attempt: Conard, *How to Know the Mosses and Liverworts*, p. 5.

p. 428 'The tropics are where you find the variety': interview with Len Ellis, Natural History Museum, London, 18 April 2002.

p. 430 he sifted through a bale of fodder sent for the ship's livestock and made new discoveries: Barber, *The Heyday of Natural History: 1820–1870*, p. 17.

p. 433 To the parts of one species of clam he gave the names: Gould, *Leonardo's Mountain of Clams and the Diet of Worms*, p. 79.

p. 433 'Love comes even to the plants. Males and females . . . hold their nuptials': quoted by Gjertsen, *The Classics of Science*, p. 237, and at University of California/UCMP Berkeley website.

p. 433 Linnaeus lopped it back to *Physalis angulata*: Kastner, *A Species of Eternity*, p. 31.

p. 434 The first edition of his great *Systema Naturae*: Gjertsen, *The Classics of Science*, p. 223.

p. 434 John Ray's three-volume *Historia Generalis Plantarum* in England: Durant, *The Age of Louis XIV*, p. 519.

p. 434 just in time to make Linnaeus a kind of father figure to British naturalists: Thomas, *Man and the Natural World*, p. 65.

p. 434 gullibly accepted from seamen and other imaginative travellers, Schwartz, *Sudden Origins*, p. 59.

p. 435 he saw that whales belonged with cows, mice and other common terrestrial animals in the order quadrupedia (later changed to mammalia): Schwartz, *Sudden Origins*, p. 59.

p. 435 other names in everyday use included *mare's fart, naked ladies, twitch-ballock, hound's piss, open arse*, and *bum-towel*: Thomas, *Man and the Natural World*, pp. 82–5.

p. 437 while Edward O. Wilson in *The Diversity of Life* puts the number at a surprisingly robust eighty-nine: Wilson, *The Diversity of Life*, p. 157.

p. 438 were transferred, amid howls, to the genus *Pelargonium*: Elliott, *The Potting-Shed Papers*, p. 18.

p. 439 Estimates range from three million to two hundred million: *Audubon*, 'Earth's Catalogue', Jan.-Feb. 2002; Wilson, *The Diversity of Life*, p. 132.

p. 439 as much as 97 per cent of the world's plant and animal species may still await discovery: *Economist*, 'A Golden Age of Discovery', 23 Dec. 1996, p. 56.

p. 439 he estimated the number of known species of all types – plants, insects, microbes, algae, everything – at 1.4 million: Wilson, *The Diversity of Life*, p. 133.

p. 439 Other authorities have put the number of known species slightly higher, at around 1.5 million to 1.8 million: *US News and World Report*, 18 Aug. 1997, p. 78.

p. 440 It took Groves four decades to untangle everything: *New Scientist*, 'Monkey Puzzle', 6 Oct. 2001, p. 54.

p. 441 only about fifteen thousand new species of all types are logged per year: *Wall Street Journal*, 'Taxonomists Unite to Catalog Every Species, Big and Small', 22 Jan. 2001.

p. 441 'It's not a biodiversity crisis, it's a taxonomist crisis!': interview with Koen Maes, National Museum, Nairobi, 2 Oct. 2002.

p. 441 'many species are being described poorly in isolated publications': *Nature*, 'Challenges for Taxonomy', 2 May 2002, p. 17.

p. 442 launched an enterprise called the All Species Foundation: *The Times*, 'The List of Life on Earth', 30

July 2001.

p. 442 your mattress is home to perhaps two million microscopic mites: Bodanis, *The Secret House*, p. 16.

p. 443 to quote the man who did the measuring, Dr John Maunder of the British Medical Entomology Centre: *New Scientist*, 'Bugs Bite Back', 17 Feb. 2001, p. 48.

p. 443 These mites have been with us since time immemorial: Bodanis, *The Secret House*, p. 15.

p. 443 Your sample will also contain perhaps a million plump yeasts: *National Geographic*, 'Bacteria', Aug. 1993, p. 39.

p. 444 'If over 9,000 microbial types exist in two pinches of substrate from two localities in Norway': Wilson, *The Diversity of Life*, p. 144.

p. 444 according to one estimate, it could be as many as 400 million: Tudge, *The Variety of Life*, p. 8.

p. 444 and discovered a thousand new species of flowering plant: Wilson, *The Diversity of Life*, p. 197.

p. 444 Overall, tropical rainforests cover only about 6 per cent of Earth's surface: Wilson, *The Diversity of Life*, p. 197.

p. 445 'over three and a half billion years of evolution': *Economist*, 'Biotech's Secret Garden', 30 May 1998, p. 75.

p. 445 one ancient bacterium was found on the wall of a country pub: Fortey, *Life*, p. 75.

p. 445 about 500 species have been identified (though other sources say 360): Ridley, *The Red Queen*, p. 54.

p. 446 Gather together all the fungi found in a typical hectare of meadowland: Attenborough, *The Private Life of Plants*, p. 177.

p. 446 it is thought the total number could be as high as 1.8 million: *National Geographic*, 'Fungi', Aug. 2000, p. 60; Leakey and Lewin, *The Sixth Extinction*, p. 117.

p. 447 The large, flightless New Zealand bird called the takahe: Flannery and Schouten, *A Gap in Nature*, p. 2.

p. 447 the horse was considered a rarity in the wider world: *New York Times*, 'A Stone-Age Horse Still Roams a Tibetan Plateau', 12 Nov. 1995.

p. 447 'a megatherium, a sort of giant ground sloth which may stand as high as a giraffe': *Economist*, 'A World to Explore', 23 Dec. 1995, p. 95.

p. 448 A single line of text in a Crampton table: Gould, *Eight Little Piggies*, pp. 32–4.

p. 448 he hiked 4,000 kilometres to assemble a collection of three hundred thousand wasps: Gould, *The Flamingo's Smile*, pp. 159–60.

Chapter 24: Cells

p. 451 you would have to miniaturize about the same number of components as are found in a Boeing 777 jetliner: *New Scientist*, 2 Dec. 2000, p. 37.

p. 451 we understand what no more than about 2 per cent of them do: Brown, *The Energy of Life*, p. 83.

p. 452 Its purpose was at first a mystery, but then scientists began to find it all over the place: Brown, *The Energy of Life*, p. 229.

p. 452 It is converted into nitric oxide in the bloodstream, relaxing the muscle linings of vessels, allowing blood to flow more freely: Alberts, et al., *Essential Cell Biology*, p. 489.

p. 452 You possess 'some few hundred' different types of cell: de Duve, *A Guided Tour of the Living Cell*, vol. 1, p. 21.

p. 452 If you are an average-sized adult you are lugging around over 2 kilograms of dead skin: Bodanis, *The Secret Family*, p. 106.

p. 453 Liver cells can survive for years: de Duve, *A Guided Tour of the Living Cell*, vol. 1, p. 68.

p. 453 not so much as a stray molecule: Bodanis, *The Secret Family*, p. 81.

p. 453 Hooke calculated that a one-inch square of cork would contain 1,259,712,000 of these tiny chambers: Nuland, *How We Live*, p. 100.

p. 455 After he reported finding 'animalcules' in a sample of pepper-water in 1676: Jardine, *Ingenious Pursuits*, p. 93.

p. 455 He calculated that there were 8,280,000 of these tiny beings in a single drop of water: Thomas, *Man and the Natural World*, p. 167.

p. 455 He called the little beings 'homunculi': Schwartz, *Sudden Origins*, p. 167.

p. 455 In one of his least successful experiments: Carey (ed.), *The Faber Book of Science*, p. 28.

p. 456 Only in 1839, however, did anyone realize that *all* living matter is cellular: Nuland, *How We Live*, p. 101.

p. 456 The cell has been compared to many things: Trefil, *101 Things You Don't Know About Science and No One Else Does Either*, p. 33; Brown, *The Energy of Life*, p. 78.

p. 456 However, scale that up and it would translate as a jolt of 20 million volts per metre: Brown, *The Energy of Life*, p. 87.

p. 457 has the approximate consistency 'of a light grade of machine oil': Nuland, *How We Live*, p. 103.

p. 457 and flying into each other up to a billion times a second: Brown, *The Energy of Life*, p. 80.

p. 458 'the molecular world must necessarily remain entirely beyond the powers of our imagination': de Duve, *A Guided Tour of the Living Cell*, vol. 2, p. 293.

p. 458 'the total is still a very minimum of 100 million protein molecules in each cell': Nuland, *How We Live*, p. 157.

p. 459 At any given moment, a typical cell in your body will have about one billion ATP molecules in it: Alberts et al., *Essential Cell Biology*, p. 110.

p. 459 Every day you produce and use up a volume of ATP equivalent to about half your body weight: *Nature*,

'Darwin's Motors', 2 May 2002, p. 25.

p. 460 On average, humans suffer one fatal malignancy for each 100 million billion cell divisions: Ridley, *Genome*, p. 237.

p. 461 what has been called 'the single best idea that anyone has ever had': Dennett, *Darwin's Dangerous Idea*, p. 21.

Chapter 25: Darwin's Singular Notion

p. 462 'Everyone is interested in pigeons': quoted in Boorstin, *Cleopatra's Nose*, p. 176.

p. 463 'You care for nothing but shooting, dogs, and rat-catching': quoted in Boorstin, *The Discoverers*, p. 467.

p. 463 The experience of witnessing an operation on an understandably distressed child: Desmond and Moore, *Darwin*, p. 27.

p. 464 some 'bordering on insanity': Hamblyn, *The Invention of Clouds*, p. 199.

p. 464 In five years . . . he had not once hinted at an attachment: Desmond and Moore, *Darwin*, p. 197.

p. 465 which suggested, not incidentally, that atolls could not form in less than a million years: Moorehead, *Darwin and the Beagle*, p. 191.

p. 465 It wasn't until the younger Darwin was back in England and read Thomas Malthus's *Essay on the Principle of Population*: Gould, *Ever since Darwin*, p. 21.

p. 465 'How stupid of me not to have thought of it!': quoted in *Sunday Telegraph*, 'The Origin of Darwin's Genius', 8 Dec. 2002.

p. 466 It was his friend the ornithologist John Gould: Desmond and Moore, *Darwin*, p. 209.

p. 466 These he expanded into a 230-page 'sketch': *Dictionary of National Biography*, vol. 5, p. 526.

p. 467 'I hate a barnacle as no man ever did before': quoted in Ferris, *Coming of Age in the Milky Way*, p. 239.

p. 467 Some wondered if Darwin himself might be the author: Barber, *The Heyday of Natural History*, p. 214.

p. 468 'he could not have made a better short abstract': *Dictionary of National Biography*, vol. 5, p. 528.

p. 468 'This summer will make the 20th year (!) since I opened my first note-book': Desmond and Moore, *Darwin*, pp. 454–5.

p. 469 'whatever it may amount to, will be smashed': Desmond and Moore, *Darwin*, p. 469.

p. 470 'all that was new in them was false, and what was true was old': quoted by Gribbin and Cherfas, *The First Chimpanzee*, p. 150.

p. 470 Much less amenable to Darwin's claim of priority was a Scottish gardener named Patrick Matthew: Gould, *The Flamingo's Smile*, p. 336.

p. 471 He referred to himself as 'the Devil's Chaplain': Cadbury, *Terrible Lizard*, p. 305.

p. 471 felt 'like confessing a murder': quoted in Desmond and Moore, *Darwin*, p. xvi.

p. 471 'The case at present must remain inexplicable': quoted by Gould, *Wonderful Life*, p. 57.

p. 472 By way of explanation he speculated: Gould, *Ever Since Darwin*, p. 126.

p. 472 'Darwin goes too far': quoted by McPhee, *In Suspect Terrain*, p. 190.

p. 472 Huxley was a saltationist: Schwartz, *Sudden Origins*, pp. 81–2.

p. 472 'The eye to this day gives me a cold shudder': quoted in Keller, *The Century of the Gene*, p. 97.

p. 473 it 'seems, I freely confess, absurd in the highest possible degree' that natural selection could produce such an instrument in gradual steps: Darwin, *On the Origin of Species* (facsimile ed.), p. 217.

p. 473 'Eventually . . . Darwin lost virtually all the support that still remained': Schwartz, *Sudden Origins*, p. 89.

p. 474 It had a library of twenty thousand books: Lewontin, *It Ain't Necessarily So*, p. 91.
p. 476 And Darwin, for his part, is known to have studied Focke's influential paper: Ridley, *Genome*, p. 44.
p. 476 Huxley had been urged to attend by Robert Chambers: Trinkaus and Shipman, *The Neandertals*, p. 79.
p. 476 bravely slogged his way through two hours of introductory remarks: Clark, *The Survival of Charles Darwin*, p. 142.
p. 478 One of his experiments was to play the piano to them: Conniff, *Spineless Wonders*, p. 147.
p. 478 Having married his own cousin: Desmond and Moore, *Darwin*, p. 575.
p. 478 Darwin was often honoured in his lifetime, but never for *On the Origin of Species*: Clark, *The Survival of Charles Darwin*, p. 148.
p. 479 Darwin's theory didn't really gain widespread acceptance until the 1930s and 1940s: Tattersall and Schwartz, *Extinct Humans*, p. 45.
p. 479 seemed set to claim Mendel's insights as his own: Schwartz, *Sudden Origins*, p. 187.

Chapter 26: The Stuff of Life

p. 481 'roughly one nucleotide base in every thousand': Sulston and Ferry, *The Common Thread*, p. 198.
p. 482 The exceptions are red blood cells, some immune system cells, and egg and sperm cells: Woolfson, *Life without Genes*, p. 12.
p. 482 'guaranteed to be unique against all conceivable odds': de Duve, *A Guided Tour of the Living Cell*, vol. 2, p. 314.
p. 482 there would be enough of it to stretch from the Earth to the Moon and back, not once or twice but again and again: Dennett, *Darwin's Dangerous Idea*, p. 151.
p. 483 you may have as much as 20 million kilometres of

DNA bundled up inside you: Gribbin and Gribbin, *Being Human*, p. 8.

p. 483 'among the most nonreactive, chemically inert molecules': Lewontin, *It Ain't Necessarily So*, p. 142.

p. 483 It was discovered as far back as 1869: Ridley, *Genome*, p. 48.

p. 484 DNA didn't do anything at all, as far as anyone could tell: Wallace et al., *Biology*, p. 211.

p. 484 The necessary complexity, it was thought, had to exist in proteins in the nucleus: de Duve, *A Guided Tour of the Living Cell*, vol. 2, p. 295.

p. 485 Working out of a small lab (which became known, inevitably, as the Fly Room): Clark, *The Survival of Charles Darwin*, p. 259.

p. 486 no consensus 'as to what the genes are – whether they are real or purely fictitious': Keller, *The Century of the Gene*, p. 2.

p. 486 we are in much the same position today in respect of mental processes such as thought and memory: Wallace et al., *Biology*, p. 211.

p. 487 Chargaff . . . suggested that Avery's discovery was worth two Nobel Prizes: Maddox, *Rosalind Franklin*, p. 327.

p. 487 including, it has been said, lobbying the authorities . . . not to give Avery a Nobel Prize: White, *Rivals*, p. 251.

p. 488 a member of a highly popular radio programme called *The Quiz Kids*: Judson, *The Eighth Day of Creation*, p. 46.

p. 489 'it was my hope that the gene might be solved without my learning any chemistry': Watson, *The Double Helix*, p. 21.

p. 489 the results of which were obtained 'fortuitously': Jardine, *Ingenious Pursuits*, p. 356.

p. 489 In a severely unflattering portrait: Watson, *The*

Double Helix, p. 17.

p. 489 'gratuitously hurtful': Jardine, *Ingenious Pursuits*, p. 354.

p. 490 To Wilkins's presumed dismay and embarrassment, in the summer of 1952 she posted a mock notice: White, *Rivals*, p. 257; Maddox, *Rosalind Franklin*, p. 185.

p. 491 'apparently without her knowledge or consent': PBS website, 'A Science Odyssey', n.d.

p. 491 Years later, Watson conceded that it 'was the key event . . . it mobilized us': quoted in Maddox, *Rosalind Franklin*, p. 317.

p. 492 a 900-word article by Watson and Crick titled 'A Structure for Deoxyribose Nucleic Acid': de Duve, *A Guided Tour of the Living Cell*, vol. 2, p. 290.

p. 492 It received a small mention in the *News Chronicle* and was ignored elsewhere: Ridley, *Genome*, p. 50.

p. 492 Franklin rarely wore a lead apron and often stepped carelessly in front of a beam: Maddox, *Rosalind Franklin*, p. 144.

p. 492 'It took over twenty-five years for our model of DNA to go from being only rather plausible, to being very plausible': Crick, *What Mad Pursuit*, p. 73–4.

p. 492 by 1968 the journal *Science* could run an article entitled 'That Was the Molecular Biology That Was': Keller, *The Century of the Gene*, p. 25.

p. 493 In this sense, they are rather like the keys of a piano, each playing a single note and nothing else: *National Geographic*, 'Secrets of the Gene', Oct. 1995, p. 55.

p. 493 Guanine, for instance, is the same stuff that abounds in, and gives its name to, guano: Pollack, *Signs of Life*, p. 22–3.

p. 495 'you could say all humans share nothing, and that would be correct, too': *Discover*, 'Bad Genes, Good Drugs', April 2002, p. 54.

p. 496 'exist for the pure and simple reason that they are

good at getting themselves duplicated': Ridley, *Genome*, p. 127.

p. 496 Altogether, almost half of human genes . . . don't do anything at all, as far as we can tell: Woolfson, *Life without Genes*, p. 18.

p. 496 Junk DNA does have a use: *National Geographic*, 'The New Science of Identity', May 1992, p. 118.

p. 497 'Empires fall, ids explode, great symphonies are written, and behind all of it is a single instinct that demands satisfaction': Nuland, *How We Live*, p. 158.

p. 497 Here were two creatures that hadn't shared a common ancestor for five hundred million years: BBC *Horizon*, 'Hopeful Monsters', first transmitted 1998.

p. 497 At least 90 per cent correlate at some level with those found in mice: *Nature*, 'Sorry, dogs – man's got a new best friend', 19–26 Dec. 2002, p. 734.

p. 497 We even have the same genes for making a tail, if only they would switch on: *Los Angeles Times* (reprinted in *Valley News*), 9 Dec. 2002.

p. 498 dubbed homeotic (from a Greek word meaning 'similar') or hox genes: BBC *Horizon*, 'Hopeful Monsters', first transmitted 1998.

p. 498 We have forty-six chromosomes, but some ferns have more than six hundred: Gribbin and Cherfas, *The First Chimpanzee*, p. 53.

p. 498 The lungfish, one of the least evolved of all complex animals, has forty times as much DNA as we have: Schopf, *Cradle of Life*, p. 240.

p. 498 Perhaps the apogee (or nadir) of this faith in biodeterminism was a study published in the journal *Science* in 1980: Lewontin, *It Ain't Necessarily So*, p. 215.

p. 499 How fast a man's beard grows . . . is partly a function of how much he thinks about sex: *Wall Street Journal*, 'What Distinguishes Us from the Chimps? Actually, Not Much', 12 April 2002, p. 1.

p. 500 'the proteome is much more complicated than the genome': *Scientific American*, 'Move Over, Human Genome', April 2002, pp. 44–5.

p. 500 Depending on mood and metabolic circumstance, they will allow themselves to be phosphorylated, glycosylated, acetylated, ubiquitinated: *The Bulletin*, 'The Human Enigma Code', 21 Aug. 2001, p. 32.

p. 501 Drink a glass of wine . . . and you materially alter the number and types of proteins at large in your system: *Scientific American*, 'Move Over, Human Genome', April 2002, pp. 44–5.

p. 501 'Anything that is true of E. coli must be true of elephants, except more so': *Nature*, 'From E. coli to Elephants', 2 May 2002, p. 22.

Chapter 27: Ice Time

p. 505 In London, *The Times* ran a small story: Williams, *Surviving Galeras*, p. 198.

p. 506 Spring never came and summer never warmed: Officer and Page, *Tales of the Earth*, pp. 3–6.

p. 506 One French naturalist named de Luc: Hallam, *Great Geological Controversies*, p. 89.

p. 507 and the other abundant clues that point to passing ice sheets: Hallam, *Great Geological Controversies*, p. 90.

p. 507 The naturalist Jean de Charpentier told the story: Hallam, *Great Geological Controversies*, p. 90.

p. 508 He lent Agassiz his notes: Hallam, *Great Geological Controversies*, pp. 92–3.

p. 508 Humboldt . . . observed that there are three stages in scientific discovery: Ferris, *The Whole Shebang*, p. 173.

p. 508 In his quest to understand the dynamics of glaciation, he went everywhere: McPhee, *In Suspect Terrain*, p. 182.

p. 509 William Hopkins, a Cambridge professor and leading member of the Geological Society: Hallam, *Great Geological*

Controversies, p. 98.

p. 511 He began to find evidence for glaciers practically everywhere: Hallam, *Great Geological Controversies*, p. 99.

p. 511 Eventually he became convinced that ice had once covered the whole Earth: Gould, *Time's Arrow*, p. 115.

p. 511 When he died in 1873 Harvard felt it necessary to appoint three professors to take his place: McPhee, *In Suspect Terrain*, p. 197.

p. 511 Less than a decade after his death: McPhee, *In Suspect Terrain*, p. 197.

p. 512 For the next twenty years, even while on holiday: Gribbin and Gribbin, *Ice Age*, p. 51.

p. 513 The cause of ice ages, Köppen decided, is to be found in cool summers, not brutal winters: Chorlton, *Ice Ages*, p. 101.

p. 513 'It is not necessarily the *amount* of snow that causes ice sheets but the fact that snow, however little, lasts': Schultz, *Ice Age Lost*, p. 72.

p. 513 'The process is self-enlarging': McPhee, *In Suspect Terrain*, p. 205.

p. 514 'you would have been hard pressed to find a geologist or meteorologist who regarded the model as being anything more than an historical curiosity': Gribbin and Gribbin, *Ice Age*, p. 60.

p. 514 The fact is, we are still very much in an ice age: Schultz, *Ice Age Lost*, p. 5.

p. 514 a situation that may be unique in the Earth's history: Gribbin and Gribbin, *Fire on Earth*, p. 147.

p. 515 it appears that we have had at least seventeen severe glacial episodes in the last 2.5 million years: Flannery, *The Eternal Frontier*, p. 148.

p. 515 about fifty more glacial episodes can be expected: McPhee, *In Suspect Terrain*, p. 4.

p. 516 Before fifty million years ago the Earth had no regular ice ages: Stevens, *The Change in the Weather*, p. 10.

p. 516 the Cryogenian, or super ice age: McGuire, *A Guide to the End of the World*, p. 69.

p. 516 The entire surface of the planet may have frozen solid: *Valley News* (from *Washington Post*), 'The Snowball Theory', 19 June 2000, p. C1.

p. 517 the wildest weather it has ever experienced: BBC *Horizon* transcript, 'Snowball Earth', broadcast 22 Feb. 2001, p. 7.

p. 518 in an event known to science as the Younger Dryas: Stevens, *The Change in the Weather*, p. 34.

p. 520 'the last thing you'd want to do is conduct a vast unsupervised experiment on it': *New Yorker*, 'Ice Memory', 7 Jan. 2002, p. 36.

p. 520 The idea is that a slight warming would enhance evaporation rates: Schultz, *Ice Age Lost*, p. 72.

p. 520 No less intriguing are the known ranges of some late dinosaurs: Drury, *Stepping Stones*, p. 268.

p. 520 In Australia – which at that time was more polar in its orientation – a retreat to warmer climes wasn't possible: Thomas H. Rich, Patricia Vickers-Rich and Roland Gangloff, 'Polar Dinosaurs', manuscript, n.d.

p. 521 there is a lot more water for them to draw on this time: Schultz, *Ice Age Lost*, p. 159.

p. 521 If so, sea levels globally would rise – and pretty quickly – by between 4.5 and 6 metres on average: Ball, *H_2O*, p. 75.

p. 521 'Did you have a good ice age?': Flannery, *The Eternal Frontier*, p. 267.

Chapter 28: The Mysterious Biped

p. 522 Just before Christmas 1887: *National Geographic*, May 1997, p. 87.

p. 523 found by railway workers in a cave at a cliff called Cro-Magnon: Tattersall and Schwartz, *Extinct Humans*, p. 149.

p. 523 The first formal description: Trinkaus and Shipman, *The Neandertals*, p. 173.

p. 523 So instead the name and credit for the discovery of the first early humans went to the Neander valley in Germany: Trinkaus and Shipman, *The Neandertals*, pp. 3–6.

p. 524 Hearing of this, T. H. Huxley in England drily observed: Trinkaus and Shipman, *The Neandertals*, p. 59.

p. 524 He did no digging himself, but instead used fifty convicts lent by the Dutch authorities: Gould, *Eight Little Piggies*, pp. 126–7.

p. 525 In fact, many anthropologists think it *is* modern, and has nothing to do with Java man: Walker and Shipman, *The Wisdom of the Bones*, p. 39.

p. 525 If it is an erectus bone, it is unlike any other found since: Trinkaus and Shipman, *The Neandertals*, p. 144.

p. 526 He also produced, with nothing but a scrap of cranium and one tooth, a model of the complete skull, which also proved uncannily accurate: Trinkaus and Shipman, *The Neandertals*, p. 154.

p. 525 To Dubois' dismay, Schwalbe thereupon produced a monograph: Walker and Shipman, *The Wisdom of the Bones*, p. 42.

p. 526 Dart could see at once that the Taung skull was not of a *Homo erectus*: Walker and Shipman, *The Wisdom of the Bones*, p. 74.

p. 526 he would sometimes bury their bodies in his back garden to dig up for study later: Trinkaus and Shipman, *The Neandertals*, p. 233.

p. 527 Dart spent five years working up a monograph, but could find no-one to publish it: Lewin, *Bones of Contention*, p. 82.

p. 527 For years, the skull . . . sat as a paperweight on a colleague's desk: Walker and Shipman, *The Wisdom of the Bones*, p. 93.

p. 527 found a single fossilized molar and on the basis of that alone quite brilliantly announced the discovery of *Sinanthropus pekinensis*: Swisher et al., *Java Man*, p. 75.

p. 528 then discovered to his horror that they had been enthusiastically smashing large pieces into small ones: Swisher et al., *Java Man*, p. 77.

p. 528 The Solo People were known variously as *Homo soloensis, Homo primigenius asiaticus*: Swisher et al., *Java Man*, p. 211.

p. 528 in 1960 F. Clark Howell of the University of Chicago, following the suggestions of Ernst Mayr and others the previous decade: Trinkaus and Shipman, *The Neandertals*, pp. 267–8.

p. 529 the whole of our understanding of human prehistory is based on the remains, often exceedingly fragmentary, of perhaps five thousand individuals: *Washington Post*, 'Skull Raises Doubts about our Ancestry', 22 March 2001.

p. 529 'You could fit it all into the back of a pickup truck if you didn't mind how much you jumbled everything up': interview with Ian Tattersall, American Museum of Natural History, New York, 6 May 2002.

p. 531 you would have to conclude that early hand tools were mostly made by antelopes: Walker and Shipman, *The Wisdom of the Bones*, p. 66.

p. 531 they show males and females evolving at different rates and in different directions: Walker and Shipman, *The Wisdom of the Bones*, p. 194.

p. 531 dismiss it as a mere 'wastebasket species': Tattersall and Schwartz, *Extinct Humans*, p. 111.

p. 531 'it is remarkable how often the first interpretations of new evidence have confirmed the preconceptions of its discoverer': quoted by Gribbin and Cherfas, *The First Chimpanzee*, p. 60.

p. 532 'And of all the disciplines in science, paleoanthropology boasts perhaps the largest share of egos': Swisher et al., *Java Man*, p. 17.

p. 532 For the first 99.99999 per cent of our history as organisms, we were in the same ancestral line as chimpanzees: Tattersall, *The Human Odyssey*, p. 60.

p. 533 'She is our earliest ancestor, the missing link between ape and human': PBS *Nova*, 'In Search of Human Origins', first broadcast Aug. 1999.

p. 533 Johanson breezily replied that he had discounted the 106 bones of the hands and feet: Walker and Shipman, *The Wisdom of the Bones*, p. 147.

p. 535 'Lucy and her kind did not locomote in anything like the modern human fashion': Tattersall, *The Monkey in the Mirror*, p. 88.

p. 535 'Only when these hominids had to travel between arboreal habitats would they find themselves walking bipedally': Tattersall and Schwartz, *Extinct Humans*, p. 91.

p. 535 'Lucy's hips and the muscular arrangement of her pelvis', he has written: *National Geographic*, 'Face-to-Face with Lucy's Family', March 1996, p. 114.

p. 536 One, discovered by Meave Leakey of the famous fossil-hunting family at Lake Turkana in Kenya: *New Scientist*, 24 March 2001, p. 5.

p. 536 making it the oldest hominid yet found – but only for a brief while: *Nature*, 'Return to the Planet of the Apes', 12 July 2001, p. 131.

p. 536 In the summer of 2002 a French team working in the Djurab Desert of Chad . . . found a hominid almost seven million years old: *Scientific American*, 'An Ancestor to Call our Own', Jan. 2003, pp. 54–63.

p. 536 Some critics believe that it was not human but an early ape: *Nature*, 'Face to Face with our Past', 19–26 Dec. 2002, p. 735.

p. 536 when you are a small, vulnerable australopithecine, with a brain about the size of an orange, the risk must have been enormous: Stevens, *The Change in the Weather*, p. 3; Drury, *Stepping Stones*, pp. 335–6.

p. 537 'but that the forests left them': Gribbin and Gribbin, *Being Human*, p. 135.

p. 537 For over three million years, Lucy and her fellow australopithecines scarcely changed at all: PBS *Nova*, 'In Search of Human Origins', first broadcast Aug. 1999.

p. 537 Absolute brain size: Gould, *Ever since Darwin*, pp. 181–3.

p. 538 yet the australopithecines never took advantage of this useful technology that was all around them: Drury, *Stepping Stones*, p. 338.

p. 538 'Perhaps,' suggests Matt Ridley, 'we ate them': Ridley, *Genome*, p. 33.

p. 539 they make up only 2 per cent of the body's mass, but devour 20 per cent of its energy: Drury, *Stepping Stones*, p. 345.

p. 539 'The body is in constant danger of being depleted by a greedy brain': Brown, *The Energy of Life*, p. 216.

p. 539 C. Loring Brace stuck doggedly to the linear concept: Gould, *Leonardo's Mountain of Clams and the Diet of Worms*, p. 204.

p. 540 *Homo erectus* is the dividing line: Swisher et al., *Java Man*, p. 131.

p. 541 It was from a boy aged between about nine and twelve who had died 1.54 million years ago: *National Geographic*, May 1997, p. 90.

p. 541 The Turkana boy was 'very emphatically one of us': Tattersall, *The Monkey in the Mirror*, p. 132.

p. 541 Someone had looked after her: Walker and Shipman, *The Wisdom of the Bones*, p. 165.

p. 542 They were unprecedentedly adventurous and spread across the globe with what seems to have been breathtaking rapidity: *Scientific American*, 'Food for Thought', Dec. 2002, pp. 108–15.

Chapter 29: The Restless Ape

p. 544 'They made them in their thousands': interview with Ian Tattersall, American Museum of Natural History, New York, 6 May 2002.

p. 547 'people may have first arrived substantially earlier than 60,000 years ago': *Proceedings of the National Academy of Sciences*, 16 Jan. 2001.

p. 547 'There's just a whole lot we don't know about the movements of people before recorded history': interview with Alan Thorne, Canberra, 20 Aug. 2001.

p. 549 'the most recent major event in human evolution – the emergence of our own species – is perhaps the most obscure of all': Tattersall, *The Human Odyssey*, p. 150.

p. 549 'whether any or all of them actually represent our species still awaits definitive clarification': Tattersall and Schwartz, *Extinct Humans*, p. 226.

p. 549 'odd, difficult-to-classify and poorly known': Trinkaus and Shipman, *The Neandertals*, p. 412.

p. 549 No Neandertal remains have ever been found in north Africa, but their tool kits turn up all over the place: Tattersall and Schwartz, *Extinct Humans*, p. 209.

p. 550 known to palaeoclimatology as the Boutellier interval: Fagan, *The Great Journey*, p. 105.

p. 550 They survived for at least a hundred thousand years, and perhaps twice that: Tattersall and Schwartz, *Extinct Humans*, p. 204.

p. 551 In 1947, while doing fieldwork in the Sahara: Trinkaus and Shipman, *The Neandertals*, p. 300.

p. 551 It is still commonly held that Neandertals lacked the intelligence or fibre to compete on equal terms with the continent's slender and more cerebrally nimble newcomers, *Homo sapiens*: *Nature*, 'Those Elusive Neanderthals', 25 Oct. 2001, p. 791.

p. 551 'Modern humans neutralized this advantage . . .

with better clothing, better fires and better shelter': Stevens, *The Change in the Weather*, p. 30.

p. 551 1.8 litres for Neandertals versus 1.4 for modern people: Flannery, *The Future Eaters*, p. 301.

p. 553 'Rhodesian man . . . lived as recently as 25,000 years ago and may have been an ancestor of the African Negroes': Canby, *The Epic of Man*, page unnoted.

p. 554 'you don't have the front end looking like a donkey and the back end looking like a horse': *Science*, 'What – or Who – Did In the Neandertals?', 14 Sept. 2001, p. 1981.

p. 555 'all present-day humans are descended from that population': Swisher et al., *Java Man*, p. 189.

p. 555 But then people began to look a little more closely at the data: *Scientific American*, 'Is Out of Africa Going Out the Door?', August 1999.

p. 555 in 1997 scientists from the University of Munich managed to extract and analyse some DNA from the arm bone of the original Neandertal man: *Proceedings of the National Academy of Sciences*, 'Ancient DNA and the Origin of Modern Humans', 16 Jan. 2001.

p. 556 suggested that all modern humans emerged from Africa within the past hundred thousand years and came from a breeding stock of no more than ten thousand individuals: *Nature*, 'A Start for Population Genomics', 7 Dec. 2000, p. 65; *Natural History*, 'What's New in Prehistory', May 2000, pp. 90–1.

p. 556 'there's more diversity in one social group of fifty-five chimps than in the entire human population': Science, 'A Glimpse of Humans' First Journey out of Africa', 12 May 2000, p. 950.

p. 556 In early 2001, Thorne and his colleagues at the Australian National University reported that they had recovered DNA from the oldest of the Mungo specimens: *Proceedings of the National Academy of Sciences*,

'Mitochondrial DNA sequences in Ancient Australians: Implications for Modern Human Origins', 16 Jan. 2001.

p. 557 'On the whole . . . the genetic record supports the out of Africa hypothesis': interview with Rosalind Harding, Institute of Biological Anthropology, Oxford, 28 Feb. 2002.

p. 559 had noted how a palaeontologist, asked by a colleague whether he thought an old skull was varnished or not, had licked its top and announced that it was: *Nature*, 27 Sept. 2001, p. 359.

p. 560 knowing of my interest in human origins for the present volume, had inserted a visit to Olorgesailie: Just for the record, the name is also commonly spelled Olorgasailie, including in some official Kenyan materials. It was this spelling that I used in a small book I wrote for CARE concerning the visit. I am now informed by Ian Tattersall that the correct spelling is with a median 'e'.

Chapter 30: Goodbye

p. 564 'unscientific voyagers, three or four oil paintings, and a few scattered osseous fragments': quoted in Gould, *Leonardo's Mountain of Clams and the Diet of Worms*, p. 237.

p. 565 Australia . . . lost no less than 95 per cent: Flannery and Schouten, *A Gap in Nature*, p. xv.

p. 566 'There's no material benefit to hunting dangerous animals more often than you need to – there are only so many mammoth steaks you can eat': *New Scientist*, 'Mammoth Mystery', 5 May 2001, p. 34.

p. 566 only four types of really hefty . . . land animals: Flannery, *The Eternal Frontier*, p. 195.

p. 566 human-caused extinction now may be running at as much as 120,000 times that level: Leakey and Lewin, *The Sixth Extinction*, p. 241.

p. 568 He set off at once for the island, but by the time he got there the cat had killed them all: Flannery, *The Future*

Eaters, pp. 62–3.

p. 569 'At each successive discharge': quoted in Matthiessen, *Wildlife in America*, pp. 114–15.

p. 569 The zoo lost it: Flannery and Schouten, *A Gap in Nature*, p. 125.

p. 570 Hugh Cuming, who became so preoccupied with accumulating objects that he built a large ocean-going ship and employed a crew to sail the world full-time picking up whatever they could find: Desmond and Moore, *Darwin*, p. 342.

p. 570 Millions of years of isolation had allowed Hawaii: *National Geographic*, 'On the Brink: Hawaii's Vanishing Species', Sept. 1995, pp. 2–37.

p. 571 The greater koa finch, an innocuous member of the honeycreeper family: Flannery and Schouten, *A Gap in Nature*, p. 84.

p. 571 a bird so sublimely rare that only one has ever been seen: Flannery and Schouten, *A Gap in Nature*, p. 76.

p. 573 By the early 1990s he had raised the figure to some six hundred per week: Easterbrook, *A Moment on the Earth*, p. 558.

p. 573 'almost certainly an underestimate': *Washington Post*, in *Valley News*, 27 Nov. 1995, 'Report Finds Growing Biodiversity Threat'.

p. 573 'One planet, one experiment': Wilson, *Diversity of Life*, p. 182.

BIBLIOGRAPHY

Aczel, Amir D., *God's Equation: Einstein, Relativity, and the Expanding Universe*. London: Piatkus Books, 2002.

Alberts, Bruce, Dennis Bray, Alexander Johnson, Julian Lewis, Martin Raff, Keith Roberts and Peter Walter, *Essential Cell Biology: An Introduction to the Molecular Biology of the Cell*. New York and London: Garland Publishing, 1998.

Allen, Oliver E., *Atmosphere*. Alexandria, Va.: Time-Life Books, 1983.

Alvarez, Walter, *T. Rex and the Crater of Doom*. Princeton, NJ: Princeton University Press, 1997.

Annan, Noel, *The Dons: Mentors, Eccentrics and Geniuses*. London: HarperCollins, 2000.

Ashcroft, Frances, *Life at the Extremes: The Science of Survival*. London: HarperCollins, 2000.

Asimov, Isaac, *The History of Physics*. New York: Walker & Co., 1966.

——*Exploring the Earth and the Cosmos: The Growth and Future of Human Knowledge*. London: Penguin Books, 1984.

——*Atom: Journey Across the Subatomic Cosmos*. New York:

Truman Talley/Dutton, 1991.

Atkins, P. W., *The Second Law*. New York: *Scientific American*, 1984.

——*Molecules*. New York: *Scientific American*, 1987.

——*The Periodic Kingdom*. London: Weidenfeld & Nicolson, 1995.

Attenborough, David, *Life on Earth: A Natural History*. London: Collins, 1979.

——*The Living Planet: A Portrait of the Earth*. London: Collins, 1984.

——*The Private Life of Plants: A Natural History of Plant Behaviour*. London: BBC Books, 1984.

Baeyer, Hans Christian von, *Taming the Atom: The Emergence of the Visible Microworld*. London: Viking, 1993.

Bakker, Robert T., *The Dinosaur Heresies: New Theories Unlocking the Mystery of the Dinosaurs and their Extinction*. New York: William Morrow, 1986.

Ball, Philip, *H_2O: A Biography of Water*. London: Phoenix/Orion, 1999.

Ballard, Robert D., *The Eternal Darkness: A Personal History of Deep-Sea Exploration*. Princeton, NJ: Princeton University Press, 2000.

Barber, Lynn, *The Heyday of Natural History: 1820–1870*. London: Jonathan Cape, 1980.

Barry, Roger G., and Richard J. Chorley, *Atmosphere, Weather and Climate*, 7th edn. London: Routledge, 1998.

Biddle, Wayne, *A Field Guide to the Invisible*. New York: Henry Holt, 1998.

Bodanis, David, *The Body Book*. London: Little, Brown, 1984.

——*The Secret House: Twenty-Four Hours in the Strange and Unexpected World in Which We Spend our Nights and Days*. New York: Simon & Schuster, 1984.

——*The Secret Family: Twenty-Four Hours Inside the Mysterious World of Our Minds and Bodies*. New York: Simon & Schuster, 1997.

——*$E = mc^2$: A Biography of the World's Most Famous Equation*. London: Macmillan, 2000.

Bolles, Edmund Blair, *The Ice Finders: How a Poet, a Professor and a Politician Discovered the Ice Age*. Washington DC: Counterpoint/Perseus, 1999.

Boorse, Henry A., Lloyd Motz and Jefferson Hane Weaver, *The Atomic Scientists: A Biographical History*. New York: John Wiley & Sons, 1989.

Boorstin, Daniel J., *The Discoverers*. London: Penguin Books, 1986.

——*Cleopatra's Nose: Essays on the Unexpected*. New York: Random House, 1994.

Bracegirdle, Brian, *A History of Microtechnique: The Evolution of the Microtome and the Development of Tissue Preparation*. London: Heinemann, 1978.

Breen, Michael, *The Koreans: Who They Are, What They Want, Where Their Future Lies*. London: Texere, 1998.

Broad, William J., *The Universe Below: Discovering the Secrets of the Deep Sea*. New York: Simon & Schuster, 1997.

Brock, William H., *The Norton History of Chemistry*. London: W. W. Norton, 1993.

Brockman, John, and Katinka Matson (eds), *How Things Are: A Science Tool-Kit for the Mind*. London: Weidenfeld & Nicolson, 1995.

Brookes, Martin, *Fly: The Unsung Hero of Twentieth-Century Science*. London: Phoenix, 2002.

Brown, Guy, *The Energy of Life*. London: Flamingo/HarperCollins, 2000.

Browne, Janet, *Charles Darwin: A Biography*, vol. 1. London: Jonathan Cape, 1995.

Burenhult, Göran (ed.), *The First Americans: Human Origins and History to 10,000 BC*. London: HarperCollins, 1993.

Cadbury, Deborah, *Terrible Lizard: The First Dinosaur Hunters and the Birth of a New Science*. New York: Henry Holt, 2000.

Calder, Nigel, *Einstein's Universe*. London: BBC Books, 1979.

——*The Comet Is Coming! The Feverish Legacy of Mr Halley*. London: BBC Books, 1980.

Canby, Courtlandt (ed.), *The Epic of Man*. New York: Time/Life, 1961.

Carey, John (ed.), *The Faber Book of Science*. London: Faber, 1995.

Chorlton, Windsor, *Ice Ages*. New York: Time-Life Books, 1983.

Christianson, Gale E., *In the Presence of the Creator: Isaac Newton and his Times*. New York: Free Press/Macmillan, 1984.

——*Edwin Hubble: Mariner of the Nebulae*. Bristol, England: Institute of Physics Publishing, 1995.

Clark, Ronald W., *The Huxleys*. London: Heinemann, 1968.

——*The Survival of Charles Darwin: A Biography of a Man and an Idea*. London: Daedalus Books, 1985.

——*Einstein: The Life and Times*. London: HarperCollins, 1971.

Coe, Michael, Dean Snow and Elizabeth Benson, *Atlas of Ancient America*. New York: Equinox/Facts on File, 1986.

Colbert, Edwin H., *The Great Dinosaur Hunters and their Discoveries*. New York: Dover Publications, 1984.

Cole, K. C., *First You Build a Cloud: And Other Reflections on Physics as a Way of Life*. San Diego: Harvest/Harcourt Brace, 1999.

Conard, Henry S., *How to Know the Mosses and Liverworts*. Dubuque, Iowa: William C. Brown Co., 1956.

Conniff, Richard, *Spineless Wonders: Strange Tales from the Invertebrate World*. London and New York: Henry Holt, 1996.

Corfield, Richard, *Architects of Eternity: The New Science of Fossils*. London: Headline, 2001.

Coveney, Peter, and Roger Highfield, *The Arrow of Time: The Quest to Solve Science's Greatest Mystery*. London: Flamingo, 1991.

Cowles, Virginia, *The Rothschilds: A Family of Fortune*. London: Futura, 1975.

Crick, Francis, *Life Itself: Its Origin and Nature*. London: Macdonald, 1982.

——*What Mad Pursuit: A Personal View of Scientific Discovery*. London: Penguin Press, 1990.

Cropper, William H., *Great Physicists: The Life and Times of Leading Physicists from Galileo to Hawking*. Oxford: Oxford University Press, 2002.

Crowther, J. G., *Scientists of the Industrial Revolution*. London: Cresset, 1962.

Darwin, Charles, *On the Origin of Species by Means of Natural Selection, or the Preservation of Favoured Races in the Struggle for Life* (facsimile edn). London: AMSPR, 1972.

Davies, Paul, *The Fifth Miracle: The Search for the Origin of Life*. London: Penguin Books, 1999.

Dawkins, Richard, *The Blind Watchmaker*. London: Penguin Books, 1988.

——*River Out of Eden: A Darwinian View of Life*. London: Phoenix, 1996.

——*Climbing Mount Improbable*. London: Viking, 1996.

Dean, Dennis R., *James Hutton and the History of Geology*. Ithaca, NY: Cornell University Press, 1992.

de Duve, Christian, *A Guided Tour of the Living Cell*, 2 vols. New York: Scientific American/Rockefeller University Press, 1984.

Dennett, Daniel C., *Darwin's Dangerous Idea: Evolution and the Meanings of Life*. London: Penguin, 1996.

Dennis, Jerry, *The Bird in the Waterfall: A Natural History of Oceans, Rivers and Lakes*. London and New York: HarperCollins, 1996.

Desmond, Adrian, and James Moore, *Darwin*. London: Penguin Books, 1992.

Dewar, Elaine, *Bones: Discovering the First Americans*. Toronto: Random House Canada, 2001.

Diamond, Jared, *Guns, Germs and Steel: The Fates of Human*

Societies. New York: Norton, 1997.

Dickinson, Matt, *The Other Side of Everest: Climbing the North Face through the Killer Storm*. New York: Times Books, 1997.

Drury, Stephen, *Stepping Stones: The Making of our Home World*. Oxford: Oxford University Press, 1999.

Durant, Will and Ariel, *The Age of Louis XIV*. New York: Simon & Schuster, 1963.

Dyson, Freeman, *Disturbing the Universe*. London and New York: Harper & Row, 1979.

Easterbrook, Gregg, *A Moment on the Earth: The Coming Age of Environmental Optimism*. London: Penguin, 1995.

Ebbing, Darrell D., *General Chemistry*. Boston: Houghton Mifflin, 1996.

Elliott, Charles, *The Potting-Shed Papers: On Gardens, Gardeners and Garden History*. Guilford, Conn.: Lyons Press, 2001.

Engel, Leonard, *The Sea*. New York: Time-Life Books, 1969.

Erickson, Jon, *Plate Tectonics: Unravelling the Mysteries of the Earth*. London and New York: Facts on File, 1992.

Fagan, Brian M., *The Great Journey: The Peopling of Ancient America*. London: Thames & Hudson, 1987.

Fell, Barry, *America B.C.: Ancient Settlers in the New World*. London: Random House, 1976.

——*Bronze Age America*. London and Boston: Little, Brown, 1982.

Ferguson, Kitty, *Measuring the Universe: The Historical Quest to Quantify Space*. London: Headline, 1999.

Ferris, Timothy, *The Mind's Sky: Human Intelligence in a Cosmic Context*. New York: Bantam Books, 1992.

——*The Whole Shebang: A State of the Universe(s) Report*. London: Phoenix, 1998.

——*Seeing in the Dark: How Backyard Stargazers Are Probing Deep Space and Guarding Earth from Interplanetary Peril*. New York: Simon & Schuster, 2002.

——*Coming of Age in the Milky Way*. London: HarperCollins, 2003.

Feynman, Richard P., *Six Easy Pieces*. London: Penguin Books, 1998.

Fisher, Richard V., Grant Heiken and Jeffrey B. Hulen, *Volcanoes: Crucibles of Change*. Princeton, NJ: Princeton University Press, 1997.

Flannery, Timothy, *The Future Eaters: An Ecological History of the Australasian Lands and People*. London: W. W. Norton, 1995.

——*The Eternal Frontier: An Ecological History of North America and its Peoples*. London: Heinemann, 2001.

——and Peter Schouten, *A Gap in Nature: Discovering the World's Extinct Animals*. London: Heinemann, 2001.

Fortey, Richard, *Life: An Unauthorised Biography*. London: Flamingo/HarperCollins, 1998.

——*Trilobite! Eyewitness to Evolution*. London: HarperCollins, 2000.

Frayn, Michael, *Copenhagen*. London: Methuen, 1998; New York: Anchor Books, 2000.

Gamow, George, and Russell Stannard, *The New World of Mr Tompkins*. Cambridge: Cambridge University Press, 2001.

Gawande, Atul, *Complications: A Surgeon's Notes on an Imperfect Science*. New York: Metropolitan Books/Henry Holt, 2002.

Giancola, Douglas C., *Physics: Principles with Applications*. London: Prentice-Hall, 1997.

Gjertsen, Derek, *The Classics of Science: A Study of Twelve Enduring Scientific Works*. New York: Lilian Barber Press, 1984.

Godfrey, Laurie R. (ed.), *Scientists Confront Creationism*. New York: W. W. Norton, 1983.

Goldsmith, Donald, *The Astronomers*. New York: St Martin's Press, 1991.

'Gordon, Mrs', *The Life and Correspondence of William*

Buckland, D.D., F.R.S. London: John Murray, 1894.

Gould, Stephen Jay, *Ever since Darwin: Reflections in Natural History*. London: Deutsch, 1978.

——*The Panda's Thumb: More Reflections in Natural History*. London and New York: W. W. Norton, 1980.

——*Hen's Teeth and Horse's Toes*. London: Penguin Books, 1984.

——*The Flamingo's Smile: Reflections in Natural History*. New York: W. W. Norton, 1985.

——*Wonderful Life: The Burgess Shale and the Nature of History*. London: Hutchinson Radius, 1990.

——*Bully for Brontosaurus: Reflections in Natural History*, London: Hutchinson Radius, 1991.

——*Time's Arrow, Time's Cycle: Myth and Metaphor in the Discovery of Geological Time*. Cambridge, Mass.: Harvard University Press, 1987.

——(ed.), *The Book of Life*. London: Ebury, 1993.

——*Eight Little Piggies: Reflections in Natural History*. London: Penguin, 1994.

——*Dinosaur in a Haystack: Reflections in Natural History*. London: Jonathan Cape, 1996.

——*Leonardo's Mountain of Clams and the Diet of Worms: Essays on Natural History*. London: Jonathan Cape, 1998.

——*The Lying Stones of Marrakech: Penultimate Reflections in Natural History*. London: Jonathan Cape, 2000.

Green, Bill, *Water, Ice and Stone: Science and Memory on the Antarctic Lakes*. New York: Harmony Books, 1995.

Gribbin, John, *In the Beginning: The Birth of the Living Universe*. London: Penguin, 1994.

——*Almost Everyone's Guide to Science: The Universe, Life and Everything*. London: Phoenix, 1998.

——and Mary Gribbin, *Being Human: Putting People in an Evolutionary Perspective*. London: Phoenix/Orion, 1993.

——*Fire on Earth: Doomsday, Dinosaurs and Humankind*. New York: St Martin's Press, 1996.

——*Ice Age*. London: Allen Lane, 2001.

——and Jeremy Cherfas, *The First Chimpanzee: In Search of Human Origins*. London: Penguin Books, 2001.

Grinspoon, David Harry, *Venus Revealed: A New Look Below the Clouds of our Mysterious Twin Planet*. Reading, Mass.: Helix/ Addison-Wesley, 1997.

Guth, Alan, *The Inflationary Universe: The Quest for a New Theory of Cosmic Origins*. London: Jonathan Cape, 1997.

Haldane, J. B. S., *Adventures of a Biologist*. New York: Harper & Brothers, 1937.

——*What is Life?* New York: Boni & Gaer, 1947.

Hallam, A., *Great Geological Controversies*, 2nd edn. Oxford: Oxford University Press, 1989.

Hamblyn, Richard, *The Invention of Clouds: How an Amateur Meteorologist Forged the Language of the Skies*. London: Picador, 2001.

Hamilton-Paterson, James, *The Great Deep: The Sea and its Thresholds*. London: Random House, 1992.

Hapgood, Charles H., *Earth's Shifting Crust: A Key to Some Basic Problems of Earth Science*. New York: Pantheon Books, 1958.

Harrington, John W., *Dance of the Continents: Adventures with Rocks and Time*. Los Angeles: J. P. Tarcher, Inc., 1983.

Harrison, Edward, *Darkness at Night: A Riddle of the Universe*. Cambridge, Mass.: Harvard University Press, 1987.

Hartmann, William K., *The History of Earth: An Illustrated Chronicle of an Evolving Planet*. London: Workman Publishing, 1991.

Hawking, Stephen, *A Brief History of Time: From the Big Bang to Black Holes*. London: Bantam Books, 1988.

——*The Universe in a Nutshell*. London: Bantam Press, 2001.

Hazen, Rombert M., and James Trefil, *Science Matters: Achieving Scientific Literacy*. New York: Doubleday, 1991.

Heiserman, David L., *Exploring Chemical Elements and their Compounds*. Blue Ridge Summit, Pa.: TAB Books/ McGraw Hill, 1992.

Hitchcock, A. S., *Manual of the Grasses of the United States*,

2nd edn. London: Peter Smith, 1971.

Holmes, Hannah, *The Secret Life of Dust*. London: John Wiley & Sons, 2001.

Holmyard, E. J., *Makers of Chemistry*. Oxford: Clarendon Press, 1931.

Horwitz, Tony, *Blue Latitudes: Boldly Going Where Captain Cook Has Gone Before*. London: Bloomsbury, 2002.

Hough, Richard, *Captain James Cook*. London: Coronet, 1995.

Jardine, Lisa, *Ingenious Pursuits: Building the Scientific Revolution*. London: Little, Brown, 1999.

Johanson, Donald, and Blake Edgar, *From Lucy to Language*. London: Weidenfeld & Nicolson, 2001.

Jolly, Alison, *Lucy's Legacy: Sex and Intelligence in Human Evolution*. Cambridge, Mass.: Harvard University Press, 1999.

Jones, Steve, *Almost Like a Whale: The Origin of Species Updated*. London: Doubleday, 1999.

Judson, Horace Freeland, *The Eighth Day of Creation: Makers of the Revolution in Biology*. London: Penguin, 1995.

Junger, Sebastian, *The Perfect Storm: A True Story of Men Against the Sea*. London: Fourth Dimension, 1997.

Jungnickel, Christa, and Russell McCormmach, *Cavendish: The Experimental Life*. Bucknell, Pa: Bucknell Press, 1999.

Kaku, Michio, *Hyperspace: A Scientific Odyssey through Parallel Universes, Time Warps, and the Tenth Dimension*. Oxford: Oxford University Press, 1999.

Kastner, Joseph, *A Species of Eternity*. New York: Knopf, 1977.

Keller, Evelyn Fox, *The Century of the Gene*. Cambridge, Mass.: Harvard University Press, 2000.

Kemp, Peter, *The Oxford Companion to Ships and the Sea*. London: Oxford University Press, 1979.

Kevles, Daniel J., *The Physicists: The History of a Scientific Community in Modern America*. London: Random House, 1978.

Kitcher, Philip, *Abusing Science: The Case against Creationism.* Cambridge, Mass.: MIT Press, 1982.

Kolata, Gina, *Flu: The Story of the Great Influenza Pandemic of 1918 and the Search for the Virus that Caused It.* London: Pan, 2001.

Krebs, Robert E., *The History and Use of our Earth's Chemical Elements.* Westport, Conn.: Greenwood, 1998.

Kunzig, Robert, *The Restless Sea: Exploring the World Beneath the Waves.* New York: W. W. Norton, 1999.

Kurlansky, Mark, *Cod: A Biography of the Fish That Changed the World.* London: Vintage, 1999.

Leakey, Richard, *The Origin of Humankind.* London: Phoenix, 1995.

——and Roger Lewin, *Origins.* New York: E. P. Dutton, 1977.

——*The Sixth Extinction: Biodiversity and its Survival.* London: Weidenfeld & Nicolson, 1996.

Leicester, Henry M., *The Historical Background of Chemistry.* New York: Dover, 1971.

Lemmon, Kenneth, *The Golden Age of Plant Hunters.* London: Phoenix House, 1968.

Lewin, Roger, *Bones of Contention: Controversies in the Search for Human Origins*, 2nd edn. Chicago: University of Chicago Press, 1997.

Lewis, Cherry, *The Dating Game: One Man's Search for the Age of the Earth.* Cambridge: Cambridge University Press, 2000.

Lewis, John S., *Rain of Iron and Ice: The Very Real Threat of Comet and Asteroid Bombardment.* Reading, Mass.: Addison-Wesley, 1996.

Lewontin, Richard, *It Ain't Necessarily So: The Dream of the Human Genome and Other Illusions.* London: Granta, 2001.

Little, Charles E., *The Dying of the Trees: The Pandemic in America's Forests.* New York: Viking, 1995.

Lynch, John, *The Weather.* Toronto: Firefly Books, 2002.

McGhee, Jr, George R., *The Late Devonian Mass Extinction:*

The Frasnian/Famennian Crisis. New York: Columbia University Press, 1996.

McGrayne, Sharon Bertsch, *Prometheans in the Lab: Chemistry and the Making of the Modern World*. London: McGraw-Hill, 2002.

McGuire, Bill, *A Guide to the End of the World: Everything You Never Wanted to Know*. Oxford: Oxford University Press, 2002.

McKibben, Bill, *The End of Nature*. London: Viking, 1990.

McPhee, John, *Basin and Range*. New York: Farrar, Straus & Giroux, 1980.

——*In Suspect Terrain*. New York: Noonday Press/Farrar, Straus & Giroux, 1983.

——*Rising from the Plains*.London: Farrar, Straus & Giroux, 1987.

——*Assembling California*. New York: Farrar, Straus & Giroux, 1993.

McSween, Harry Y., Jr, *Stardust to Planets: A Geological Tour of the Solar System*. New York: St Martin's Press, 1993.

Maddox, Brenda, *Rosalind Franklin: The Dark Lady of DNA*. London: HarperCollins, 2002.

Margulis, Lynn, and Dorion Sagan, *Microcosmos: Four Billion Years of Evolution from Our Microbial Ancestors*. London: HarperCollins, 2002.

Marshall, Nina L., *Mosses and Lichens*. New York: Doubleday, Page & Co., 1908.

Matthiessen, Peter, *Wildlife in America*. London: Penguin Books, 1995.

Moore, Patrick, *Fireside Astronomy: An Anecdotal Tour through the History and Lore of Astronomy*. Chichester: John Wiley & Sons, 1992.

Moorehead, Alan, *Darwin and the Beagle*. London: Hamish Hamilton, 1969.

Morowitz, Harold J., *The Thermodynamics of Pizza*. New Brunswick, NJ: Rutgers University Press, 1991.

Musgrave, Toby, Chris Gardner and Will Musgrave, *The*

Plant Hunters: Two Hundred Years of Adventure and Discovery around the World. London: Ward Lock, 1999.

Norton, Trevor, *Stars beneath the Sea: The Extraordinary Lives of the Pioneers of Diving*. London: Arrow Books, 2000.

Novacek, Michael, *Time Traveler: In Search of Dinosaurs and Other Fossils from Montana to Mongolia*. New York: Farrar, Straus & Giroux, 2001.

Nuland, Sherwin B., *How We Live: The Wisdom of the Body*. London: Vintage, 1998.

Officer, Charles, and Jake Page, *Tales of the Earth: Paroxysms and Perturbations of the Blue Planet*. New York: Oxford University Press, 1993.

Oldroyd, David R., *Thinking about the Earth: A History of Ideas in Geology*. London: Athlone, 1996.

Oldstone, Michael B. A., *Viruses, Plagues and History*. New York: Oxford University Press, 1998.

Overbye, Dennis, *Lonely Hearts of the Cosmos: The Scientific Quest for the Secret of the Universe*. London: Macmillan, 1991.

Ozima, Minoru, *The Earth: Its Birth and Growth*. Cambridge: Cambridge University Press, 1981.

Parker, Ronald B., *Inscrutable Earth: Explorations in the Science of Earth*. New York: Charles Scribner's Sons, 1984.

Pearson, John, *Serpents and Stags: The Story of the House of Cavendish and the Dukes of Devonshire*. London: Macmillan, 1983.

Peebles, Curtis, *Asteroids: A History*. Washington: Smithsonian Institution Press, 2000.

Plummer, Charles C., and David McGeary, *Physical Geology*. London: McGraw-Hill Education, 1997.

Pollack, Robert, *Signs of Life: The Language and Meanings of DNA*. London: Penguin Books, 1995.

Powell, James Lawrence, *Night Comes to the Cretaceous: Dinosaur Extinction and the Transformation of Modern Geology*. New York: W. H. Freeman, 1998.

——*Mysteries of Terra Firma: The Age and Evolution of the*

Earth. New York: Free Press/Simon & Schuster, 2001.
Psihoyos, Louie, with John Knoebber, *Hunting Dinosaurs*. London: Cassell Illustrated, 1995.
Putnam, William Lowell, *The Worst Weather on Earth*. London: Mountaineers Books, 1991.
Quammen, David, *The Song of the Dodo*. London: Hutchinson, 1996.
——*The Boilerplate Rhino: Nature in the Eye of the Beholder*. London: Touchstone, 2001.
——*Monster of God*. New York: W. W. Norton, 2003.
Rees, Martin, *Just Six Numbers: The Deep Forces that Shape the Universe*. London: Phoenix/Orion, 2000.
Ridley, Matt, *The Red Queen: Sex and the Evolution of Human Nature*. London: Penguin, 1994.
——*Genome: The Autobiography of a Species*. London: Fourth Estate, 1999.
Ritchie, David, *Superquake! Why Earthquakes Occur and When the Big One Will Hit Southern California*. London: Random House, 1989.
Rose, Steven, *Lifelines: Biology, Freedom, Determinism*. London: Penguin, 1997.
Rudwick, Martin J. S., *The Great Devonian Controversy: The Shaping of Scientific Knowledge among Gentlemanly Specialists*. Chicago: University of Chicago Press, 1985.
Sacks, Oliver, *An Anthropologist on Mars: Seven Paradoxical Tales*. London: Picador, 1996.
——*Oaxaca Journal*. London: National Geographic, 2002.
Sagan, Carl, *Cosmos*. London: Random House, 1980.
——and Ann Druyan, *Comet*. London: Random House, 1985.
Sagan, Dorion, and Lynn Margulis, *Garden of Microbial Delights: A Practical Guide to the Subvisible World*. Boston: J. Harcourt Brace Jovanovich, 1988.
Sayre, Anne, *Rosalind Franklin and DNA*. London: W. W. Norton, 2002.
Schneer, Cecil J. (ed.), *Toward a History of Geology*. London:

MIT Press, 1970.

Schopf, J. William, *Cradle of Life: The Discovery of Earth's Earliest Fossils*. Princeton, NJ: Princeton University Press, 1999.

Schultz, Gwen, *Ice Age Lost*. Garden City, NY: Anchor Press/Doubleday, 1974.

Schwartz, Jeffrey H., *Sudden Origins: Fossils, Genes and the Emergence of Species*. New York: John Wiley & Sons, 1999.

Semonin, Paul, *American Monster: How the Nation's First Prehistoric Creature Became a Symbol of National Identity*. New York: New York University Press, 2000.

Shore, William H. (ed.), *Mysteries of Life and the Universe*. San Diego: Harvest/Harcourt Brace & Co., 1992.

Silver, Brian, *The Ascent of Science*. New York: Solomon/Oxford University Press, 1998.

Simpson, George Gaylord, *Fossils and the History of Life*. New York: Scientific American, 1983.

Smith, Anthony, *The Weather: The Truth about the Health of our Planet*. London: Hutchinson, 2000.

Smith, Robert B., and Lee J. Siegel, *Windows into the Earth: The Geologic Story of Yellowstone and Grand Teton National Parks*. Oxford: Oxford University Press, 2002.

Snow, C. P., *Variety of Men*. London: Macmillan, 1967.

——*The Physicists*. London: House of Stratus, 1979.

Snyder, Carl H., *The Extraordinary Chemistry of Ordinary Things*. London: John Wiley & Sons, 1995.

Stalcup, Brenda (ed.), *Endangered Species: Opposing Viewpoints*. San Diego: Greenhaven, 1996.

Stanley, Steven M., *Extinction*. New York: *Scientific American*, 1987.

Stark, Peter, *Last Breath: Cautionary Tales from the Limits of Human Endurance*. New York: Ballantine Books, 2001.

Stephen, Sir Leslie, and Sir Sidney Lee (eds), *Dictionary of National Biography*. Oxford: Oxford University Press, 1973.

Stevens, William K., *The Change in the Weather: People,*

Weather, and the Science of Climate. New York: Delacorte, 1999.

Stewart, Ian, *Nature's Numbers: Discovering Order and Pattern in the Universe*. London: Phoenix, 1995.

Strathern, Paul, *Mendeleyev's Dream: The Quest for the Elements*. London: Penguin Books, 2001.

Sullivan, Walter, *Landprints*. New York: Times Books, 1984.

Sulston, John, and Georgina Ferry, *The Common Thread: A Story of Science, Politics, Ethics and the Human Genome*. London: Bantam Press, 2002.

Swisher III, Carl C., Garniss H. Curtis and Roger Lewin, *Java Man: How Two Geologists' Dramatic Discoveries Changed our Understanding of the Evolutionary Path to Modern Humans*. London: Little, Brown, 2001.

Sykes, Bryan, *The Seven Daughters of Eve*. London: Bantam Press, 2001.

Tattersall, Ian, *The Human Odyssey: Four Million Years of Human Evolution*. New York: Prentice-Hall, 1993.

——*The Monkey in the Mirror: Essays on the Science of What Makes Us Human*. Oxford: Oxford University Press, 2002.

——and Jeffrey Schwartz, *Extinct Humans*. Boulder, Colo.: Westview/Perseus, 2001.

Thackray, John, and Bob Press, *The Natural History Museum: Nature's Treasurehouse*. London: Natural History Museum, 2001.

Thomas, Gordon, and Max Morgan Witts, *The San Francisco Earthquake*. London: Souvenir, 1971.

Thomas, Keith, *Man and the Natural World: Changing Attitudes in England, 1500–1800*. London: Penguin Books, 1984.

Thompson, Dick, *Volcano Cowboys: The Rocky Evolution of a Dangerous Science*. New York: St Martin's Press, 2000.

Thorne, Kip S., *Black Holes and Time Warps: Einstein's Outrageous Legacy*. London: Picador, 1994.

Tortora, Gerard J., and Sandra Reynolds Grabowski,

Principles of Anatomy and Physiology. London: John Wiley & Sons, 1999.

Trefil, James, *The Unexpected Vista: A Physicist's View of Nature*. New York: Charles Scribner's Sons, 1983.

——*Meditations at Sunset: A Scientist Looks at the Sky*. New York: Charles Scribner's Sons, 1987.

——*Meditations at 10,000 Feet: A Scientist in the Mountains*. New York: Charles Scribner's Sons, 1987.

——*101 Things You Don't Know About Science and No One Else Does Either*. London: Cassell Illustrated, 1997.

Trinkaus, Erik, and Pat Shipman, *The Neandertals: Changing the Image of Mankind*. London: Pimlico, 1994.

Tudge, Colin, *The Time before History: Five Million Years of Human Impact*. New York: Touchstone/Simon & Schuster, 1996.

——*The Variety of Life: A Survey and a Celebration of All the Creatures that Have Ever Lived*. Oxford: Oxford University Press, 2002.

Vernon, Ron, *Beneath our Feet: The Rocks of Planet Earth*. Cambridge: Cambridge University Press, 2000.

Vogel, Shawna, *Naked Earth: The New Geophysics*. New York: Dutton, 1995.

Walker, Alan, and Pat Shipman, *The Wisdom of the Bones: In Search of Human Origins*. London: Weidenfeld & Nicolson, 1996.

Wallace, Robert A., Jack L. King and Gerald P. Sanders, *Biology: The Science of Life*, 2nd edn. Glenview, Ill.: Scott, Foresman & Co., 1986.

Ward, Peter D., and Donald Brownlee, *Rare Earth: Why Complex Life Is Uncommon in the Universe*. New York: Copernicus, 1999.

Watson, James D., *The Double Helix: A Personal Account of the Discovery of the Structure of DNA*. London: Penguin Books, 1999.

Weinberg, Samantha, *A Fish Caught in Time: The Search for the Coelacanth*. London: Fourth Estate, 1999.

Weinberg, Steven, *The Discovery of Subatomic Particles*. London: W. H. Freeman, 1990.
——*Dreams of a Final Theory*. London: Vintage, 1993.
Whitaker, Richard (ed.), *Weather*. London: Warner Books, 1996.
White, Michael, *Isaac Newton: The Last Sorcerer*. London: Fourth Dimension, 1997.
——*Rivals: Conflict as the Fuel of Science*. London: Vintage, 2001.
Wilford, John Noble, *The Mapmakers*. London: Random House, 1981.
——*The Riddle of the Dinosaur*. London: Faber, 1986.
Williams, E. T., and C. S. Nicholls (eds), *Dictionary of National Biography, 1961–1970*. Oxford: Oxford University Press, 1981.
Williams, Stanley, and Fen Montaigne, *Surviving Galeras*. Boston: Houghton Mifflin, 2001.
Wilson, David, *Rutherford: Simple Genius*. London: Hodder, 1984.
Wilson, Edward O., *The Diversity of Life*. London: Allen Lane/Penguin Press, 1993.
Winchester, Simon, *The Map That Changed the World: The Tale of William Smith and the Birth of a Science*. London: Viking, 2001.
Woolfson, Adrian, *Life without Genes: The History and Future of Genomes*. London: Flamingo, 2000.

INDEX